AF906277

Current Advances in Biotechnological Production of Xylitol

Maria das Graças de Almeida Felipe ·
Anuj Kumar Chandel
Editors

Current Advances in Biotechnological Production of Xylitol

Fermentative Production of Xylitol

Springer

Editors
Maria das Graças de Almeida Felipe
Department of Biotechnology
Engineering School of Lorena (EEL)
University of São Paulo (USP)
Lorena, São Paulo, Brazil

Anuj Kumar Chandel
Department of Biotechnology
Engineering School of Lorena (EEL)
University of São Paulo (USP)
Lorena, São Paulo, Brazil

ISBN 978-3-031-04941-5 ISBN 978-3-031-04942-2 (eBook)
https://doi.org/10.1007/978-3-031-04942-2

© The Editor(s) (if applicable) and The Author(s), under exclusive license to Springer Nature Switzerland AG 2022
Chapter 9 is licensed under the terms of the Creative Commons Attribution 4.0 International License (http://creativecommons.org/licenses/by/4.0/). For further details see license information in the chapter.
This work is subject to copyright. All rights are solely and exclusively licensed by the Publisher, whether the whole or part of the material is concerned, specifically the rights of translation, reprinting, reuse of illustrations, recitation, broadcasting, reproduction on microfilms or in any other physical way, and transmission or information storage and retrieval, electronic adaptation, computer software, or by similar or dissimilar methodology now known or hereafter developed.
The use of general descriptive names, registered names, trademarks, service marks, etc. in this publication does not imply, even in the absence of a specific statement, that such names are exempt from the relevant protective laws and regulations and therefore free for general use.
The publisher, the authors, and the editors are safe to assume that the advice and information in this book are believed to be true and accurate at the date of publication. Neither the publisher nor the authors or the editors give a warranty, expressed or implied, with respect to the material contained herein or for any errors or omissions that may have been made. The publisher remains neutral with regard to jurisdictional claims in published maps and institutional affiliations.

This Springer imprint is published by the registered company Springer Nature Switzerland AG
The registered company address is: Gewerbestrasse 11, 6330 Cham, Switzerland

Preface

Biotransformation of lignocellulosic biomass into biofuels and biochemicals offers numerous strategic and environmental benefits. One of the valuable compounds which can be made from lignocellulosic biomass is Xylitol. It is a commercially viable sugar alcohol that has several applications in the food, confectionaries, odontological, and pharmaceutical sectors. Based on the growing demand for xylitol, the US Department of Energy (USDOE) has included xylitol in the top 12 value-added chemicals produced from lignocellulosic feedstock [1]. Xylitol production from the hemicellulose hydrolysate is considered more profitable than ethanol or co-production of ethanol and furfural [2].

Xylitol is a white crystalline powder, soluble in water, and similar in taste to table sugar with a lower calorific content, and it has anticariogenicity and antimicrobial properties, among other properties. Besides the common use of xylitol as a low calorific sweetener, it can be considered as a platform chemical for the chemical synthesis of xylaric acid, ethylene glycol, glycerol, etc. [3]. Therefore, in recent years, the commercial demand for xylitol has increased significantly. Currently, xylitol has a market of US$823.6 million which is expected to rise approximately to US$ 1.37 billion by 2025 [4]. Currently, this increased demand worldwide is being met through the chemical conversion of hemicellulosic hydrolysate. However, chemical synthesis has several environmental and sustainable challenges like high chemical requirements, high carbon footprints, high water usage, and reactor corrosions. In order to replace the chemical transformations, biotechnological methods of xylitol production have the advantages of being eco-friendly and sustainable [5]. Xylitol production by microbial fermentation offers several unique advantages because it has a very low amount of inhibitors compared to the chemical synthesis. The fermentation broth after separation of biomass can be directly used for xylitol recovery and concentration using minimum steps of purification. The concentrated xylitol solution can also be directly used in the confectionaries or in toothpaste formulations avoiding the requirement of crystallization.

Recent developments made in lignocellulosic feedstock pretreatment for efficient xylose recovery, fermentation methods and economic media formulations, and strain improvement for enhanced xylose conversion and xylitol recovery have shown the potential of microbial xylitol production to carter its growing demands in the market. For this, the requirement of high-quality concentrated xylose sugar syrup without the inhibitors is very important. A considerable research progress has been made for the best-quality xylose sugar recovery as this is the primary feedstock for xylitol production followed by strain improvement, fermentation of xylose solution, and downstream processing of microbial xylitol production. The major barrier to the microbial xylitol production at the commercial scale is the requirement of more time in the fermentation step in turn reducing the xylitol productivity. This book is aimed to bring the recent research output in addition to some important advancements made in the xylose recovery, fermentative production of xylitol and strain improvement, etc. Besides the improvements made in xylose production, the book will also disseminate cutting-edge information on the demand for xylitol from various commercial sectors and its applications.

This book has compiled **11** specialized chapters written by experts from various universities/institutes/industries. Chapter 1 of **de Almeida et al**. presents the recent progress and future perspectives on hemicellulose destruction aiming to xylose recovery with desired yields with a minimum amount of plant cell wall-derived inhibitors. These inhibitors include weak acids, phenolics, furans, and others. These inhibitors influence the microorganism's growth and xylitol productivities. This is necessary to remove these inhibitors from the hydrolysate prior to the fermentation of hydrolysates in order to obtain the optimum productivity of xylitol. Chapter 2 by **Velmurugan et al.** provides a critical assessment of various detoxification strategies applied to the lignocellulose hydrolysates conditioning and to remove the inhibitors. Various new wild xylitol-producing microbial strains have been found in nature. These strains offer the important characteristics of growth and xylitol production profile. These strains display the different growth profile and substrate utilization rates eventually showing different xylitol production profiles. So, Chap. 3 by **Queiroz and colleagues** summarizes the important features of xylitol-producing microorganisms and growth characteristics employing various substrates. In order to achieve the desired product yield and productivity, continuous efforts are being made to develop the superior microorganisms by classical and new molecular biology techniques. In this sense, Chap. 4 by **Gupta et al.** appraises the various strain improvement methods for enhanced xylitol production. Various fermentation strategies have been assessed for xylitol production from a variety of lignocellulose hydrolysates. These techniques include batch, fed-batch, semi-continuous, and continuous cultivation. Recycling of microbial cells in free or immobilized form has also been attempted to make the xylitol production process robust and economic. Besides the development made in fermentation methods, optimization of media and fermentation parameters have been studied in detail. Optimization of fermentation media using economic carbon and nitrogen sources is essential to obtain the economic xylitol production. These insights are important for the xylitol production at a large scale. Therefore, Chap. 5 by **Prado et al.** provides a comprehensive review on a variety of fermentation methods and optimization of fermentation parameters and media ingredients. For the commercial viability of microbial xylitol production, this is necessary to adopt integrated approaches in sugarcane, wood, or corn processing mills in order to reduce the capital and operational expenses. This will save the biomass handling cost, steam and electricity savings and reduce the biomass handling and processing cost. Taking this into consideration, **Martinez et al.,** in Chap. 6, elaborate on the critical elements for xylitol production in integrated sugarcane and corn processing mills. After the fermentation of hemicellulosic hydrolysate, downstream processing for xylitol recovery is done by precipitation, purification, and crystallization steps. Chapter 7 by **Durán-Aranguren and co-authors** is aimed to present a critical analysis on downstream processing methodologies for xylitol recovery. Techno-economics analysis of any bio-based process is utmost important to have the key insights into the fermentative production of xylitol. This will enable the economic viability of the xylitol production process considering various process variables in different case scenarios. These insights are valuable for xylitol production at a commercial scale. Therefore, Chap. 8 by **Piedrahita-Rodríguez and Alzate** provides a detailed information on techno-economics analysis of xylitol production covering important scenarios for xylitol production in standalone and integrated biorefineries. Even though a significant progress has been seen toward the fermentative production of xylitol in research institutes, there is still a big gap that exists for xylitol production at a commercial scale via fermentation routes. Chemical synthesis of xylitol is the a primary source of growing xylitol demand in society. There are several reasons which hinder the large-scale production of xylitol by microbial routes. These issues and their possible solution have been highlighted in Chap. 9 by **Rao and colleagues.** In fact, this chapter presents a closer perspective on the commercial production of xylitol by microbial routes.

In the last two or three decades, there is a continuous rise in demand for xylitol in health, food, and medical sectors because of the growing health and food habit concerns. Recently, several new potential applications of xylitol have been found in medical applications in the form of low calorific sweetener, odontological applications, otitis media, antimicrobial property, and osteoporosis. Chapter 10 by **Arruda and co-workers** essentially provides the key and valuable information on the applications of xylitol in health and medical applications, confectionaries, and toothpaste, among others.

Finally, Chap. 11 by **Hans et al.** provides an important information on the market, consumption trends, and commercial status of xylitol production. This information is pivotal for the commercial-scale production of xylitol in the biorefineries.

In a nutshell, this book provides a comprehensive review on the current state and future perspectives of the technological developments in microbial production of xylitol. The book features the authoritative contribution from leading researchers from both academia and industry on xylitol production and its applications. Editors would like to thank Padma Subbaiyan, Werner Hermens, and Anthony Doyle from Springer Nature for their continuous encouragement and support to edit this important book.

Among the editors, **M. G. A. Felipe** is grateful to the National Council for Scientific and Technological Development—CNPq, the Research Council for the State of São Paulo—FAPESP Brazil, and Coordenação de Aperfeiçoamento de Pessoal de Nível Superior—CAPES for providing necessary research infrastructure and support for conducting research on various aspects of microbial conversion of lignocellulose biomass into xylitol and other bioproducts. **Anuj K. Chandel** is also grateful to the Research Council for the State of São Paulo (FAPESP), Brazil, for providing financial assistance (Process No. 2020/12559-6) through *jovem pesqvisador* program and the National Council for Scientific and Technological Development (CNPq) for the fellowship of scientific productivity (PQ: 1-309214/2021-1) in the area of biomass valorization into low-carbon fuels and chemicals.

The editors are confident that the book will serve the purpose to provide the colossus information to the readers of different disciplines who are keen to see the advancement in lignocellulose biotechnology, xylitol production, and sustainability.

Lorena, Brazil

Maria das Graças de Almeida Felipe
mgafelipe@usp.br

Anuj Kumar Chandel
anuj10@usp.br

References

1. U.S. Department of Energy. U.S. billion-ton update: biomass supply for a bioenergy and bioproducts industry. In: Perlack RD, Stokes BJ, editors. Agricultural and biosystems engineering technical reports and white papers. ORNL/TM- 2011/224. Oak Ridge (TN); 2011. p. 227
2. Jong Ed, Stichnothe H, Bell G et al: Bio-based chemicals: a 2020 update. 2020
3. Silva SS, Chandel AK. D-Xylitol. Fermentative production, application and commercialization. Berlin (Germany): Springer-Verlag; 2012
4. Grand View Research, Inc. Xylitol building block mar- ket size, application analysis, regional outlook, com- petitive strategies and forecasts, 2014 To 2020. 2014. Available from: https://www.grandviewresearch.com/industry-analysis/xylitol-building-block-market
5. Hernández-Pérez AF, de Arruda PV, Sene L, Silva SS, handel AK, Felipe MGA (2019) Xylitol bioproduction: State-of-the-art, industrial paradigm shift and opportunities for integrated biorefineries. Critical Reviews in Biotechnology. 39: 924-943

Contents

About the Editors

Prof. Dr. de Almeida Felipe Maria das Graças is a Full Professor in the Department of Biotechnology at the School of Engineering of Lorena—EEL, University of São Paulo—USP since 2020. She is working for almost 34 years at EEL-USP teaching in the area of applied microbiology and conducting research on the sustainable production of xylitol and other biomolecules such as ethanol from lignocellulosic biomass. Her line of research is advancing the scientific and technological knowledge to limit the bottlenecks in the biotechnological use of plant biomass concurrently imparting the training to undergraduate, and post-graduate students. Prof. Felipe in the Scopus database (March, 2022) figures as the second author with the largest number of publications in the world on the scientific work on "xylitol production" with the H-index of 33, while in Web of Science database she had H-index of 30, and 43 according to The Scholar Google database. Prof. Felipe has published 111 peer reviewed journals and 25 book chapters in the area of lignocellulose biotechnology, in a context of circular bioeconomy.

Prof. Chandel Anuj Kumar has 22 years research experience in industrial biotechnology working with Dalas Biotech Ltd, Bhiwadi India, University of Delhi, Celestial Biolabs Ltd, Hyderabad, India and Sugarcane Technology Centre-Piracicaba, Brazil. Dr. Chandel has also awarded MBA in agri-business having specialization in Biorefinery commercialization. Anuj has worked as a post doctorate fellow at University of Stellenbosch, South Africa, EEL-University of São Paulo, Brazil and University of Arkansas, Fayetteville, USA for 6 years. Anuj has worked at EEL-University of São Paulo-Lorena as a visiting professor for 3 years. Currently, he is working as a professor at USP-Lorena. He is an editor of 12 books on D-xylitol, lignocellulose degradation and Brazilian Biofuels Development, Biogas production, amongst others. Dr. Chandel has published 98 articles in peer-reviewed journals, 44 book chapters and recorded one Brazilian Patent. His contributions span industrial biotechnology, the circular carbon economy and policy domains. Dr. Chandel is an active consultant to some Biotech companies and startups in biomass conversion and biorefinery and industrial enzymes production.

Chapter 1
Methods for Hemicellulose Deconstruction Aiming to Xylose Recovery: Recent Progress and Future Perspectives

Sâmilla G. C. de Almeida, Veronica T. F. Silva, Jonas P. de Souza, Cleiton D. Prado, Débora K. S. Oliveira, Débora D. V. Silva, and Kelly J. Dussán

Abstract Lignocellulosic biomass currently represents the most significant potential to produce biofuels and biochemical compounds because of its abundance and cost savings. For using these materials, pretreatment is one of the essential steps in the biomass conversion process. However, the physical and chemical barriers from the main constituents (cellulose, hemicellulose, lignin) and their interactions form a hardheaded structure, creating a barrier to recover the fermentable sugars. Hemicellulose hydrolysis results in a xylose-rich hydrolysate that can be converted to high add-value chemicals, like xylitol. Each type of pretreatment will affect most downstream processes and represent a portion of the costs of the bioprocess. In this way, the selection of one pretreatment cost-effective and viable is a significant

S. G. C. de Almeida · J. P. de Souza · D. K. S. Oliveira · K. J. Dussán (✉)
Department of Engineering, Physics and Mathematics, Institute of Chemistry, São Paulo State University-UNESP, Araraquara, São Paulo, Brazil
e-mail: kelly.medina@unesp.br

V. T. F. Silva
Department of Biotechnology, School of Engineering of Lorena (EEL), University of São Paulo, Lorena, São Paulo, Brazil

C. D. Prado
Department of Genetics and Evolution, Health and Biological Sciences Center, Federal University of São Carlos, São Carlos, Brazil

D. D. V. Silva
Department of Chemistry and Biological Sciences, Institute of Biosciences, São Paulo State University-Unesp, Botucatu, São Paulo, Brazil

K. J. Dussán
Center for Monitoring and Research of the Quality of Fuels, Biofuels, Crude Oil, and Derivatives - CEMPEQC, Institute of Chemistry (UNESP), Araraquara, São Paulo, Brazil

Bioenergy Research Institute (IPBEN), São Paulo State University (Unesp), Araraquara, São Paulo, Brazil

© The Author(s), under exclusive license to Springer Nature Switzerland AG 2022

M. das G. de Almeida Felipe and A. K. Chandel (eds.), *Current Advances in Biotechnological Production of Xylitol*,
https://doi.org/10.1007/978-3-031-04942-2_1

challenge. Moreover, each specific pretreatment will act differently in the lignocellulosic matrix; therefore, the choice must consider the configuration of the process used and the characteristics of the subsequent fermentation. This chapter aims to provide an overview and critically about hemicellulose deconstruction techniques aiming at xylose recovery, approach the perspectives, and show promising methods for disrupting the lignocellulosic matrix.

Keywords Lignocellulosic biomass · Pretreatment · Xylitol · Fermentation

1.1 Introduction

Lignocellulosic biomass has been targeted as a non-fossil, renewable, and low-cost feedstock for many different areas of application, such as fuels, energy, and biobased chemicals (Mankar et al. 2021; Choi et al. 2019; Arruda et al. 2021). The main aspects that characterize this biomass as an attractive alternative as feedstock for industrial purposes are its composition in carbohydrates and phenolic fraction, being relatively abundant, and its renewable and sustainable nature. Furthermore, applying the concept of biorefinery consists of processing a base material in a sustainable, economical, and environmental way to generate energy and bioproducts. The utilization of lignocellulosic biomass enables the obtention of a wide range of bioproducts such as biofuels (bioethanol, biodiesel) and chemicals with high add-value, namely xylitol, mannitol, furfural, among others (Arevalo-Gallegos et al. 2017; Romaní et al. 2020).

As many advantages as it has, lignocellulosic biomass presents a significant issue, a natural resistance to physical, chemical, and biological modification, limiting its usage. As a result of this resistance, also defined as recalcitrant, the extraction of cellulose, hemicellulose, and lignin fractions are quite complicated. Therefore, the application of pretreatment is required, as an essential step, to overcome this negative aspect of lignocellulosic biomass.

Pretreatments are classified in physical, chemical, and biological, and a combination of these techniques can be made to increase the efficiency of lignocellulosic deconstruction (Kim 2018; Kumar and Sharma 2017). However, selecting the ideal pretreatment must be low energy intake, selective toward the desirable fraction to be extracted, generating limited amounts of inhibitors products, and generally, the pretreatment must be feasible in terms of overall cost and sustainability (Mankar et al. 2021; Liao et al. 2020; Haldar and Purkait 2021).

The deconstruction process of lignocellulosic biomass enables the fractionation of cellulose, hemicellulose, and lignin and its subsequent transformation into high added-value products. Cellulose is a well-known base material on an industrial scale, and lignin and hemicellulose have been gaining attention regarding their high potential as starting materials.

Here it will highlight some characteristics of hemicellulose structure and methodologies and processes to its deconstruction aiming xylose recovery. Hemicellulose fraction consists mainly of a pentose (C5), as xylose. The sugars that constitute this fraction can be converted into bioethanol, antioxidant products (xylooligosaccharides), biofilms, xylitol, furfural, and gelling agents (Geng et al. 2019; Zhang et al. 2018; Jofre et al. 2021). Among these bioproducts, xylitol production has been extensively investigated due to its beneficial features in various industrial sectors, such as food, dental, pharmaceutical, and cosmetics (Jofre et al. 2021; Saha and Kennedy 2020). Xylitol is a high-value product whose price has been rising as reported by the global market, with an estimated value of US$1.3 billion up to 2027, with a 5.5% Compound Annual Growth Rate (CAGR) from 2020 to 2027 (Research and Market Reports, Xylitol - Global Market Trajectory Analytics 2021). This market expansion justifies the search for methods that enable the implementation of its production through a biotechnological route in biorefineries.

1.2 Factors Responsible for the Recalcitrance of Lignocellulosic Biomass

The complex and entangled structure of lignocellulosic biomass confers to this material a natural resistance against physical, chemical, and biological deconstruction. This natural resistance, so-called recalcitrant, has many sources within the constitution of lignocellulosic biomass. First, its highly organized structure (Fig. 1.1), where the cellulose microfibrils are embedded in an amorphous matrix made up of hemicellulose and lignin. The lignin by itself works as a physical barrier, protecting the carbohydrate fraction (cellulose and hemicellulose) from physical and chemical damages and biological attacks. The presence of lignin also provides adhesion and cohesion among the biomass fractions (Kim 2018; Brienzo et al. 2016; Ek et al. 2009). Besides the physical resistance conferred by its organized structure, the recalcitrance of lignocellulosic biomass to some extent is also due to the physical and chemical composition of each fraction of the biomass (Yoo et al. 2020).

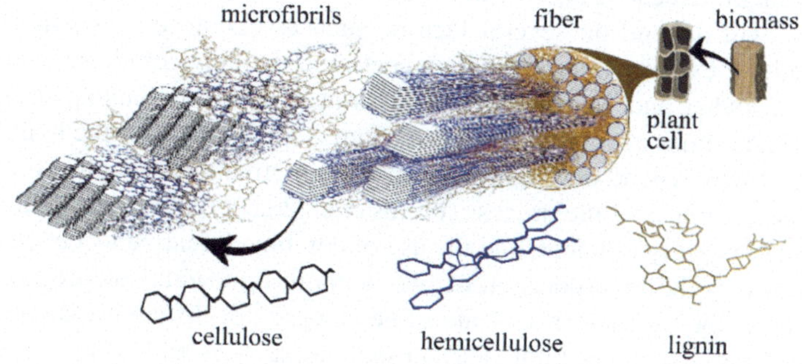

Fig. 1.1 Model proposed for the arrangement of the components in lignocellulosic biomass

Hemicellulose is a physical barrier that surrounds cellulose fibers, protecting them (Gírio et al. 2010). It is a heterogeneous and amorphous polysaccharide that has its chain built up from more than one type of monosaccharide such as pentoses (xylose and arabinose); hexoses (glucose, mannose, galactose), and other sugars (fucose, rhamnose). The hemicellulosic chain may also contain 4-O-methyl-glucuronic acid and be esterified with acetyl and feruloyl groups (Liao et al. 2020). The hemicellulose backbone is formed by the $\beta(1 \rightarrow 4)$ bond between the building blocks carbohydrates. In nature, hemicellulosic can differ between plant species and tissue types. Thus, it is possible to find different hemicelluloses, such as glucuronoxylan and glucomannan in hardwood, galacto-glucomannan, and arabino-glucuronoxylan and xylan, in hardwood grasses and softwoods (Fengel and Wegener 1984; Biely et al. 2016). However, the xylans are the most relevant and abundant hemicellulose type, and their hydrolysis releases monomers (mainly xylose) and oligomers (i.e., xylooligosaccharides) (Hilpmann et al. 2016).

Depending on the kind of biomass (sugarcane bagasse, sweet sorghum bagasse, barley straw, for example), the amount of hemicellulose can vary from around 18 to 30% (Dussán et al. 2016; Camargo et al. 2019; Moraes et al. 2020). Therefore, converting most hemicellulose into soluble sugars is essential to use this material in xylitol production efficiently.

1.3 Methods for the Deconstruction of Hemicellulose

Different methods can be employed to deconstruct the lignocellulosic biomass cell wall and the solubilization of sugars present in its fractions (hemicellulosic and cellulosic). There are physical, chemical, and physicochemical methods (Arora et al. 2020; Alokika et al. 2020). The choice of the technique must consider that maximum recovery of polymers from biomass is achieved, which will allow obtaining a hydrolysate rich in sugar and with a low content of toxic compounds, especially when it comes to hemicellulose hydrolysate (Arora et al. 2020; Alokika et al. 2020).

Considering the recalcitrance of lignocellulosic biomass, the efficiency in matrix degradation will depend on several factors, such as cellulose crystallinity, lignin content, and the existing interconnections between the hemicellulosic fraction and lignin (Bussemaker and Zhang 2013; Ravindran et al. 2017; Santos, et al. 2020). Therefore, sometimes a single pretreatment step is not enough for the hydrolysis of monomers; it will depend on the limitation of the pretreatment employed and other intrinsic disadvantages of the process (Flores et al. 2021). Thus, none treatment can be assumed to be fully efficient, and the use of combined treatments becomes pertinent; as it allows the incorporation of two or more lignocellulosic deconstruction techniques; maybe mechanisms of the same category or different (Mussatto et al. 2016), and allows to achieve high yields in sugar monomers with low severity conditions (Liu et al. 2017). Moreover, in some cases, providing the recovery of each component of lignocellulosic biomass in single streams, facilitating the subsequent conversion and valorization processes (Lomovsky et al. 2016; Moreno, et al. 2019).

Usually, the combination of pretreatment steps is closely associated with physical treatments (methods that are not modified chemically or biologically the lignocellulose structure). For example, the widest physicals pretreatment reported are milling, grinding, extrusion, and irradiation using ultrasound and microwave (Esfahani and Azin 2011; Nanda et al. 2016).

The use of unique chemical or mechanical pretreatment steps may require extreme conditions to deconstruct lignocellulosic biomass, which favors unsatisfactory monomer yields and a high generation of degradation products. Thus, efforts have been needed to combine treatments to perform well in mild processing conditions (Liu et al. 2017).

Each pretreatment method, despite its nature, has advantages and disadvantages, such as production cost, production of inhibitors, use of harsh chemicals. Thus, to select the best pretreatment for hemicellulosic extraction from lignocellulosic biomass aiming xylose production, it is crucial to follow some criteria, such as being able to reduce biomass recalcitrance by mitigating its causes, being efficient in pretreating different lignocellulosic materials, producing low content of inhibitors of microorganisms and be as sustainable and environmental-friendly as possible (Mankar et al. 2021).

This section will discuss the main methods of pretreatment of biomass for hemicellulose extraction aiming at xylose recovery. Future perspectives will be addressed as well in terms of new and advanced techniques.

1.3.1 Chemical Pre-treatments

Deconstruction of hemicellulosic fraction using acid, organic solvents, or the combination of them with other methods of pretreatment possibilities obtain hydrolysates rich in xylose that can be used for xylitol production. There are various possible pretreatments, but here it will be discussed those more common or preferred chemical pretreatment practices.

1.3.1.1 Acid Hydrolysis

Initially, acid hydrolysis was developed to separate or reduce the hemicellulose fraction of biomass, exposing the cellulose fibers to enzymatic action for bioethanol production. However, it was found that the large content of xylose in the C5 fraction should be considered as a raw material to produce both ethanol and other bioproducts (Moraes et al. 2020; Raj et al. 2021a; Yu et al. 2021). This fact has increased the interest in developing pretreatment methods that improve the efficiency of cellulosic hydrolysis and provide a hemicellulosic hydrolysate with a high content of xylose and the lowest possible content of inhibitor compounds, favoring its use as another raw material from lignocellulosic biomass.

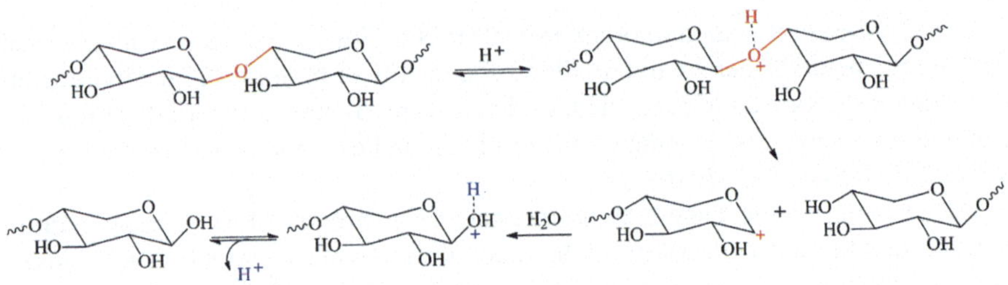

Fig. 1.2 General representation of acid hydrolyses in polysaccharides

In acidic pretreatments, depolymerization of carbohydrates also takes place, where the acid catalysis occurs in a few steps: a rapid electrophilic attack from proton (H^+) to the oxygen in the glycosidic bond (C–O–C), the generation of a carbocation ion that results from the mobility of the positive charge to the carbon one and the slow split of the glycosidic linkage (Fig. 1.2). As a result of these steps, a polysaccharide residue is released in the medium, such as oligosaccharides or monosaccharides. Finally, the remaining carbocation ion reacts with a molecule of water from the medium, and it is stabilized, and a proton (H^+) is released (Ek et al. 2009; Fengel and Wegener 1984).

For the reduction of hemicellulose polysaccharide in its monomers, acid hydrolysis has been conducted using organic (citric acid, acetic acid, formic acid) or inorganic acids (H_2SO_4; HCl, H_3PO_4), which are more effective (Camargo et al. 2019; Lavarack et al. 2002). Depending on the kind of biomass and the acid, it is possible to choose either concentrated (>10%) acid hydrolysis at a low temperature (<100 °C) or dilutes (<10%) acid hydrolysis at a high temperature (>100 °C) (Hilpmann et al. 2016; Lavarack et al. 2002). Other parameters such as solid-to-liquid ratio and the reaction time also are considered important for high hydrolysis efficiency (Moraes et al. 2020; Wijaya et al. 2014).

Acid hydrolysis of lignocellulosic biomass (pretreatment) promotes the extraction of pentoses and hexoses present in hemicellulosic fraction with efficiency over 80%, which is desirable to use them as raw material for xylitol bioconversion (Lavarack et al. 2002; Canilha et al. 2011). Silva et al. (2020) evaluated xylitol production from bagasse and straw and obtained 83% hydrolysis efficiency using diluted H_2SO_4 (100 mg/g dry matter, 1:10 solid–liquid ratio, 121 °C, 20 min), with low content of toxic compounds. Moraes et al. (2020) reported a diluted-acid pretreatment efficiency of 99% with minimal inhibitors content and energy consumption of 8.41 KW/Kg xylose for barley straw hydrolysis with 1.0% H_2SO_4 (w/v) at 120 °C, 40 min. Moreover, according to Dalli et al. (2017), the hydrolysis of poplar wood with diluted H_2SO_4 (1.75% (w/w) and 120 °C) resulted in a hydrolysate with three times higher xylose content and less by-product formation, improving xylitol production. Table 1.1 summarizes the xylose solubilization rates after diluting acid pretreatment with different feedstocks, acids, and other hydrolysis conditions.

As noted in Table 1.1, phosphoric acid has been less used than sulfuric acid, which can be 15 times more expensive than sulfuric acid (Siripong et al. 2016), resulting in

Table 1.1 Different acid hydrolysis of hemicellulose for xylose recovery

Feedstock	Hydrolysis parameters	Xylose recovery	References
Pinewood	4.49% phosphoric acid (H_3PO_4), 106.7 °C, 4.57 h, solution-to-feed ratio of (mL/g) 12.51	90.95%*	Cao et al. (2018)
Rice husk	2.60% sulfuric acid (H_2SO_4), 127 °C, 60 min, solid/liquid ratio (w/v) of 10	87.7%*	Temiz and Akpinar (2017)
Quercus mongolica (hardwood)	4.0% sulfuric acid (H_2SO_4), 121 °C, 105 min solid/liquid ratio (w/v) of 7	83.0%*	Jang et al. (2018)
Corn stover	0.7% hydrochloric acid (HCl) followed by wet milling could, 120 °C, 40 min, solid/liquid ratio (w/v) of 10	81.0%*	Liu et al. (2016)
Wheat straw	0.1 mol/L maleic acid ($C_4H_4O_4$), 150 °C, 40 min, 70 g/L solid/liquid ratio (w/v)	77.12%*	Liu et al. (2021)
Bioenergy sorghum	2% sulfuric acid (H_2SO_4), 120 °C, 5 min, 10% solids loading, after hydrothermal pretreatment and disk milling	64.9%*	Cheng et al. (2020)
Pine tree wood	5% sulfuric acid (H_2SO_4), 121 °C, 60 min, 1:10 solid/liquid ratio	92.0%*	Gonzales et al. (2016)
Empty palm fruit bunch	5% sulfuric acid (H_2SO_4), 121 °C, 30 min, 1:10 solid/liquid ratio	93.4%*	Gonzales et al. (2016)
Sugarcane straw	1% sulfuric acid (H_2SO_4), 121 °C, 20 min, 1:10 solid/liquid ratio	18.6 g/L	Hernández-Pérez et al. (2016)
Sugarcane bagasse and straw (1:1)	1% sulfuric acid (H_2SO_4), 121 °C, 20 min, 1:10 solid/liquid ratio	71.6 g/L**	Jofre et al. (2021)
Sugarcane bagasse and straw (1:1)	100 mg of H2SO4 per gram of dry matter, 1:10 solid–liquid ratio, at 121 °C for 20 min	18.76 g/L	Silva et al. (2020)
Rapeseed straw	2% sulfuric acid (H_2SO_4), 130 °C, 1 h, (10% w/v feedstock load)	10.30 g/L	López-Linares et al. (2018)

(continued)

Table 1.1 (continued)

Feedstock	Hydrolysis parameters	Xylose recovery	References
Corn cobs	0.5% nitric acid (HNO_3) 121 °C, 30 min, 1:10 of solid/liquid ratio	15 g/L	Kumar et al. (2018)

*Based on xylan
**After vacuum evaporation

a higher cost of the process. Nonetheless, its use is investigated for being associated with a lower generation of degradation compounds and minor corrosion in the reactors. On the other hand, Sulfuric acid is the most used in acid hydrolysis. It is the main method of effective depolymerization and enhancement of C5 fraction from lignocellulosic biomass due to its easy operation and low cost. However, its main disadvantage is the consumption of alkalis to neutralize the hydrolysate obtained (Loow et al. 2761; Girisuta et al. 2013; Norrahim et al. 2021; Akhlisah et al. 2021). Wan et al. (Wan et al. 2021) evaluated to overcome this problem by employing hydrolysis with solid acid sulfated zirconia. However, detoxification of hydrolysate continued to be recommended for its use as a fermentation medium for xylitol production. Acidic solids catalyzed hydrolysis has shown promise, emphasizing the use of heteropoly-acids, zeolites, mesoporous materials, sulfonated or modified with acid oxide derived from tungsten or zirconium (Amarasekara and Wiredu 2012; Guo et al. 2012; Lu et al. 2021). Their utilization has been gaining notability, as it provides easy separation from the final product, with the perspective of recovery and reuse of the catalyst (Suganuma et al. 2008; Lanzafame et al. 2012; Ye et al. 2021).

Detoxification of hydrolysate is necessary because during acid hydrolysis, even in low temperature and short time, the reactions release sugars, and other compounds, which are inhibitory to fermentative processes. Such compounds can be classified into four main groups: by-products derived from sugars (furfural originated from the degradation of pentoses and hydroxymethylfurfural, formed in the degradation of glucose), acetic acid (from acetyl groups present in hemicellulose fraction), lignin degradation products that include a wide variety of aromatic and polyaromatic compounds. Furthermore, the pretreatment using strong acids, high temperature, and pressure promotes the wear of the equipment and the presence of metal and minerals that can be inhibitors for microorganism metabolism (Palmqvist and Hahn-Hägerdal 2000; Silva et al. 2004, 2014; Alvira et al. 2010).

1.3.1.2 Organosolv

Organosolv is a pretreatment method that uses a solvent system for the deconstruction of lignocellulosic biomass using organic aqueous solvent (i.e., methanol, ethanol, and acetone), in a temperature range of 100–250 °C and must occur under controlled conditions due to the associated risk of volatile solvents (Raj et al. 2021a; Espinoza-Acosta et al. 2014; Zhao et al. 2009). As a result, monomers from hemicellulose and lignin are obtained dissolving in the liquid fraction, and the solid fraction obtained is rich in cellulose (Zhao et al. 2009; Teramura et al. 2018; Batista Meneses et al. 2020). In addition, high purity xylose and lignin are obtained during posterior treatment of liquid fraction (Zhang et al. 2016).

When alcohol is used as an organic solvent in lignocellulosic biomass at high temperatures and pressure, the hemicellulose and lignin will be fragmented and solubilized in the liquor (Wildschut et al. 2013). As a result, the availability of sugar monomers and degradation compounds may occur. Therefore, filtration is used to separate the cellulosic fraction from the black liquor. For solvent recovery, the liquid is evaporated, the concentrated fraction is mixed with acidic water so that the lignin is precipitated. A further filtration and distillation step are employed to separate components such as furfural, acetic acid, and xylose; from the hydrolysis of hemicellulose (Zhang et al. 2016). Thus, obtaining at the end of the process a pulp consisting mainly of cellulose, high-purity lignin, and an aqueous fraction containing the constituent monomers of hemicellulose and their degradation products (Wildschut et al. 2013; Rio et al. 2012).

The ethanol process was applied to fractionate wheat bran, evaluating the temperature (160–200 °C) and ethanol concentration (30–60%w/w) for 30 min, and conversion of hemicelluloses to xylose 60% was obtained (Reisinger et al. 2014). Previous research on wheat straw used a three-step aqueous pretreatment biorefining approach to pre-hydrolysis hemicellulose to sugars, (2) organosolv delignification, and (3) enzymatic hydrolysis of cellulose to glucose. Fractionating wheat bran using 60% ethanol, 10/1 (v/w) liquor/solid ratio, 220 °C and 60 min, 67% xylose yield xylose was obtained (Huijgen et al. 2012).

Organosolv pretreatment is commonly performed at harsh conditions (high concentrations of organic solvents (>50%) and high temperatures) and may be necessary to use a catalyst to obtain a higher xylose yield (Zhang et al. 2016; Huijgen et al. 2012; Meng et al. 2020). In addition, mineral acids such as hydrochloric acid, sulfuric acid, and phosphoric acid can accelerate the process of delignification and degradation of xylan. Still, organic acids (formic acid, oxalic, acetylsalicylic, salicylic acid) can also be used as catalysts (Zhao et al. 2009; Huijgen et al. 2012).

Teramura et al. (2016), aiming to reduce processing cost, evaluated fractioning sorghum bagasse with low solvent concentration. The authors reported effective biomass fractionation using 12.5% 1-butanol or 1-pentanol as the solvent, 1% H_2SO_4, 180 °C, 45 min. Teramura et al. (2018) also evaluated the solvent recovery using a nanofiltration membrane. They reported that this step could be beneficial for using the hemicellulosic hydrolysate in fermentative processes.

According to Zhang et al. (2016) and Salapa et al. (2017), especially in a process with not very high temperatures (<195 °C), ethanol and methanol could be chosen as a solvent, since they are more easily recovered, preferably ethanol, due to lack of toxicity. After treatment, the liquor (black liquor) is evaporated and condensed, recovering the alcohol solvent, and additional filtration is applied to recovering xylose.

Other possibilities also have been explored. For example, Huijgen et al. (2010) reported 82% hydrolysis of hemicellulose from wheat straw employing 50:50%

(w/w) acetone–water for 60 min at 205 °C. Furthermore, Yu et al. (2015) showed that 94.1% of xylan was converted in xylose from sugarcane bagasse pretreated with green liquor-ethanol combined with H_2O_2 without using xylanase.

1.3.2 *Hydrothermal Technologies*

Hydrothermal pretreatments are the most promising fractionation technologies for lignocellulosic biomass-based biorefineries (Díaz et al. 2021). They consist of a combination of physical and chemical principles operating in an aqueous medium under different temperature and pressure ranges. In addition, hydrothermal pretreatment is an effective technique for extracting hemicellulose fraction highlights: steam explosion and autohydrolysis (Beig et al. 2020; Sarker et al. 2021).

1.3.2.1 Steam Explosion

Steam explosion (SE) is a widely used method for fractionating lignocellulosic biomass and disrupts xylan structure. This method combines thermal, chemical, and mechanical effects on biomass. As a result, it is possible to obtain hemicellulosic hydrolysate, change cellulose crystallinity, and induce lignin transformations (Duque et al. 2016; Cebreiros et al. 2021).

The steam explosion system comprises a steam generator that feeds a reactor, which uses high pressure saturated steam at temperatures 160 and 260 °C and is maintained from seconds to several minutes (Biswas et al. 2015; Bhutto et al. 2017). Subsequently, a sudden depressurization occurs, the material is ejected from the reactor and recovered in the explosion tank (Jacquet et al. 2015).

The steam explosion occurs without a catalyst, promoting acetyl groups' liberation present in the biomass (Carvalheiro et al. 2008). As a result, cleavage of glycosidic bonds and hemicellulose solubilization occurs at lower temperatures (190 °C, 10 min) or short residence time (270 °C, 1 min). As a result, a liquid fraction and insoluble solids from the pretreated material. The liquid, or hydrolyzed, contains mainly hemicellulose sugars solubilized and almost also degradation products generated during pretreatment (Carvalheiro et al. 2008; Mosier et al. 2005).

Some factors can influence the efficiency of the steam explosion, the main ones being particle size, temperature, moisture, and residence time (Duque et al. 2016). The high residence time favors complete hydrolysis of the hemicellulosic fraction and contributes to fermentation (Jacquet et al. 2015). However, the oligosaccharides and monosaccharides generated may undergo dehydration, fragmentation, and condensation reactions. The products obtained from these reactions are furfural, hydroxymethylfurfural, levulinic and formic acid, and aromatic compounds and are generated due to increased retention time (Jacquet et al. 2015). All these degradation products are fermentation inhibitors, so they must be minimized. Thus, high temperatures (220 °C) decrease overall carbohydrate yield due to the parallel reactions

(Ruiz et al. 2008). In addition, particle size and raw material moisture influence heat transfer into of reactor. For example, small biomass particles have better and faster heat transfer, avoiding surface overheating, production of degradation products, and incomplete hydrolysis (Duque et al. 2016; Brownell et al. 1986).

Bonfiglio et al. (2021) reported that to obtain hydrolysates of switchgrass and *Eucalyptus globulus* biomasses rich in sugars, avoiding a subsequent concentration step before fermentation, the better conditions were 200 °C and 10 min residence time, and the hydrolysates obtained were efficient for xylitol production.

An alternative to improve the solubilization of hemicellulose is to use a combination of acid catalysts such as SO_2, H_2SO_4, H_3PO_4, and steam explosion process (Duque et al. 2016; Devi et al. 2021). Furthermore, when the steam explosion is made with softwood, it is essential to combine this pretreatment with the impregnation of biomass with acids. This combination helps to reduce the residence time and temperature of the process and produce low degradation products (Devi et al. 2021; Behera et al. 2014).

Wang et al. (Wang et al. 2015) proposed a pretreatment of corn straw combining impregnation of biomass with 1% H_2SO_4 followed by a steam explosion at 180 °C for 9 min and obtained a hydrolysate xylose-rich and with low content of furfural and 5-HMF, that was fermented, without detoxification step, with effective xylitol production by *Candida tropicalis*.

Walker et al. (2018) also evaluated the use of steam explosion pretreatment (12 bar(g), 3 min, 1.2% H_3PO_4, and 500 g substrate) for different feedstock (wheat straw, corn stover, Miscanthus, and willow). They yielded up to 94% xylose release with minimal fermentation inhibitor production, allowing xylitol production without detoxification, by *Scheffersomyces shehatae*. Still, the hydrolysates were concentrated to increase xylose concentration.

The use of H_3PO_4, although less common, has been pointed out more advantage over the H_2SO_4 as a catalyst for SE, to be less toxic and corrosive, contributing to reducing the capital cost of the plant (Duque et al. 2016; Geddes et al. 2011). Furthermore, it is possible to recover phosphoric acid and use it as a source of nutrients (López-Linares et al. 2013).

1.3.2.2 Auto-Hydrolysis

Auto-hydrolysis or hot water pretreatment (LHW) is a pretreatment process without using chemicals, in which compressed liquid water is used to hydrolyze the hemicellulose fraction. This process has a mechanism similar to acid hydrolysis, and the reaction conditions and decomposition mechanism have been widely studied (Yu et al. 2012; Zhuang et al. 2012). In this method, the self-ionization of water takes place, releasing hydrogen ions, promoting the depolymerization of hemicellulose fraction. Besides, the acetyl groups, present in lignocellulosic biomass, also act as catalysts, improving the reaction mechanism (Yu et al. 2012; Zhuang et al. 2012; Gullon et al. 2012; Serna-Loaiza et al. 2021; Carvalheiro et al. 2009).

Autohydrolysis allows the selective hemicellulose extraction at high temperatures, without catalysts, and without lignin solubilization, obtaining a liquid fraction rich in soluble saccharides and a solid fraction rich in cellulose and lignin.

However, studies show that hemicellulosic hydrolysates obtained by autohydrolysis of biomass are complex. Such hydrolysates contain a mixture of oligosaccharides (60–80%), monosaccharides (especially pentoses, 10–15%). They can also present inhibitory compounds of microbial metabolisms, such as acetic acid and some furans (10–15%), such as furfural and 5-hydroxymethylfurfural, which can further develop decompose, producing formic and levulinic acid (Yu et al. 2012; Zhuang et al. 2012; Carvalheiro et al. 2009; Kuhad et al. 2010; Amendola et al. 2012), requiring a further detoxification step.

The factors that affect autohydrolysis are temperature associated with heating and cooling profile, reaction time, and liquid–solid ratio. The most commonly used temperatures are in the ranges of 150–250 °C, under high pressures to keep water in a liquid state, with S/L ratios of 2–100 (w/w) and reaction times of seconds to hours. Under these conditions, the extraction processes of the hemicellulosic fraction present 55–84% recovery (Gírio et al. 2010; Gurgel et al. 2014; Garrote et al. 1999; Walsum 2001).

This procedure is economical and environmentally correct when comparing autohydrolysis with other biomass pretreatment methods (Carvalheiro et al. 2009). In addition, LHW also has other advantages such as low energy requirement and high yields, few corrosive problems due to its moderate pH, and can be used in a variety of lignocellulosic biomasses, such as sugarcane bagasse, grape stalks, corn cobs, sweet sorghum bagasse. Therefore, it could be used in biorefineries, contributing to less waste generation, and promoting the efficient use of raw materials, minimizing environmental impacts (Carvalheiro et al. 2008; Amendola et al. 2012; Gurgel et al. 2014; Garrote et al. 1999; Amiri and Karimi 2015; Khalili and Amiri 2020; Rivas et al. 2002).

Some authors have reported the combination of LHW with other pretreatments, mainly using acid, trying to increase xylose extraction. Brandenburg et al. (2016) pretreated birch wood chips with hot water extraction (165 °C, 90 min) followed by acid hydrolysis (4% H_2SO_4, 120 °C, 60 min) and obtained a hydrolysate with 45.06 g/L xylose, 0.46 g/L glucose, 13.07 g/L acetic acid, and 4.70 g/L furfural. Kim et al. (2015) reported the use of acetic acid in the hydrothermal fractionation for method sugar recovery from empty fruit bunches, achieving 50.7% sugars (xylose, mannose, and galactose) recovery yield (and 1.0 g/L furfural) with 6.9 wt.% acetic acid, at 170 °C, 18 min.

Lyu et al. (2018) also evaluated LHW but used mixed acid (lactic acid + acetic acid) to treat wheat straw and corn straw. The authors reported a mixed acid synergistic catalysis (3 wt%) on strengthening LHW pretreatment at 180 °C for 60 min, obtaining a C5 sugars yield of 93.83%.

Such results encourage new studies to adjust this pretreatment to become a viable alternative in biorefineries.

1.3.3 *Physical Pretreatments*

Physical pretreatments include mechanical operations, irradiations, and ultrasonic processes (Bhutto et al. 2017; Subhedar and Gogate 2016; Sun et al. 2016). However, these methods alone cannot obtain oligomers and monomers to be hydrolyzed and used as substrates in biotechnological processes, so they are usually combined with another type of pretreatment.

1.3.3.1 Ultrasound

The association of conventional pretreatments with the application of ultrasonic irradiation is an excellent approach to intensify the deconstruction of lignocellulosic biomass. Furthermore, studies have addressed that the application of ultrasound associated with another pretreatment significantly improves the yield of polysaccharides from biomass (Flores et al. 2021; Minjares-Fuentes et al. 2016).

The treatment is based on the formation of cavities/bubbles micro size, which results from the propagation of ultrasound waves in a liquid medium. This phenomenon is called cavitation, which results in physical and chemical effects in liquid solutions (Ravindran et al. 2017; Zheng et al. 2014; Leong et al. 2011). The physical impact caused is a collapse of the cavitation bubbles formed and intensified the fluids transport and furthered the dissolution of solid particles. In addition, this result will produce chemical changes through the formation of free radicals (Flores et al. 2021; Gallo et al. 2018). The combination of these effects generates impact and destruction of the plant cell wall.

The result of these cavities results in high temperatures and pressure gradients in specific regions. This entire process lasts for a short period, creating a hot-spot effect in the liquid (Kentish and Ashokkumar 2011). In addition, the high energy generated in the cavitation regions results in a significant collapse in the lignocellulose matrix, which leads to substantial changes in morphology, with a high contact surface, being advantageous for the intensification of other pretreatments due to the greater surface area that results in better mass transfer (Yan et al. 2021).

Based on that, the energy generated by the ultrasound process has been reported as an efficient method for enhancing several types of pretreatment, which is associated with the turbulence of the process, making the oligomers that compose the biomass more available for subsequent operations, and a decrease in lignin content (Bussemaker and Zhang 2013). Besides, ultrasound has the advantage of removing inorganic materials associated with biomass, especially when using low operating frequencies (Yunus et al. 2010).

Thus, the associated use of ultrasound with other technologies, such as chemical treatment to fractionate the biomass and obtain the hemicellulosic fraction, can increase yields and generate less waste and toxic by-products.

The application of ultrasound beyond the cavitation zones results in intense turbulence, shock waves, and shear forces with high pressures and temperatures for a short period due to the bubble collapse (Subhedar and Gogate 2016; Kentish and Ashokkumar 2011). Furthermore, cavitation results in a mechanical effect called microjets, one of the main mechanical effects. Bubbles will collapse asymmetrically in the solid surface, and a formed liquid jet implodes on the particle's surface, decreasing the size of these particles (Leong et al. 2011; Singh et al. 2014). On the other hand, micro-streaming is a consequence of cavitation in the presence of suspended powder. Thus, it can be assumed that these mechanical effects act in synergy, whereas microjets reduce particle size and disperse particles facilitating the dissolution.

Hence, the technique employed results in a mechanochemical effect (Rooze et al. 2013) that will accelerate the extraction of organic compounds from the lignocellulosic biomass by causing a kind of cell wall rupture, increasing the contact surface and mass transfer cell content (Yuan et al. 2010; Luo et al. 2014). Firstly, the radicals formed during cavitation will cleavage the process of lignin and xylan networks, together with the shear forces of the ultrasonic mixture (Ong and Wu 2020).

This combination of factors leads to significant physical and morphological changes, like an increase in surface area, a reduction in the degree of polymerization, which will result in a greater degree of intensification of the associated treatment, bringing more expressive yields as well as greater use of lignocellulosic biomass (Yunus et al. 2010; Sun et al. 2021). Table 1.2 shows various ultrasound-assisted pretreatment for the many lignocellulosic biomass.

A study by Yunus et al. (2010) studied the effect of dilute sulfuric acid with ultrasonic pretreatment of oil palm empty fruit bunch. The ultrasound treatment showed to be more effective for recovery xylose (58%) of the hemicellulosic fraction when compared with the condition in the absence of ultrasonic (22%). Yang and Wang (2019) also used the combination of acid treatment with ultrasound to improve biomass pretreatment efficiency for fermentable sugars recovery. This ultrasound-assisted dilute acid pretreatment also reduces the required amount of acid for the pretreatment, which is more environmentally attractive.

Minjares-Fuentes et al. (2016) realized an ultrasound-assisted alkaline treatment to extract pentoses from grape pomace; the maximum yield extract was 8.6%. Thus, the author concluded that it is possible to obtain the hemicellulose fraction using low concentrations of KOH and short extraction times when associated with ultrasound; however, it is still a low performance when considered a large-scale application.

The frequency used in ultrasound is one of the main parameters influencing the physicochemical changes in biomass fiber. It is generally used in the range of 16–100 kHz; the increased frequency is necessary to create the cavitation zones, in addition to causing the flow of liquid required and turbulence (Luo et al. 2014). Another parameter is the time that the ultrasonic is applied. Thus, it can be established that the processing time will depend on the biomass recalcitrance used, mass transfer resistances, and the desired fractionation (He et al. 2017). Studies used for the recovery of hemicellulose state that, in addition to a time established as optimal, excessive severity in the use of ultrasound can result in an unchanged fractionation or even be harmful to the pretreatment process, and consequently results in an aggregate of particles reducing the surface area of the biomass (Loow et al. 2016; Ong and

Table 1.2 Overview of ultrasound-assisted pretreatment techniques combined for recovery xylose

Application	Ultrassonic Configuration	Catalyst	Main observation	References
Ultrasound-assisted alkaline pretreatment for the extraction of hemicelluloses from grape pomace	37 kHz, 140 W, 20 ± 2 °C	KOH	Ultrasound treatment yielded 8.6% hemicelluloses, and reaction time is reduced from 15 to 4 h	Minjares-Fuentes et al. (2016)
Ultrasound alkali pretreatment of sugarcane bagasse and ultrasound acid hydrolysis of solid fraction from the production of sugar monomers	24 kHz, 200 W, 50 °C	H_2SO_4	7.34 g/L of xylose was observed in the liquid fraction	Velmurugan and Muthukumar (2011)
Ultrasound-assisted dilute-acid pretreatment to the grass	260 W, 30 min	HCl	Xylose and arabinose concentrations were improved to 93.7%	Yang and Wang (2019)
Acid hydrolysis following ultrasound pretreatment of oil palm empty fruit bunch	2 kW 20 kHz, 45 min (90% power)	H_2SO_4	Maximum xylose yield of 58% compared to 22% without ultrasound. Ultrasound removed the silica on the surface of the fibers	Yunus et al. (2010)
Sequential ultrasound-assisted deep eutectic solvent pretreatment of oil palm fronds	70% amplitude, 30 min	Deep eutectic solvent	Reduce lignin content to 14.01% while improving xylose recovery by 58%	Ong et al. (2019)
Eucalyptus grandis was delignified with sodium chlorite in an acidic solution adjusted by acetic acid combined with KOH solution under ultrasound	21 kHz, 180 W, 30 min	5% KOH at 50 °C	The highest yield of hemicelluloses extraction (95.2%) with 83% of xylose	Xu et al. (2018)

Wu 2020). Considering the economic factors associated with this type of treatment, it is important to understand the optimized conditions of ultrasonic irradiation. The increase in time can result in no advantage for the process and lead to energy waste due to a large amount of energy consumed (García et al. 2011; Qian et al. 2021).

It is important to emphasize that ultrasonic techniques generate energy and the liquid medium undergoes an increase in temperature, requiring strict control of this parameter. (Louis and Venkatachalam 2020). At higher temperatures, the cavitational intensity will decrease due to the solvent vapor pressure, which will increase; the bubbles generated during the process are less intense (Subhedar and Gogate 2016), thus reducing the desired effect. Based on that, optimizing the temperature is also a necessary process to obtain satisfactory results. In general, the influence of temperature on the techniques in which the ultrasonic method is used will depend on the type of associated pretreatment and other process parameters (Flores et al. 2021).

The technique associated with ultrasound brings the advantages of potentializing the pretreatment, higher yields at lower process temperatures, a shorter operating time when compared to conventional pretreatment, and the possibility of working using a reactor operating at atmospheric pressure (Flores et al. 2021; Minjares-Fuentes et al. 2016; Ong and Wu 2020). In this way, promoting a more ecological and safer process, which requires lower operating temperature and pressure, and energy demand (Bhutto et al. 2017).

Moreover, it has been discussed that the association of pretreatment steps with ultrasonic irradiation (Minjares-Fuentes et al. 2016; Ong and Wu 2020; Ríos-González et al. 2021) significantly improves the yield of hemicellulose extraction during the deconstruction of lignocellulose biomass compared to treatment without the use of ultrasound, moreover to lower temperatures and shorter incubation time.

1.3.3.2 Ultrasound-Assisted Dilute Acid Treatment

As said previously, dilute acid pretreatment is a widely used method for fractionating lignocellulosic biomass; however, an ultrasound-assisted approach is still very limited (Loow et al. 2016). The most evident aspect of these associated pretreatment steps is the significant improvement in the xylan recovery rate and yield and reduced extraction time compared to conventional treatment (Flores et al. 2021). Combined pretreatment results in inhibitors production, for example, acetic acid that comes from acetyl groups linked to a hemicellulosic fraction. The accumulation of acetic acid above 0.95 g/L (Velmurugan and Muthukumar 2011) is an inhibitor of microbial growth. It acts by entering through cell membranes and a consequent decrease intracellular pH, thus resulting in metabolic alteration. Therefore, the highest production of acetic acid is generally higher with an increasing concentration of acid employed.

As regards furfural, it is derived from dehydration reactions of xylose (Lee et al. 2019). Therefore, high process temperatures and prolonged reaction time positively contribute to the accumulation of furfural, affecting the growth and gene expression of microorganisms (Velmurugan and Muthukumar 2011; Rao et al. 2016).

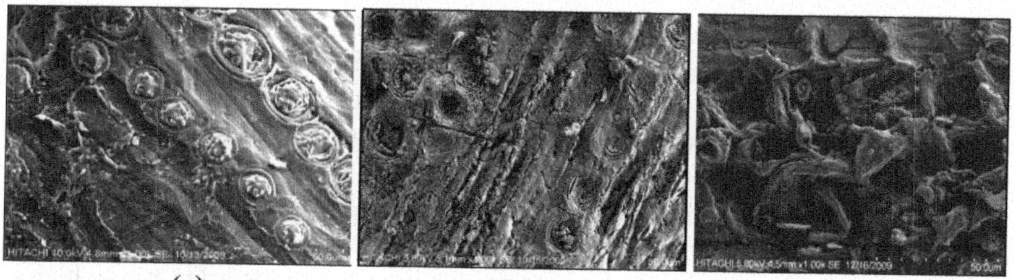

(a) (b) (c)

Fig. 1.3 Scanning Electron Microscopy images of untreated OPEFB Fibre, **a** acid hydrolysis at 140 °C without ultrasonic pretreatment **b** and acid hydrolysis at 140 °C with ultrasonic pretreatment (**c**). Reprinted from Bioresource Technology, 101/24, Yunus, R., Salleh, S. F., Abdullah, N., Biak, D. R. A., Effect of ultrasonic pretreatment on low-temperature acid hydrolysis of oil palm empty fruit bunch, 9792–9796, Copyright (2010), with permission from Elsevier (Yunus et al. 2010)

Based on that, the possibility of using milder conditions arising from the association with ultrasound will positively contribute to a lower formation of furfural, hydroxymethylfurfural, and acetic acid.

Yunus et al. (2010) studied the effect of ultrasonication on the oil palm empty fruit bunch (OPEFB) fiber before acid hydrolysis. In the absence of ultrasonic pretreatment, the yield of xylose was 22%. With ultrasonic acid hydrolysis, the yield was 58% at 140 °C. Regarding the morphological changes observed on the surface of the fibers (Fig. 1.3), it is possible to detect the presence of round shaped silica on the surface of the sample untreated, beyond an organized and more uniform appearance when compared to the others (Fig. 1.3a). Even with the acid hydrolysis step without ultrasound, it is possible to observe silica bodies are still intact, with the presence of cracks and holes, besides an interrupted appearance and a rough surface (Fig. 1.3b); on the other hand, the combination of acid hydrolysis with the ultrasonic application shows a positive and evident impact on the morphology of the fibers and the yield of sugar due to higher surface area and porosity. Furthermore, the release of silica bodies allows exposure of the hemicelluloses with enhanced acid accessibility.

The use of diluted acid is more common when compared to concentrated acid as they are less corrosive to the reactor used. The ultrasound-assisted acid hydrolysis approach typically uses inorganic acid catalysts that hydrolysis the ether linkages present in the biomass. In addition, it has attracted interest in reporting the lower release of inhibitors generated in the process (Ong and Wu 2020).

1.3.3.3 Cold Alkaline Extraction (CAE) Treatment Assisted Ultrasound

Many pretreatments have been employed to isolate the hemicellulosic streams of plant biomass; among these, the use of alkalis can be easily implemented and offers an attractive cost–benefit ratio (Xu et al. 2018; Bian et al. 2012). The cold alkaline extraction (CAE) or low-temperature alkali treatment (LTA) is a pretreatment that has been proposed and is based on the principle of extracting the hemicellulosic fraction at low temperatures, between 20 and 50 °C (Minjares-Fuentes et al. 2016; Carvalho et al. 2016), to avoid a high degree of delignification and is used different alkali concentrations. At the end of the process, hemicelluloses are regenerated, which can be hydrolyzed into their constituent monomers and widely used to synthesize or produce biomaterials and biochemicals (Subhedar and Gogate 2016).

The treatment in an alkaline medium causes swelling in the cell wall, which leads to breakage of the hydrogen bonds present between cellulose and hemicellulose and the removal of the hemicellulose fraction for the liquor in polymeric form, which is an advantage without substantial degradation of oligomers constituents (Bian et al. 2012; Carvalho et al. 2016). Usually, the higher the alkaline charge used, the greater the removal of the xylan. This pretreatment is advantageous due to the relatively milder operational conditions since at low temperatures, the use of alkali does not act as a delignifying agent, but for xylan recovery (Quintero et al. 2021), the generation of a smaller amount of degradation products that affect the fermentation process besides breaking up the macromolecular structure of lignin (Sindhu et al. 2015; Li et al. 2018), showing a potential strategy for pretreatment of different biomass.

One factor that makes alkali pretreatment attractive is lower process temperature and pressure (Bian et al. 2012) compared to other techniques. These are important parameters when large-scale implementation is desired since the use of high temperatures increases processing costs.

Based on that, Xu et al. (Xu et al. 2019) used a pretreatment step ultrasonic-assisted NaOH to deconstruct sugarcane bagasse. The hydrolysate was employed to produce xylitol, and a yield of 62.9 g/L was achieved.

Furthermore, the alkaline extraction of hemicellulose can fractionate the heteropolysaccharides of biomass into three main streams with low physical and chemical change, which allow their use in several supply chains (García et al. 2013). For example, the residue from the process composed of a more significant amount of cellulose can be hydrolyzed and used for the production of C6 derivatives (Sun et al. 2016; López-Linares et al. 2020), alkali-soluble lignin can be used to produce phenolic and burned to provide energy (Radhakrishnan et al. 2021), and isolated hemicellulose can be used to obtain several products derived from this fraction, such as xylose, xylitol, and furfural (Hernández-Pérez et al. 2016; Alonso et al. 2013). On the other hand, a disadvantage of the process is the generation of salt residues that are not recoverable or can be incorporated as salts in the biomass during pretreatment (Moraes et al. 2020; Peleteiro et al. 2016).

Some biomass is more resistant to CAE than grasses (Bian et al. 2012), for example. Thus, it is essential to associate it with other additional pretreatments, such as ultrasound and comminution mechanical (Minjares-Fuentes et al. 2016; Xu et al. 2018; Yuan et al. 2015); therefore, the synergistic behavior allows better performance with greater extraction of sugars.

Xu et al. (2018) evaluated the effect of ultrasound and physicochemical properties of the hemicelluloses from *Eucalyptus grandis*. The process used in the first stage sodium chlorite at pH 3.8 (adjusted with acetic acid) and neutralized and precipitated with ethanol after 5% KOH solution under ultrasound. The application of ultrasound led to an increase in the yield of the hemicellulosic fraction obtained from 75.6 to

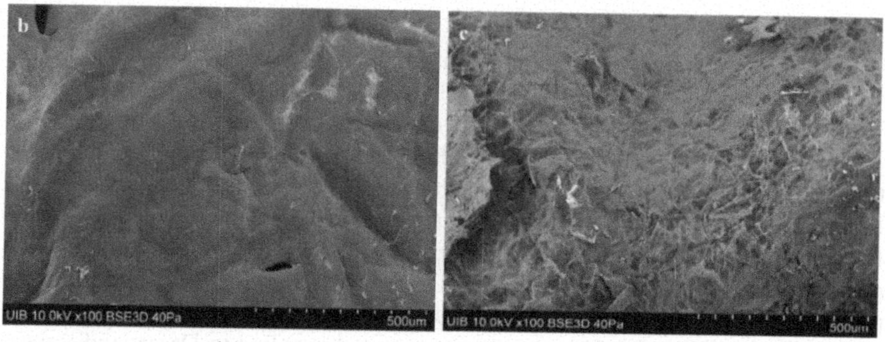

(a) (b)

Fig. 1.4 Scanning Electron Microscopy images of biomass treated with KOH 2 M during 60 min at 20 °C without ultrasound **a** and using ultrasound assistance (**b**). Reprinted from Carbohydrate Polymers, 138/15, R. Minjares-Fuentes, A. Femenia, M. C. Garau, M. G. Candelas-Cadillo, S. Simal, C. Rosselló, Ultrasound-assisted extraction of hemicelluloses from grape pomace using response surface methodology, 180–191, Copyright (2016), with permission from Elsevier (Minjares-Fuentes et al. 2016)

95.2%; such behavior was attributed to the cavitation phenomenon that intensified the mass transfer and improved the access of the alkaline solution to the hemicellulosic fraction.

Minjares-Fuentes et al. (2016) also observed that the ultrasound-assisted alkaline treatment was significantly more efficient in removing hemicellulose than conventional treatment. The maximum yields obtained with 2 M and 4 M KOH were ~7.8% and ~8.6%, respectively, were obtained after four h; while similar results using the same conditions, but without ultrasound application, were only obtained after 15 h. Figure 1.4 revealed that the surface morphology the surface of grape pomace before and after ultrasonic-assisted alkaline treatment. It is observed that disruption of the surface significantly liberates fermentable sugars of the hemicellulosic fraction.

1.3.3.4 Mechanical Comminution

Comminuting lignocellulosic biomass is a physical pretreatment technology used to reduce particle size, reduce polymerization, increase transfer efficiency of mass and heat during subsequent processes, and the specific surface area of biomass, improving raw material handling and next treatment of the polymers comprising cell walls (Cheng and Timilsina 2011; Bychkov et al. 2012; Rodiahwati and Sriariyanun 2016). Usually applied before other pretreatment methods or combined with other pretreatment technologies, synergistic interactions with increased process efficiency and energy and cost savings.

The process of mechanical comminution combines steps of chipping, grinding, and milling, which will reduce the size of the raw material, in the case of chipping between 10–30 mm and 0.2–2 mm after milling or grinding (Zhao et al. 2012).

Due to the high energy consumption associated with the comminution step, one must consider the variety of techniques that can be used, classified based on the type of mechanical action on the particles, where most devices used to involve different mechanisms and the properties of the raw material should be considered.

Mechanical comminution is considered an environmentally friendly technology. It does not require the use of chemical catalysts, and the resulting pretreated material is free of degradation compounds that are inhibitory in the fermentation process (Brodeur et al. 2011; Zhang et al. 2019). Besides, it is crucial for converting biomass into a biobased product through subsequent treatments (Sun et al. 2016). The increase in the surface area resulting from particle size reduction will directly influence the hydrolysis efficiency of biomass constituents. According to Dasari and Berson (Dasari and Berson 2007), reducing that particle size 590–850 to 33–75 μm doubled sugar conversion rates.

The major impacts associated with the comminution method refer to the disturbance generated by the inherent ultrastructure of lignocellulosic biomass, considerable surface increase, an improvement that will facilitate the access of catalysts and other pretreatment agents, a decrease in the degree of polymerization of the material, reducing the crystallinity. Mills machines commonly used for such purposes include ball, vibro-mill, hammer, knife, two-roll, and attrition mills (Bhutto et al. 2017).

The main disadvantage associated with this pretreatment is the energy demanded, which is intensified depending on the final particle size and the biomass characteristics; due to this, it is considered unfeasible in a large-scale implementation (Moreno, et al. 2019; Alvira et al. 2010; Bhutto et al. 2017). Furthermore, from an economic viewpoint, the mechanical energy demanded in physical processes is expensive energy, requiring the development of combined technologies for are used efficiently and allow the recovery of fermentable sugars.

The comminution is typically combined with chemical processes, thus allowing the hydrolysis of the constituent oligomers of the biomass. Using these combined treatments can be called mechanochemical. However, there are still called activator mills, which allow carrying out chemical reactions directly during mechanical treatment (Lomovsky et al. 2016).

Choosing the equipment that will perform the technique, the moisture content of the biomass in question, and the type of biomass must be considered. Lignin, for example, is an important constituent that will determine the mechanical aspects of plant fibers and tissues in plants (Bychkov et al. 2012). Table 1.3 summarizes the parameters of some processes, advantages, and disadvantages in different technologies.

For comminating dry biomass (10–15% wet basis) (Table 1.3), commonly used two-roll, hammer, attrition, and knife mills. In comparison, colloid mills and extruders are used for comminating wet materials with moisture contents of more than 20% (wet basis). And in the case of dry or wet materials, the ball and vibrio mills are used (Rodiahwati and Sriariyanun 2016).

Table 1.3 Summary of parameters of the process, advantages, and disadvantages in different pretreatment of milling devices

Attribute	Milling devices		
	Hammermill	Ball mills	Vibrocentrifugal mill
Biomass moisture	Moisture contents of up to 10–15% (wet basis) (Zheng et al. 2014; Fu et al. 2018)	Irrespective of moisture content of biomass (Zheng et al. 2014; Fu et al. 2018)	Used for either dry or wet materials (Zheng et al. 2014)
Mechanical stress	Crushing, collision, and inter-particle impacts (Fu et al. 2018)	Low rotation: abrasion High rotation: impact action (Lomovsky et al. 2016)	Splitting and delamination of the large fiber bundles along with and across the fiber direction (Fu et al. 2018)
Advantages	Relatively inexpensive, easy to operate, and produce a wide range of particle sizes (Bhutto et al. 2017). High-efficiency milling for the initial size reduction (Fu et al. 2018)	Ball milling changes the ultrastructural properties of materials like crystallinity and particle size (Liu et al. 2017)	Destroys the fibrous structure and results in the formation of amorphous microparticles (Fu et al. 2018). Effectively break chemical bonds between lignin and hemicellulose if decreasing particle size (Liu et al. 2017)
Disadvantages	High specific energy consumption (90–130 kWh t^{-1}) (Fu et al. 2018) and the sharp decrease in their productivity with a reduction of the mesh size of the screen (Lomovsky et al. 2016). Besides, lengthy biomass may not be directly fed into hammer mills (Bhutto et al. 2017)	Energy-intensive and low productivity compared to the other milling processes (Lomovsky, et al. 2016; Fu et al. 2018)	High temperature in the milling zone leads to the decomposition of organic substances (Lomovsky, et al. 2016)

1.4 Challenges and Prospects

One of the significant challenges in biomass valorization for its integral implementation in biorefineries is pretreatment. Pretreatment is well recognized as an indispensable unit operation to recover the xylose-rich fraction. It will directly influence the total cost of the bioprocess and determine the toxic degradation products from this step, which affect the performance of the microorganism and other process variables. Furthermore, pretreatment of biomass is one of the main bottlenecks that can enable an economical and competitive bioprocess with a chemical process to produce xylitol.

Regarding the economic aspects, several factors, directly and indirectly, will influence this one. For example, energy requirements are a limiting factor for the bioprocess to be commercialized. The choice of raw material also affects this aspect, as it will determine the energy required for processing. There is a marginal variation in the cost of various types of pretreatments, and it will impact the cost of bioprocess, whether conventional or hybrid. For example, the lower operational cost of the pretreatment step can be compensated in the recovery and purification step (Sun et al. 2016).

The inhibitors produced during the pretreatment stages represent a problem for the downstream and fermentation steps. These components combine with hydrolysis products and negatively impact the bioprocess (Alvira et al. 2010; Teramura et al. 2018; Beig et al. 2020).

An alternative to reduce these shortcomings is that the combination of steps arises from filling an isolated treatment's deficiency. However, it is worth noting that the effects obtained in the pretreatment steps and physicochemical changes will depend on the type of biomass investigated.

A new possibility that has been evaluated is the use of enzymatic hydrolysis of hemicellulose. Jamaldheen et al. (2019) reported the enzymatic hydrolysis of hemicellulose from Finger millet straw pretreated with NaOH (1% (w/v) combined with oven heating (120 °C, 20 min) for production of xylose. Endo-1,4-β-xylanase (55 °C) and exo-1,4-β-xylosidase (37 °C) were used, with a 24.7% conversion of xylan to xylose.

Ontañon et al. (2019) evaluated the extraction of xylose and xylooligosaccharides from hemicellulose of barley straw using an enzymatic cocktail composed of extra and intracellular xylanases of Cellulomonas strains. The authors concluded that xylanases were efficient in the deconstruction of hemicellulose. The hydrolysis results can boost this new technology for high-value products production, such as xylitol.

Another possibility that needs to be further studied is the one reported by Silveira et al. (2018), in which steam explosion is combined to enzymatic hydrolysis. These authors evaluated different steam explosion conditions for sugarcane bagasse pretreatment. The hydrolysate obtained was submitted for enzymatic hydrolysis using three commercial enzymes in different ratios, which resulted in 85% yield sugars and acetic acid as the main inhibitory component.

Of course, that multiple factors influence this process and need to be elucidated, as mentioned by Dutta and Chakraborty (2019, 2016), that the use of enzyme cocktail for hemicellulose hydrolysis could improve sugar yield and reduce energy costs, as long as synergistic effects between enzymes are taken into account because such effects could affect the timescales of the various transport and reaction processes.

According to Ostadjoo et al. (2019), using xylanases is a promisor and innovative way to hydrolyze hemicellulosic fractions from biomass. When this is achieved more efficiently, there will be no need for chemical pretreatment or bulk water, reducing waste.

Exploring enzymatic hydrolysis of hemicellulose certainly provides novel insights into the best combination of pretreatment that can overcome the problems due to recalcitrant cell wall of the biomass, allowing a high xylose recovery, improving factory design and industrial bioconversion processes into integrated biorefineries.

Another important aspect that must be considered, especially when one intends to think that xylitol production must be an environmental-friendly process, is about the generation of wastes and carbon and water use footprint.

When the combination of different pretreatment steps occur, the increase in unit operations, in addition to making the process more expensive due to the demand for resources, results in large amounts of wastewater, formation of toxic compounds that need detoxification steps, and neutralization of effluents, processes that can lead to loss of soluble sugars. Therefore, it is essential to study new technologies that will result in a high xylose recovery that provides high xylitol production, identifying hazard potential and environmental performance, to become the entire process economically and environmentally attractive.

References

Akhlisah ZN, Yunus R, Abidin ZZ, Lim BY, Kania D (2021) Pretreatment methods for an effective conversion of oil palm biomass into sugars and high-value chemicals. Biomass Bioenergy 144:1

Alokika A, Kumar A, Kumar V, Singh B (2020) Cellulosic and hemicellulosic fractions of sugarcane bagasse: potential, challenges and future perspective. Int J Biol Macromol 169:564

Alonso DM, Wettstein SG, Mellmer MA, Gurbuz EI, Dumesic JA (2013) Integrated conversion of hemicellulose and cellulose from lignocellulosic biomass. Energy Environ Sci 6:76

Alvira P, Tomás-Pejó E, Ballesteros M, Negro MJ (2010) Pre-treatment technologies for an efficient bioethanol production process based on enzymatic hydrolysis: a review. Bioresour Technol 101:4851

Amarasekara AS, Wiredu B (2012) Aryl sulfonic acid catalyzed hydrolysis of cellulose in water. Appl Catal A-Gen 417:259

Amendola D, De Faveri DM, Egües I, Serrano L, Labidi J, Spigno G (2012) Autohydrolysis and organosolv process for recovery of hemicelluloses, phenolic compounds and lignin from grape stalks. Bioresour Technol 107:267

Amiri H, Karimi K (2015) Autohydrolysis: A promising pretreatment for the improvement of acetone, butanol, and ethanol production from woody materials. Chem Eng Sci 137:722

Arevalo-Gallegos A, Ahmad Z, Asgher M, Parra-Saldivar R, Iqbal HMN (2021) Lignocellulose: a sustainable material to produce value-added products with a zero waste approach—A review. Int J Biol Macromol 99:308

Arora A, Nandal P, Singh J, Verma ML (2020) Nanobiotechnological advancements in lignocellulosic biomass pretreatment. Mater Sci Energy Technol 3:308

Batista Meneses D, Montes de Oca-Vásquez G, Vega-Baudrit JR, Rojas-Álvarez M, Corrales-Castillo J, Murillo-Araya LC (2020) Pretreatment methods of lignocellulosic wastes into value-added products: recent advances and possibilities. Biomass Convers. Biorefinery

Behera S, Arora R, Nandhagopal N, Kumar S (2014) Importance of chemical pretreatment for bioconversion of lignocellulosic biomass. Renew Sustain Energy Rev 36:91

Beig B, Riaz M, Naqvi RS, Hassan M, Zheng Z, Karimi K, Pugazhendhi A, Atabani AE, Chi TL (2020) Current challenges and innovative developments in pre-treatment of lignocellulosic residues for biofuel production: a review. Fuel 287:1

Bhutto AW, Qureshi K, Harijan K, Abro R, Abbas T, Bazmi AA, Karim S, Yu G (2017) Insight into progress in pre-treatment of lignocellulosic biomass. Energy 122:724
Bian J, Peng F, Peng XP, Xu F, Sun RC, Kennedy JF (2012) Isolation of hemicelluloses from sugarcane bagasse at different temperatures: structure and properties. Carbohydr Polym 88:638
Biely P, Singh S, Puchart V (2016) Towards enzymatic breakdown of complex plant xylan structures: state of the art. Biotechnol Adv 34:1260
Biswas R, Uellendahl H, Ahring BK (2015) Wet explosion: a universal and efficient pretreatment process for lignocellulosic biorefineries. Bioenergy Res 8:1101
Bonfiglio F, Cagno M, Yamakawa CK, Mussatto SI (2021) Production of xylitol and carotenoids from switchgrass and Eucalyptus globulus hydrolysates obtained by intensified steam explosion pretreatment. Ind Crops Prod 170:113800
Brandenburg J, Blomqvist J, Pickova J, Bonturi N, Sandgren M, Passoth V (2016) Lipid production from hemicellulose with Lipomyces starkeyi in a pH regulated fed-batch cultivation. Yeast 33:451
Brienzo M, Carvalho AFA et al (2016) Sugarcane bagasse hemicellulose properties, extraction technologies and xylooligosaccharides production. In: Riley GL (ed) Food waste. Nova Science Publishers, Inc., pp 156–188
Brodeur G, Yau E, Badal K, Collier J, Ramachandran KB, Ramakrishnan S (2011) Chemical and physicochemical pre-treatment of lignocellulosic biomass: a review. Enzyme Res 1
Brownell HH, Yu EKC, Saddler JN (1986) Steam-explosion pretreatment of wood: effect of chip size, acid, moisture content and pressure drop. Biotechnol Bioeng 28:792
Bussemaker MJ, Zhang D (2013) Effect of ultrasound on lignocellulosic biomass as a pre-treatment for biorefinery and biofuel applications. Ind Eng Chem Res 52:3563
Bychkov AL, Ryabchikova EI, Korolev KG, Lomovsky OI (2012) Ultrastructural changes of cell walls under intense mechanical treatment of selective plant raw material. Biomass Bioenergy 47:260
Camargo D, Sydney EB, Leonel LV, Pintro TC, Sene L (2019) Dilute acid hydrolysis of sweet sorghum bagasse and fermentability of the hemicellulosic hydrolysate. Braz J Chem Eng 36:143
Canilha L, Santos VT, Rocha GJ, Silva JBA, Giulietti M, Silva SS, Felipe MGA, Ferraz A, Milagres AMF, Carvalho W (2011) A study on the pretreatment of a sugarcane bagasse sample with dilute sulfuric acid. J Ind Microbiol 38:1467
Cao L, Chen H, Tsang DCW, Luo G, Hao S, Zhang S, Chen J (2018) Optimizing xylose production from pinewood sawdust through dilute-phosphoric-acid hydrolysis by response surface methodology. J Clean Prod 178:572–579
Carvalheiro F, Duarte LC, Gírio FM (2008) Hemicellulose biorefineries: a review on biomass pretreatments. J Sci Ind Res 67:849
Carvalheiro F, Silva-Fernandes T, Duarte LC, Girio FM (2009) Wheat straw auto-hydrolysis: process optimization and products characterization. Appl Biochem Biotechnol 153:84
Carvalho DM, Sevastyanova O, Queiroz JH, Colodette JL (2016) Cold alkaline extraction as a pretreatment for bioethanol production from eucalyptus, sugarcane bagasse and sugarcane straw. Energy Convers Manag 124:315
Cebreiros F, Risso F, Cagno M, Cabrera MN, Rochón E, Jauregui G, Boix E, Bothig S, Ferrari MD, Lareo C (2021) Enhanced production of butanol and xylosaccharides from Eucalyptus grandis wood using steam explosion in a semi-continuous pre-pilot reactor. Fuel 290:1
Cheng JJ, Timilsina GR (2011) Status and barriers of advanced biofuel technologies: a review. Renew Energy 36:3541
Cheng MH, Sun L, Jin YS, Dien B, Singh V (2020) Production of xylose enriched hydrolysate from bioenergy sorghum and its conversion to β-carotene using an engineered Saccharomyces cerevisiae. Bioresour Technol 308:1
Choi JH, Park SY, Kim JH, Cho SM, Jang SK, Hong C, Choi IG (2019) Selective deconstruction of hemicellulose and lignin with producing derivatives by sequential pre-treatment process for biorefining concept. Bioresour Technol 291:1

Dalli SS, Patel M, Rakshit SK (2017) Development and evaluation of poplar hemicellulose prehydrolysate upstream processes for the enhanced fermentative production of xylitol. Biomass Bioenergy 105:402

Dasari RK, Berson RE (2007) The effect of particle size on hydrolysis reaction rates and rheological properties in cellulosic slurries. Appl Biochem Biotechnol 137:289

De Arruda PV, Dos Santos JC, Rodrigues RCLB, Da Silva DDV, Yamakawa CK, Rocha GJM, Júnior JN, Pradella JGC, Rossel CEV, Felipe MGA (2021) Scale up of xylitol production from sugarcane bagasse hemicellulosic hydrolysate by Candida guilliermondii FTI 20037. J Ind Eng Chem 47:297

Del Rio LF, Chandra RP, Saddler JN (2012) Fibre size does not appear to influence the ease of enzymatic hydrolysis of organosolv-pretreated softwoods. Bioresour Technol 107:235

Devi A, Singh A, Bajar S, Pant D, Din ZU (2021) Ethanol from lignocellulosic biomass: an in-depth analysis of pre-treatment methods, fermentation approaches and detoxification processes. J Environ Chem Eng 9:105798

Díaz S, Ortega Z, Benítez AN, Costa D, Carvalheiro F, Fernandes MC, Duarte LC (2021) Assessment of the effect of autohydrolysis treatment in banana's pseudostem pulp. Waste Manag 119:306

dos Santos MG, Pires LO et al (2020) The role of heterogeneous catalysts in converting cellulose to platform chemicals. In: Inamuddin AMA (ed) Nanotechnology in the life sciences. Springer, pp 305–328

Duque A, Manzanares P et al (2016) Steam explosion as lignocellulosic biomass pretreatment. In: Solange IM (ed) Biomass fractionation technologies for a lignocellulosic feedstock based biorefinery. Elsevier, pp 349–368

Dussán KJ, Silva DD, Perez VH, Silva SS (2016) Evaluation of oxygen availability on ethanol production from sugarcane bagasse hydrolysate in a batch bioreactor using two strains of xylose-fermenting yeast. Renew Energy 87:703

Dutta SK, Chakraborty S (2016) Pore-scale dynamics of enzyme adsorption, swelling and reactive dissolution determine sugar yield in hemicellulose hydrolysis for biofuel production. Sci Rep 6:1

Dutta SK, Chakraborty S (2019) Multiscale dynamics of hemicellulose hydrolysis for biofuel production. Ind Eng Chem Res 58:8963

Ek M, Gellerstedt G, Henriksson G (eds) (2009) Pulp and paper chemistry and technology, wood chemistry and wood biotechnology. De Gruyrer, Berlin, pp 1–320

Esfahani MR, Azin M (2011) Pre-treatment of sugarcane bagasse by ultrasound energy and dilute acid. Asia-Pacific J Chem Eng 7:274

Espinoza-Acosta JL, Torres-Chávez PI, Carvajal-Millán E, Ramírez-Wong B, Bello-Pérez LA, Montaño-Leyva B (2014) Ionic liquids and organic solvents for recovering lignin from lignocellulosic biomass. Bioresources 9:3660

Fengel D, Wegener G (1984) Wood: chemistry, ultrastructure, reactions. Walter de Inc., New York, pp 1–566

Flores EMM, Cravotto G, Bizzi CA, Santos D, Lop GD (2021) Ultrasound-assisted biomass valorization to industrial interesting products: state-of-the-art, perspectives and challenges. Ultrason Sonochem 72:1

Fu Y, Gu BJ, Wang J, Gao J, Ganjyal GM, Wolcott MP (2018) Novel micronized woody biomass process for production of cost-effective clean fermentable sugars. Bioresour Technol 260:311

Gallo M, Ferrara L, Naviglio D (2018) Application of ultrasound in food science and technology: a perspective. Foods 7:1

García A, Alriols MG, Llano-Ponte R, Labidi J (2011) Ultrasound-assisted fractionation of the lignocellulosic material. Bioresour Technol 102:6326

García JC, Díaz MJ, Garcia MT, Feria MJ, Goméz DM, Lopéz (2013) Search for optimum conditions of wheat straw hemicelluloses cold alkaline extraction process. Biochem Eng J 71:127

Garrote G, Dominguez H, Parajo JC (1999) Hydrothermal processing of lignocellulosic materials. Eur J Wood Wood Prod 57:191

Geddes CC, Mullinnix MT, Nieves IU, Peterson JJ, Hoffman RW, York SW, Yomano LP, Miller EN, Shanmugam KT, Ingram LO (2011) Simplified process for ethanol production from sugarcane bagasse using hydrolysate-resistant Escherichia coli strain MM160. Bioresour Technol 102:2702

Geng W, Venditti RA, Pawlak JJ, De Assis T, Gonzalez RW, Phillips RB, Chang H-M (2019) Techno-economic analysis of hemicellulose extraction from different types of lignocellulosic feedstocks and strategies for cost optimization. Biofuel Bioprod Biorefin 14:225

Gírio FM, Fonseca C, Carvalheiro F, Duarte LC, Marques S, Bogel-Łukasik R (2010) Hemicelluloses for fuel ethanol: a review. Bioresour Technol 101:4775

Girisuta B, Dussan K, Haverty D, Leahy JJ, Hayes MHB (2013) A kinetic study of acid catalysed hydrolysis of sugar cane bagasse to levulinic acid. Chem Eng Sci 217:61

Gonzales RR, Sivagurunathan P, Kim SH (2016) Effect of severity on dilute acid pretreatment of lignocellulosic biomass and the following hydrogen fermentation. Int J Hydrogen Energy 41:21678

Gullon P, Romani A, Vila C, Garrote G, Parajo JC (2012) Potential of hydrothermal treatments in lignocellulose biorefineries. Biofuels Bioprod Biorefining 6:219

Guo F, Fang Z, Xu CC, Smith RL Jr (2012) Solid acid mediated hydrolysis of biomass for producing biofuels. Prog Energy Combust Sci 38:672

Gurgel LVA, Pimenta MTB, Curvelo AAS (2014) Enhancing liquid hot water (LHW) pretreatment of sugarcane bagasse by high pressure carbon dioxide (HP-CO_2). Ind Crop Prod 57:141

Haldar D, Purkait MK (2021) A review on the environment-friendly emerging techniques for pretreatment of lignocellulosic biomass: mechanistic insight and advancements. Chemosphere 264:1

He Z, Wang Z, Zhao Z, Yi S, Mu J, Wang X (2017) Influence of ultrasound pre-treatment on wood physiochemical structure. Ultrason Sonochem 34:136

Hernández-Pérez AF, Costa IAL, Silva DDV, Dussán KJ, Villela TR, Canettieri EV, Carvalho JA, Neto TGS, Felipe MGA (2016) Biochemical conversion of sugarcane straw hemicellulosic hydrolyzate supplemented with co-substrates for xylitol production. Bioresour Technol 200:1085

Hilpmann G, Becher N, Pahner FA, Kusema B, Mäki-Arvela P, Lange R, Murzin DY, Salmi T (2016) Acid hydrolysis of xylan. Catal Today 259:376

Huijgen WJJ, Reith JH, den Uil H (2010) Pretreatment and fractionation of wheat straw by an acetone-based organosolv process. Ind Eng Chem Res 49:10132

Huijgen WJJ, Smit AT, de Wild PJ, den Uil H (2012) Fractionation of wheat straw by prehydrolysis, organosolv delignification and enzymatic hydrolysis for production of sugars and lignin. Bioresour Technol 114:389

Jacquet N, Maniet G, Vanderghem C, Delvigne F, Richel A (2015) Application of steam explosion as pretreatment on lignocellulosic material: a review. Ind Eng Chem Res 54:2593

Jamaldheen SB, Thakur A, Moholkar VS, Goyal A (2019) Enzymatic hydrolysis of hemicellulose from pretreated Finger millet (Eleusine coracana) straw by recombinant endo-1, 4-β-xylanase and exo-1, 4-β-xylosidase. Int J Biol Macromol 135:1098

Jang SK, Kim JH, Jeong H, Choi JH, Lee SM, Choi IG (2018) Investigation of conditions for dilute acid pretreatment for improving xylose solubilization and glucose production by supercritical water hydrolysis from Quercus mongolica. Renew Energy 117:150

Jofre FM, Hernández-Pérez AF, Dos Santos JC, Felipe MGA (2021) Use of dry yeast biomass as a new approach for detoxification of hemicellulosic hydrolysates aiming to xylitol production. Ind Crops Prod 170:1

Kentish S, Ashokkumar M (2011) The physical and chemical effects of ultrasound. In: Ultrasound technologies for food and bioprocessing. Springer, pp 1–12

Khalili F, Amiri H (2020) Integrated processes for production of cellulosic and hemicellulosic biobutanol from sweet sorghum bagasse using autohydrolysis. Ind Crops Prod 145:1

Kim DY, Um BH, Oh KK (2015) Acetic acid-assisted hydrothermal fractionation of empty fruit bunches for high hemicellulosic sugar recovery with low byproducts. Appl Biochem Biotechnol 176:1445

Kim D (2018) Physico-chemical conversion of lignocellulose: inhibitor effects and detoxification strategies: a mini review. Molecules 23:1

Kuhad RC, Gupta R, Khasa YP, Singh A (2010) Bioethanol production from Lantana camara (red sage): pretreatment, saccharification and fermentation. Bioresour Technol 101:8348
Kumar V, Krishania M, Sandhu PP, Ahluwalia V, Gnansounou E, Sangwan RS (2018) Efficient detoxification of corn cob hydrolysate with ion-exchange resins for enhanced xylitol production by Candida tropicalis MTCC 6192. Bioresour Technol 251:416
Kumar AK, Sharma S (2017) Recent updates on different methods of pre-treatment of lignocellulosic feedstocks: a review. Bioresour Bioprocess 4:1
Lanzafame P, Temi DM, Perathoner S, Spadaro AN, Centi G (2012) Direct conversion of cellulose to glucose and valuable intermediates in mild reaction conditions over solid acid catalysts. Catal Today 179:178
Lavarack BP, Griffin GJ, Rodman D (2002) The acid hydrolysis of sugarcane bagasse hemicellulose to produce xylose, arabinose, glucose and other products. Biomass Bioenergy 23:367
Lee CBTL, Wu TY, Ting CH, Tan JK, Siow LF, Cheng CK, Jahim JM, Mohammad AW (2019) One-pot furfural production using choline chloride-dicarboxylic acid based deep eutectic solvents under mild conditions. Bioresour Technol 278:486
Leong T, Ashokkumar M, Sandra K (2011) The fundamentals of power ultrasound—A review. Acoust Aust 39:54
Li HY, Chen X, Li YJ, Cao XF, Sun SN, Sun RC (2018) The effect of ionic liquids pre-treatment on the distribution and structure of alkali-soluble hemicelluloses from Eucalyptus. Sep Purif Technol 191:364
Liao Y, Beeck BOD, Thielemans K, Ennaert T, Snelders J, Dusselier M, Courtin CM, Sels J (2020) The role of pre-treatment in the catalytic valorization of cellulose. Mol Catal 487:1
Liu Q, Li W, Ma Q, An S, Li M, Jameel H, Chang HM (2016) Pretreatment of corn stover for sugar production using a two-stage dilute acid followed by wet-milling pretreatment process. Bioresour Technol 211:435
Liu Y, Wang J, Wolcott MP (2017) Evaluating the effect of wood ultrastructural changes from mechanical treatment on kinetics of monomeric sugars and chemicals production in acid bisulfite treatment. Bioresour Technol 226:24
Liu Z, Shi E, Ma F, Jiang K (2021) An integrated biorefinery process for co-production of xylose and glucose using maleic acid as efficient catalyst. Bioresour Technol 325:1
Lomovsky O, Bychkov A et al (2016) Mechanical Pre-treatment. In: Mussato SI (ed) Biomass fractionation technologies for a lignocellulosic feedstock based biorefinery. Elsevier Inc., pp 23–55
Loow YL, Wu TY, Yang GH, Jahim JM, Teoh WH, Mohammad AW (2016) Role of energy irradiation in aiding pre-treatment of lignocellulosic biomass for improving reducing sugar recovery. Cellulose 23:2761
López-Linares JC, Cara C, Moya M, Ruiz E, Castro E, Romero I (2013) Fermentable sugar production from rapeseed straw by dilute phosphoric acid pretreatment. Ind Crops Prod 50:525
López-Linares JC, Romero I, Cara C, Castro E, Mussatto SI (2018) Xylitol production by Debaryomyces hansenii and Candida guilliermondii from rapeseed straw hemicellulosic hydrolysate. Bioresour Technol 247:736
López-Linares JC, García-Cubero MT, Lucas S, Coca M (2020) Integral valorization of cellulosic and hemicellulosic sugars for biobutanol production: ABE fermentation of the whole slurry from microwave pretreated brewer's spent grain. Biomass Bioenergy 135:1
Louis ACF, Venkatachalam S (2020) Energy efficient process for valorization of corn cob as a source for nanocrystalline cellulose and hemicellulose production. Int J Biol Macromol 163:260
Lu S, Wang Q, Liang Z, Wang W, Liang C, Wang Z, Yuan Z, Lan P, Qi W (2021) Saccharification of sugarcane bagasse by magnetic carbon-based solid acid pre-treatment and enzymatic hydrolysis. Ind Crops Prod 1
Luo J, Fang Z, Smith RL (2014) Ultrasound-enhanced conversion of biomass to biofuels. Prog Energy Combust Sci 41:56
Lyu H, Zhou J, Lyu C, Meng F, Geng Z, Zhang M (2018) Exploration and optimization of mixed acid synergistic catalysis pretreatment for maximum C5 sugars. Bioresour Technol 260:53

Mankar AR, Pandey A, Modak A, Pant KK (2021) Pre-treatment of lignocellulosic biomass: a review on recent advances. Bioresour Technol 334:1
Meng X, Bhagia S, Wang Y, Zhou Y, Pu Y, Dunlap JR, Shuai L, Ragauskas AJ, Yoo CG (2020) Effects of the advanced organosolv pretreatment strategies on structural properties of woody biomass. Ind Crops Prod 146:112
Minjares-Fuentes R, Femenia A, Garau MC, Candellas-Cadillo MG, Simal S, Roselló C (2016) Ultrasound-assisted extraction of hemicelluloses from grape pomace using response surface methodology. Carbohydr Polym 138:180
Moraes EJC, Silva DDV, Dussán KJ, Tesche LZ, Silva JBA, Rai M, Felipe MGA (2020) Xylitol-sweetener production from barley straw: optimization of acid hydrolysis condition with the energy consumption simulation. Waste Biomass Valorization 11:1837
Moreno AD, Tomás-Pejó E et al (2019) Pre-treatment technologies for lignocellulosic biomass deconstruction within a biorefinery perspective. In: Pandey A (ed) Biofuels: alternative feedstocks and conversion processes for the production of liquid and gaseous biofuels, pp 379–399
Mosier N, Wyman C, Dale B, Elander R, Lee YY, Holtzapple M, Ladisch M (2005) Features of promising technologies for pretreatment of lignocellulosic biomass. Bioresour Technol 96:673
Mussatto SI, Bikaki N (2016) Technoeconomic considerations for biomass fractionation in a biorefinery context. In: Mussatto SI (ed) Biomass fractionation technologies for a lignocellulosic feedstock based biorefinery. Elsevier Inc., pp 587–610
Nanda S, Kozinski JA, Dalai AK (2016) Lignocellulosic biomass: a review of conversion technologies and fuel products. Curr Biochem Eng 3:24
Norrrahim MNF, Ilyas RA, Nurazzi NM, Rani MSA, Atikah MSN, Shazleen SS (2021) Chemical pre-treatment of lignocellulosic biomass for the production of bioproducts: an overview. Appl Sci Eng Prog 14:588
Ong VZ, Wu TY (2020) An application of ultrasonication in lignocellulosic biomass valorisation into bio-energy and bio-based products. Renew Sustain Energy Rev 132:1
Ong VZ, Wu TY, Lee CBTL, Cheong NWR, Shak KPY (2019) Sequential ultrasonication and deep eutectic solvent pre-treatment to remove lignin and recover xylose from oil palm fronds. Ultrason Sonochem 58:1
Ontañon OM, Bedő S, Ghio S, Garrido MM, Topalian J, Jahola D, Fehér A, Valacco MP, Campos E, Fehér C (2021) Optimisation of xylanases production by two Cellulomonas strains and their use for biomass deconstruction. Appl Microbiol Biotechnol 105:4577
Ostadjoo S, Hammerer F, Dietrich K, Dumont MJ, Friscic T, Auclair K (2019) Efficient Enzymatic Hydrolysis of Biomass Hemicellulose in the Absence of Bulk Water. Molecules 24:1
Palmqvist E, Hahn-Hägerdal B (2000) Fermentation of lignocellulosic hydrolysates. I: inhibition and detoxification. Bioresour Technol 74:17
Peleteiro S, Rivas S, Alonso JL, Santos V, Parajó JC (2016) Furfural production using ionic liquids: a review. Bioresour Technol 202:181
Qian J, Zhao F, Gao J, Qu L, He Z, Yi S (2021) Characterization of the structural and dynamic changes of cell wall obtained by ultrasound-water and ultrasound-alkali treatments. Ultrason Sonochem 77:1
Quintero LP, Souza NPQ, Milagres AMF (2021) The effect of xylan removal on the high-solid enzymatic hydrolysis of sugarcane bagasse. Bioenergy Res
Radhakrishnan R, Patra P, Das M, Ghosh A (2021) Recent advancements in the ionic liquid mediated lignin valorization for the production of renewable materials and value-added chemicals. Renew Sustain Energy Rev 149:1
Raj T, Chandrasekhar K, Banu R, Yoon JJ, Kumar G, Kim SH (2021) Synthesis of γ-valerolactone (GVL) and their applications for lignocellulosic deconstruction for sustainable green biorefineries. Fuel 303:1
Research and Market Reports, Xylitol - Global Market Trajectory & Analytics (2021). https://www.researchandmarkets.com/reports/5140553/xylitol-global-market-trajectory-and-analytics
Rao LV, Goli JK, Gentela J, Koti S (2016) Bioconversion of lignocellulosic biomass to xylitol: an overview. Bioresour Technol 213:299

Ravindran R, Jaiswal S, Abu-Ghannam N, Jaiswal AK (2017) Evaluation of ultrasound assisted potassium permanganate pre-treatment of spent coffee waste. Bioresour Technol 224:680
Reisinger M, Tirpanalan Ö, Huber F, Kneifel W, Novalin S (2014) Investigations on a wheat bran biorefinery involving organosolv fractionation and enzymatic treatment. Bioresour Technol 170:53
Ríos-González LJ, Medina-Morales MA, Rodríguez-De la Garza JA, Romero-Galarza A, Medina DD, Moralez-Martínez TK (2021) Comparison of dilute acid pre-treatment of agave assisted by microwave versus ultrasound to enhance enzymatic hydrolysis. Bioresour Technol 319:1
Rivas B, Domınguez JM, Domınguez H, Parajó JC (2002) Bioconversion of posthydrolysed autohydrolysis liquors: an alternative for xylitol production from corn cobs. Enzyme Microb Technol 31:431
Rodiahwati W, Sriariyanun M (2016) Lignocellulosic biomass to biofuel production: integration of chemical and extrusion (Screw Press) pre-treatment. Int J Appl Sci Technol 9:289
Romaní A, Morais ES, Soares PO, Freire MG, Freire CSR, Silvestre AJD, Domingues L (2020) Aqueous solutions of deep eutectic systems as reaction media for the saccharification and fermentation of hardwood xylan into xylitol. Bioresour Technol 311:1
Rooze J, Rebrov EV, Schouten JC, Keurentjes JTF (2013) Dissolved gas and ultrasonic cavitation—A review. Ultrason Sonochem 20:1
Ruiz E, Cara C, Manzanares P, Ballesteros M, Castro E (2008) Evaluation of steam explosion pre-treatment for enzymatic hydrolysis of sunflower stalks. Enzyme Microb Technol 42:160
Saha BC, Kennedy GJ (2020) Production of xylitol from mixed sugars of xylose and arabinose without co-producing arabitol. Biocatal Agric Biotechnol 29:101786
Salapa I, Katsimpouras C, Topakas E, Sidiras D (2017) Organosolv pretreatment of wheat straw for efficient ethanol production using various solvents. Biomass Bioenergy 100:10
Sarker TR, Pattnaik F, Nanda S, Dalai AK, Meda V, Naik S (2021) Hydrothermal pre-treatment technologies for lignocellulosic biomass: a review of steam explosion and subcritical water hydrolysis. Chemosphere 284:1
Serna-Loaiza S, Zikeli F, Adamcyk J, Friedl A (2021) Towards a wheat straw biorefinery: combination of Organosolv and Liquid Hot Water for the improved production of sugars from hemicellulose and lignin hydrolysis. Bioresour Technol Rep 14:1
Silva DD, Felipe MG, Mancilha IM, Luchese RH, Silva SS (2004) Inhibitory effect of acetic acid on bioconversion of xylose in xylitol by Candida guilliermondii in sugarcane bagasse hydrolysate. Braz J Microbiol 35:248
Silva DD, Arruda PV, Dussán KJ, Felipe MG (2014) Adaptation of Scheffersomyces stipitis cells as a strategy to the improvement of ethanol production from sugarcane bagasse hemicellulosic hydrolysate. Chem Eng 38:1
Silva DD, Dussán KJ, Idarraga A, Grangeiro L, Silva SS, Cardona CA, Quintero J, Felipe MGA (2020) Production and purification of xylitol by *Scheffersomyces amazonenses* via sugarcane hemicellulosic hydrolysate. Biofuels Bioprod Biorefining 14:344
Silveira MHL, Chandel AK, Vanelli BA, Sacilotto KS, Cardoso EB (2018) Production of hemicellulosic sugars from sugarcane bagasse via steam explosion employing industrially feasible conditions: pilot scale study. Bioresour Technol Rep 3:138
Sindhu R, Pandey A, Binod P (2015) Alkaline treatment. In: Pandey A, Negi S, Binod P, Larroche C (eds) Pre-treatment of biomass: processes and technologies. Elsevier, pp 51–60
Singh S, Bharadwaja STP, Yadav PK, Moholkar VS, Goyal A (2014) Mechanistic Investigation in ultrasound-assisted (Alkaline) delignification of parthenium hysterophorus biomass. Ind Eng Chem Res 53:14241
Siripong P, Duangporn P, Takata E, Tsutsumi Y (2016) Phosphoric acid pretreatment of Achyranthes aspera and Sida acuta weed biomass to improve enzymatic hydrolysis. Bioresour Technol 203:303
Subhedar PB, Gogate PR (2016) Use of ultrasound for pre-treatment of biomass and subsequent hydrolysis and fermentation. In: Mussato SI (ed) Biomass fractionation technologies for a lignocellulosic feedstock based biorefinery. Elsevier Inc., pp 127–149

Suganuma S, Nakajima K, Kitano M, Yamaguchi D, Kato H, Hayashi S, Hara M (2008) Hydrolysis of cellulose by amorphous carbon bearing SO_3H, COOH, and OH groups. J Am Chem Soc 130:12787
Sun S, Sun S, Cao X, Sun R (2016) The role of pre-treatment in improving the enzymatic hydrolysis of lignocellulosic materials. Bioresour Technol 199:49
Sun SF, Yang HY, Yang J, Wang DW, Shi ZJ (2021) Integrated treatment of perennial ryegrass: structural characterization of hemicelluloses and improvement of enzymatic hydrolysis of cellulose. Carbohydr Polym 254:1
Temiz E, Akpinar O (2017) The effect of severity factor on the release of xylose and phenolics from rice husk and rice straw. Waste Biomass Valorization 8:505
Teramura H, Sasaki K, Oshima T, Matsuda F, Okamoto M, Shirai T, Kawaguchi H, Ogino C, Hirano K, Sazuka T, Kitano H, Kikuchi J, Kondo A (2016) Organosolv pretreatment of sorghum bagasse using a low concentration of hydrophobic solvents such as 1-butanol or 1-pentanol. Biotechnol Biofuels 9:1
Teramura H, Sasaki K, Oshima T, Kawaguchi H, Ogino C, Sazuka T, Kondo A (2018) Effective usage of sorghum bagasse: optimization of organosolv pretreatment using 25% 1-butanol and subsequent nanofiltration membrane separation. Bioresour Technol 252:157
Van Walsum GP (2001) Severity function describing the hydrolysis of xylan using carbonic acid. Biotechnol Appl Biochem 317:91
Velmurugan R, Muthukumar K (2011) Utilization of sugarcane bagasse for bioethanol production: Sono-assisted acid hydrolysis approach. Bioresour Technol 102:7119
Walker DJ, Gallagher J, Winters A, Somani A, Ravella SR, Bryant DN (2018) Process optimization of steam explosion parameters on multiple lignocellulosic biomass using Taguchi method-a critical appraisal. Front. Energy Res. 6:1
Wan L, Gao Z, Wu B, Cao F, Jiang M, Wei P, Jia H (2021) Hydrolysis of corncob hemicellulose by solid acid sulfated zirconia and its evaluation in xylitol production. Biotechnol Appl Biochem 193:205
Wang W, Ling H, Zhao H (2015) Steam explosion pretreatment of corn straw on xylose recovery and xylitol production using hydrolysate without detoxification. Process Biochem 50:1623
Wijaya YP, Putra RDD, Widyaya VT, Ha JM, Suh DJ, Kim CS (2014) Comparative study on two-step concentrated acid hydrolysis for the extraction of sugars from lignocellulosic biomass. Bioresour Technol 164:221
Wildschut J, Smit AT, Reith JH, Huijgen WJJ (2013) Ethanol-based organosolv fractionation of wheat straw for the production of lignin and enzymatically digestible cellulose. Bioresour Technol 135:58
Xu JY, Yuan TQ, Xiao L, Sun RC (2018) Effect of ultrasonic time on the structural and physico-chemical properties of hemicelluloses from Eucalyptus grandis. Carbohydr Polym 195:114
Xu L, Liu L, Li S, Zheng W, Cui Y, Liu R, Sun W (2019) Xylitol production by Candida tropicalis 31949 from sugarcane bagasse hydrolysate. Sugar Tech 21:341
Yan D, Ji Q, Yu X, Li M, Fakayode OA, Yagoub AEA, Zhou C (2021) Multimode-ultrasound and microwave assisted natural ternary deep eutectic solvent sequential pre-treatments for corn straw biomass deconstruction under mild conditions. Ultrason Sonochem 72:1
Yang G, Wang J (2019) Ultrasound combined with dilute acid pre-treatment of grass for improvement of fermentative hydrogen production. Bioresour Technol 275:10
Ye K, Liu Y, Wu S, Zhuang J (2021) A review for lignin valorization: challenges and perspectives in catalytic hydrogenolysis. Ind Crops Prod 172:1
Yoo CG, Meng X, Pu Y, Ragauskas AJ (2020) The critical role of lignin in lignocellulosic biomass conversion and recent pre-treatment strategies: a comprehensive review. Bioresour Technol 301:1
Yu Q, Zhuang X, Wang Q, Qi W, Tan X, Yuan Z (2012) Hydrolysis of sweet sorghum bagasse and eucalyptus wood chips with liquid hot water. Bioresour Technol 116:220
Yu H, You Y, Lei F, Liu Z, Zhang W, Jiang J (2015) Comparative study of alkaline hydrogen peroxide and organosolv pretreatments of sugarcane bagasse to improve the overall sugar yield. Bioresour Technol 187:161

Yu Q, Baroutian S, Xie J (2021) Hydrothermal co-hydrolysis of corncob/sugarcane bagasse/Broussonetia papyrifera blends: kinetics, thermodynamics and fermentation. Bioresour Technol 342:1

Yuan TQ, Xu F, He J, Sun RC (2010) Structural and physico-chemical characterization of hemicelluloses from ultrasound-assisted extractions of partially delignified fast-growing poplar wood through organic solvent and alkaline solutions. Biotechnol Adv 28:583

Yuan Z, Long J, Wang T, Shu R, Zhang Q, Ma L (2015) Process intensification effect of ball milling on the hydrothermal pre-treatment for corn straw enzymolysis. Energy Convers Manag 101:481

Yunus R, Salleh SF, Abdullah N, Biak DRA (2010) Effect of ultrasonic pre-treatment on low temperature acid hydrolysis of oil palm empty fruit bunch. Bioresour Technol 101:9792

Zhang K, Pei Z, Wang D (2016) Organic solvent pretreatment of lignocellulosic biomass for biofuels and biochemicals: a review. Bioresour Technol 199:21

Zhang J, Wang YHW, Wei QY, Du XJ, Qu YS (2018) Investigating desorption during ethanol elution to improve the quality and antioxidant activity of xylo-oligosaccharides from corn stalk. Bioresour Technol 264:1

Zhang J, Zhou H et al (2019) Pre-treatment of lignocellulosic biomass for efficient enzymatic saccharification of cellulose. In: Yousuf A, Pirozzi D, Sannino F (eds) Lignocellulosic biomass to liquid biofuels. Academic Press, pp 17–65

Zhao X, Cheng K, Liu D (2009) Organosolv pretreatment of lignocellulosic biomass for enzymatic hydrolysis. Appl Microbiol Biotechnol 82:815

Zhao X, Zhang L, Liu D (2012) Biomass recalcitrance. Part II: fundamentals of different pre-treatments to increase the enzymatic digestibility of lignocellulose Xuebing. Biofuels Bioprod Biorefining 561

Zheng Y, Zhao J, Xu F, Li Y (2014) Pre-treatment of lignocellulosic biomass for enhanced biogas production. Prog Energy Combust Sci 42:35

Zhuang X, Yu Q, Wang W, Qi W, Wang Q, Tan X, Yuan Z (2012) Decomposition behavior of hemicellulose and lignin in the step-change flow rate liquid hot water. Biotechnol Appl Biochem 168:206

Chapter 2
Detoxification of Hemicellulosic Hydrolysates for Improved Xylitol Production

Rajendran Velmurugan, Ramesh Chander Kuhad, Arumugam Dhanesh Gandhi, Ranganathan Babujanarthanam, and Aran Incharoensakdi

Abstract Xylitol, a naturally occurring sugar alcohol, is used in various medical and non-medical applications. The microbial production of xylitol from lignocellulosic biomass has gained much attention as the process is involved in greener synthesis. The production of xylitol from lignocellulosic biomass consists of disintegration of lignin–hemicellulose meshwork, hydrolysis to produce xylose, and microbial fermentation to convert xylose to xylitol. In microbial conversion, the generation of inhibitory products in hydrolysis is the rate limiting factor in xylose to xylitol conversion. The major inhibitory products include organic acids from lignin and furfural from cellulose and hydroxymethyl furfural from hemicellulose, respectively. As the inhibitors have their unique characteristic feature, the development of sugar-free, inhibitor specific detoxification process is difficult. Various bioprocesses including evaporation, neutralization, overliming, adsorption have been developed in which the calcium hydroxide, activated charcoal, ion-exchange resins and enzymes have been used so-far for detoxification. Recent progress on developing hybrid methods is interesting as the process undergoes two different principles aiming detoxification of chemically distinct inhibitors. In this chapter, the route of inhibitor generation, complications in biomass to xylitol conversion, methods used for detoxification and recent perspectives have been discussed.

Keywords Lignocellulose · Hemicellulose hydrolysis · Detoxification · Xylitol

---

R. Velmurugan · A. Incharoensakdi (✉)
Cyanobacterial Biotechnology Laboratory, Department of Biochemistry, Faculty of Science, Chulalongkorn University, Bangkok 10330, Thailand
e-mail: aran.i@chula.ac.th

R. C. Kuhad
Lignocellulose Biotechnology Laboratory, Department of Microbiology, University of Delhi South Campus, New Delhi 110021, India

A. D. Gandhi · R. Babujanarthanam
Department of Biotechnology, Thiruvalluvar University, Vellore 632115, India

A. Incharoensakdi
Academy of Science, Royal Society of Thailand, Bangkok 10300, Thailand

© The Author(s), under exclusive license to Springer Nature Switzerland AG 2022

M. das G. de Almeida Felipe and A. K. Chandel (eds.), *Current Advances in Biotechnological Production of Xylitol*,
https://doi.org/10.1007/978-3-031-04942-2_2

2.1 Introduction

Lignocellulosic biomass from agricultural and industrial processes is considered as renewable and has been recognized as a potential low-cost source for various biomolecules production. The utilization of lignocellulosic biomass as a source of industrial feedstock is difficult, because of its complex nature. Lignocellulose is composed of three main fractions: cellulose, hemicellulose and lignin (Wiselogel et al. 1996).

The cellulose molecule is characterized by β-1,4-glucosidic linkages between sequential glucose units. On the other hand, the hemicellulose is composed of heteropolymer such as xylan (polymer of xylose), arabinan (polymer of arabinose), and oligosaccharides. The extraction of carbohydrate from lignocellulosic biomass is not easy as it is meshed with lignin fraction. Therefore, the hydrolysis step is required to efficiently hydrolyse these polysaccharides for the successful production of fermentable sugar (Katahira et al. 2006). The theoretical maximum yield is 1.136 kg of pentose per kg of C_5 polysaccharides (Eq. 2.1):

$$n\, C_5H_8O_4 + n\, H_2O \rightarrow n\, C_5H_{10}O_5 \tag{2.1}$$

The conversion of xylan to xylose consisted of two steps viz, protonation followed by hydrolysis (Fig. 2.1). Various processes were developed for the conversion of xylan into xylose using enzymes or acids. Enzymatic hydrolysis is more expensive due to its cost associated with the isolation and purification of enzymes. Even though acid hydrolysis is less expensive, the generation of inhibitory compound is a major problem. Furthermore, the hydrolysis of hemicellulose is more complicated than that of cellulose, due to the presence of noncarbohydrate fractions. The acid hydrolysis of hemicellulose under thermal or acid produced mainly sugar monomers. Among the hemicellulose components, the xylan is more susceptible to dilute acid hydrolysis, due to its amorphous nature compared to cellulose (Rahman et al. 2007). Acids attack the β-1,4-linkage and branched glucuronoxylan and arabinoxylan. During acid hydrolysis, xylose is degraded rapidly to hydroxy methyl furfural and other condensation by-products and these degradation products inhibit microbial growth. The inhibitory effect of different compounds like furfural, 5-hydroxymethyl furfural (HMF), acetate and other degradative products on yeast growth is well reported (Rao et al. 2006).

The removal of such inhibitory components has been carried out in detoxification process by various physico-chemical methods including evaporation, solvent extraction and membrane filtration; chemical methods include neutralization, overliming, activated charcoal adsorption and ion-exchange resins; biological methods include enzymes (laccase, peroxidases) and direct use of microorganisms (Mussato and Roberto 2004; Chandel et al. 2011).

An effective detoxification method mostly removes charged inhibitory components (furans and organic acids) and traces of sugar monomers. Depending on the raw material, the rest of the monosaccharides in hydrolysate can be xylose, glucose,

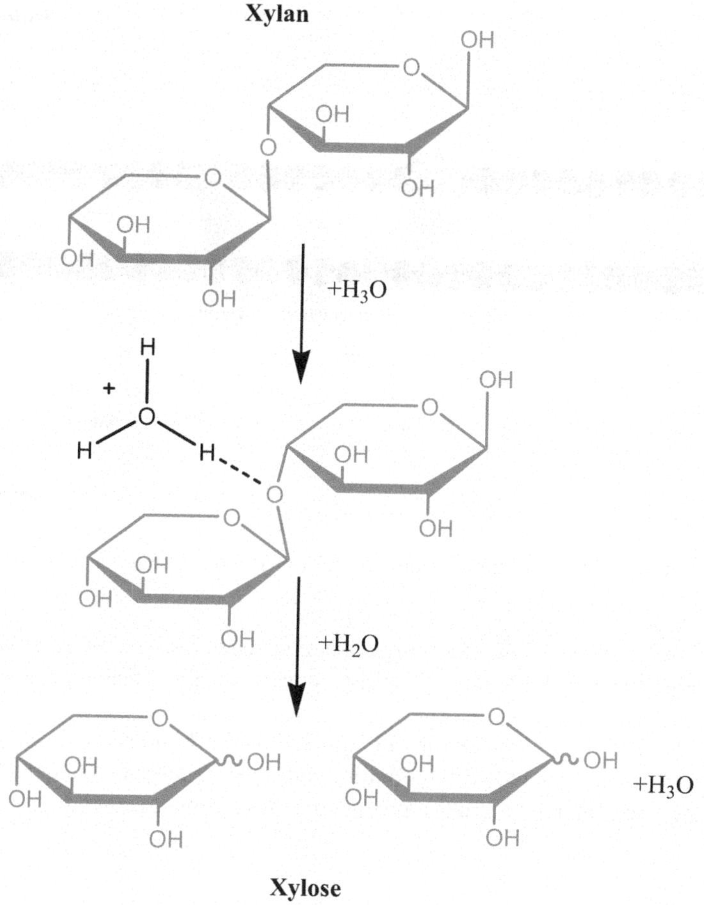

Fig. 2.1 The reaction involved in xylan to xylose conversion

mannose, galactose, arabinose, and oligosaccharides. Although the xylose is the source of xylitol production, the microorganism should utilize rest of the monosaccharides to maintain the cellular growth. According to the reactions, the theoretical maximum yield is 1.01 kg xylitol per kg of xylose (Eq. 2.2):

$$n\ C_5H_{10}O_5 + n\ NADPH + H \rightarrow n\ C_6H_{12}O_6 + NADP \tag{2.2}$$

The xylitol producing microorganisms, that are most promising for industrial exploitation include *Pachysolen tannophilus* NRRL Y-7426 (Converti et al. 2000), *Candida tropicalis* (Zhuang et al. 2012), *Candida guilliermondii* FTI 20,037 (Rodrigues et al. 2001). Natural xylose utilizing yeast, such as *Candida tropicalis* (Jeon et al. 2013) and *Saccharomyces cerevisiae* (Oh et al. 2013), and bacteria such as *Escherichia coli* (Nair and Zhao 2010) and *Candida glutamicum* (Sasaki et al.

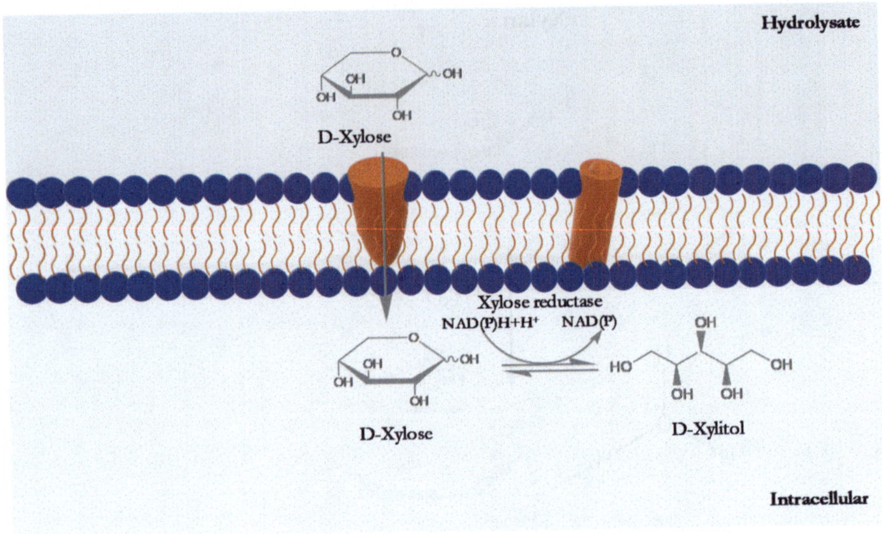

Fig. 2.2 The utilization of extracellular xylose for microbial conversion of xylose to xylitol

2010). On the other hand, the microorganisms has been engineered to enhance the xylitol production via the action of xylose reductase (XR) (Katahira et al. 2006) (Fig. 2.2).

In this chapter, the characteristics of lignocellulosic biomass, current issues in xylitol production associated with inhibitory components, methods in detoxification of inhibitory components and future prospective have been discussed.

2.2 Lignocellulose—Components

Lignocellulose biomass, the most abundant structural component of plants, consisted of cellulose, hemicellulose and lignin. The cellulose is a structural component of plants which is associated with hemicelluloses and lignin. The denser and crystalline domains of cellulose with hindered access to the ether bonds causes complication in enzymatic and chemical access, while hemicelluloses have amorphous structure allowing easy chemical access (Lynd et al. 2002). The monosaccharides extracted from plant materials are the actual raw material for current industrial applications. Glucose is used in various applications including food and fermentative production of various biochemicals. Xylose is another major pentose sugar, which is hydrogenated to xylitol for food applications and also for the production of furfural by acid hydrolysis.

2.2.1 Cellulose

Cellulose is a linear polymer of glucose, composed of thousands of molecules of anhydroglucose linked by β(1,4)-glycosidic bonds. The secondary and tertiary conformation of cellulose, as well as its close association with lignin and hemicellulose, makes cellulose resistant to hydrolysis. However, cellulose can be hydrolysed using acid or enzymes.

2.2.2 Hemicellulose

Hemicellulose is a branched heteropolymers containing sugar residues such as pentoses (D-xylose, L-arabinose) and hexoses (D-galactose, L-galactose, D-mannose, L-rhamnose, L-fucose). They also contain smaller amounts of acetyl groups, which are non-sugars (Lynd 1996). The composition of hemicellulose depends on the source of the raw material (Wiselogel et al. 1996). There are different enzymes available for the hydrolysis of hemicellulose such as endoxylanase, exoxylase, xylosidase, arabinosidase, glucoronisidase and acetyl xylan esterase.

2.2.3 Lignin

Lignin (10–30%) is a complex, hydrophobic and cross-linked aromatic polymer in nature. Lignins are polymers of phenylpropane units, consisting of coniferyl alcohol and sinapyl alcohol, with minor quantity of p-coumaryl alcohol units (Kirk et al. 1977). The complex of these components is cross-linked together through carbon–carbon, ester and ether linkages.

2.3 Complications in Biomass to Xylitol Conversion

2.3.1 Structural Integrity of Cellulose, Hemicellulose and Lignin

The complexity of lignocellulosic biomass is considered as the major drawback for their use in value-added product production. In order to utilize xylose, an efficient deconstruction step is a prerequisite for removal of lignin. For effective xylitol production, the hydrolysate should be free from lignin degradatives.

2.3.2 Depolymerization

As the hemicellulose is composed of multiple sugars, the hydrolysis is complicated in obtaining xylose monomer. The non-specific hydrolysis of hemicellulose under acidic condition is easy, however the generation of furan derivatives is a main drawback. On the other hand, the enzymatic method is reported to not generate such inhibitory products. However, the requirement of multiple enzymes makes the process not feasible.

2.3.3 Presence of Non-carbohydrate Fraction

Apart from hexoses and pentoses, the lignocellulosic biomass carries organic acids such as acetic and uronic acids as a structural component. Although the quantity is very less when compared to sugar units, those organic acids are remarked to inhibit the fermenting organism.

2.3.4 Degradation of Sugars upon Acid Hydrolysis

The acid hydrolysis is a well-recognized method for hydrolysis of hemicellulose. However, as a secondary reaction, the acids degrade the hexoses and pentoses into hydroxymethyl furfural and furfural, respectively. The development of moderate conditions that hydrolyses sugars and not degrade further are essential for a potential hydrolysis.

2.3.5 Loss of Sugars During Detoxification

The detoxification is a promising technique to remove the inhibitory compounds from hydrolysate. However, the adsorption of sugar monomer is also reported in detoxification process. To avoid the sugar monomer losses, an efficient method has to be applied to selectively remove inhibitory compounds.

2.4 Hydrolysis Methods—Sugars and Inhibitors Generation

2.4.1 Chemical

This is the oldest method for converting cellulose into fermentable sugars. In dilute acid hydrolysis, the hemicellulose fraction is depolymerized at a lower temperature than the cellulosic fraction. Dilute H_2SO_4 is mixed with lignocelluosic biomass to hydrolyse hemicellulose to xylose and other sugars (Chandel et al. 2007). Most dilute acid processes are limited to a sugar recovery efficiency of around 50% (Badger 2002). The primary challenge is to increase the sugar yield to 70%, to make the process economically viable. The main advantage of the dilute acid processes is a higher reaction rate, and the main disadvantage is a low sugar yield.

The concentrated acid process provides a complete and rapid conversion of polysaccharides into monosaccharides with little degradation of sugars (Fig. 2.3). The concentrated acid process uses relatively mild pressures and temperatures. This process is carried out with the acid concentration of 70% at 313–323 K for 2–4 h in a reactor. The primary advantage of the concentrated acid process is higher sugar recovery (Demirbas 2005). Concentrated H_2SO_4 or HCl is difficult to work with, and the acids present must be recovered and reconcentrated in order to make the process cost effective and eco-friendly. The concentrated acid process offers more potential for cost reduction than the dilute acid process. However, the time requirement, corrosion problems and the generation of inhibitors are the main drawbacks (Wyman 1996).

2.4.2 Biological

Enzymatic hydrolysis utilizes hydrolysing enzymes for the hydrolysis of polysaccharides. The most widely used organism for the production of cellulase and hemicellulase enzymes is *Trichoderma reesei* (Sheehan and Himmel 1999). Cellulases and hemicellulases are also produced by a wide variety of bacteria, actinomycetes, fungi, etc. Bacteria belonging to *Clostridium, Cellulomonas, Bacillus, Thermomonospora, Ruminococcus, Bacteriodes, Erwinia, Acetovibrio, Microbispora*, and *Streptomyces* produce cellulases (Sun and Tomkinson 2003). Filamentous fungi are the major source of cellulases and hemicellulases (Gusakov et al. 2007). The wild type and mutant strains of *Trichoderma* sp. (*T. viride, T. reesei* and *T. longibrachiatum*) and *Aspergillus* sp. have also been considered to be the most productive and powerful destroyers of crystalline cellulose.

The most important enzyme activities required for the hydrolysis of cellulose and hemicellulose are presented in Table 2.1 (Carvalheiro et al. 2008). At least, three major types of cellulase enzymatic activities are believed to be involved in

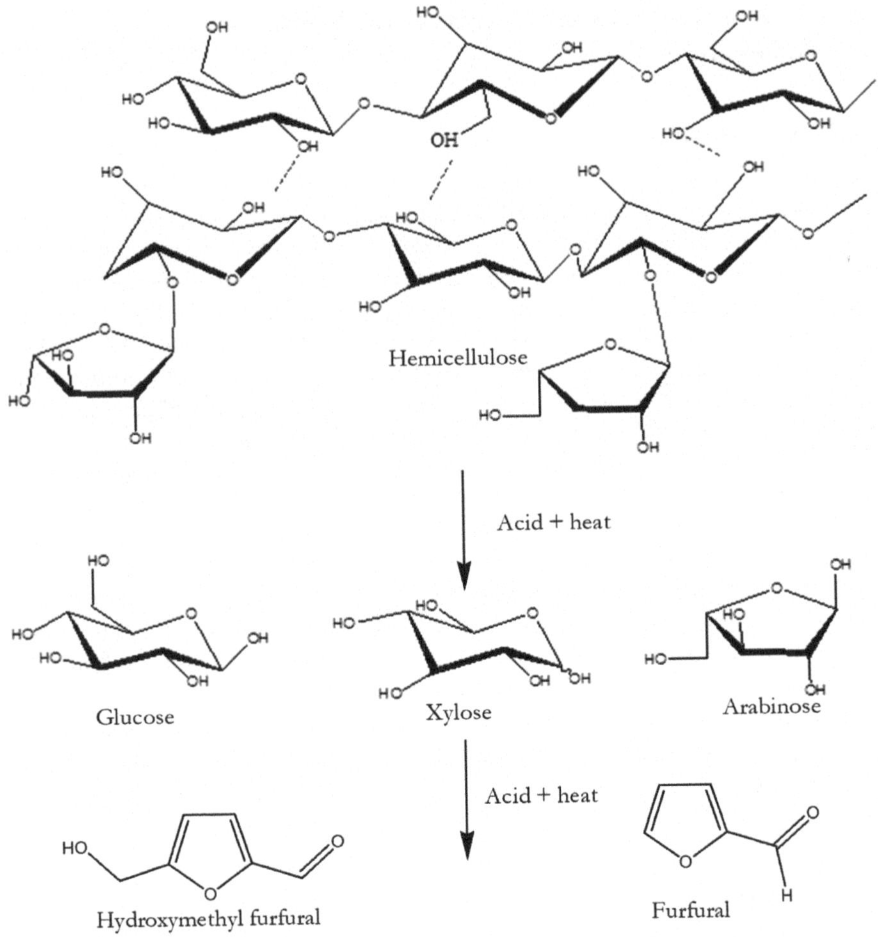

Fig. 2.3 Conversion of hemicellulose into hexose and pentoses and further conversion into hydroxymethyl furfural and furfural under acidic condition

cellulose hydrolysis, based on their structural properties: endoglucanases or 1,4-β-D-glucan-4-glucanohydrolases (EC 3.2.1.4), exoglucanases, including 1,4-β-D-glucan glucanohydrolases (also known as cellodextrinases) (EC 3.2.1.74) and 1,4-β-D-glucan cellobiohydrolases (cellobiohydrolases) (EC 3.2.1.91), and β-glucosidases or β-glucoside glucohydrolasess (EC 3.2.1.21) (Lynd et al. 2002). Endoglucanase plays an important role in cellulose hydrolysis by cleaving cellulose chains randomly. Endoglucanase hydrolyses accessible intramolecular β-1,4-glucosidic bonds of cellulose chains randomly, to produce new chain ends; exoglucanases processively cleave cellulose chains to release soluble cellobiose or glucose; and β-glucosidases hydrolyse cellobiose to glucose in order to eliminate cellobiose inhibition (Zhang

Table 2.1 Types of enzymes and its mode of action on hydrolysis

Enzyme	EC number	Mode of action	Main product
Endoxylanase	3.2.1.8	Hydrolyses internal β-1,4-xylan linkages of main chain	Oligosaccharides
Exoxylanase	n.c	Hydrolyses terminal β-1,4-xylose linkages	Xylose, xylobionse
β-Xylosidase	3.2.1.37	Hydrolyses xylobiose and terminal β-1,4-linkages	Xylose
α-Arabinosidase	3.2.1.55	Hydrolyses terminal nonreducing α-arabinofuranose from arabinoxylans	Arabinose
α-Glucuronisidase	3.2.1.139	Hydrolyses glucuronoxylans	Methyl-glucuronic acids
Acetyl xylan esterase	3.1.1.72	Hydrolyses acetylester bonds in acetyl xylans	Acetic acid
Feruloyl esterases	3.1.1.73	Hydrolyses feruloylester bonds in xylans	Ferulic acid

n.c—Not classified

et al. 2006). β-glucosidases complete the hydrolysis process by catalyzing the hydrolysis of cellobiose to glucose.

Xylan degrading enzymes are produced by a wide variety of fungi and bacteria which include *Trichodrema* sp. (Wong and Saddler 1992; Haltrich et al. 1996), *Penicillium* sp. (Filho et al. 1991; Jorgensen et al. 2003), *Talaromyces* sp. (Filho et al. 1991; Tuohy et al. 1993), *Aspergillus* sp. (Reis et al. 2003), and *Bacillus* sp. (Virupakshi et al. 2005). The enzymatic hydrolysis of xylan involves a multi-enzyme system, including endoxylanase, exoxylanase, β-xylosidase, α-arabinofuranosidase, α-glucoronisidase, acetyl xylan esterase, and ferulic acid esterase (Saha 2003). The endoxylanase attacks the main chains of xylans, and the β-xylosidase hydrolyses xylooligosaccharides to xylose. The α-arabinofuranosidase and α-glucuronidase remove the arabinose and 4-0-methyl glucuronic acid substituents, respectively, from the xylan backbone (Saha 2003). Hemicellulolytic esterases include acetyl esterases, which hydrolyse the acetyl substitutions on xylose moieties, and feruloyl esterases which hydrolyse the ester bond between the arabinose substitutions and ferulic acid. Feruloyl esterases aid the release of hemicellulose from lignin and render the free polysaccharide product more amenable to degradation by the other hemicellulases (Howard et al. 2003).

2.5 Inhibitors

2.5.1 *Sugar Degradative Products*

The acid hydrolysis is known to produce inhibitory products such as furfural and hydroxymethyl furfural upon dehydration of pentose and hexose, respectively. Although the furans can further be converted to other products such as furoic, levulinic, formic and hydroxyvaleric acid, the furfural and hydroxymethyl furfural are mostly found under hemicellulose hydrolysis conditions.

2.5.2 *Sugar-Bound Organic Acids*

The hemicellulose is a heteropolymer which carries not only sugar monomers, but also organic acids such as acetic and uronic acids in trace amount. During acid or enzymatic hydrolysis, the organic acids are released from the solid biomass along with the pentose sugar which results in inhibition of xylitol production.

2.5.3 *Lignin Degradatives*

Lignin is a complex organic polymer composed of cross-linked phenolic precursors. The conferyl alcohol, sinaphyl alcohol and paracoumaryl alcohol are the major components of lignin (Velmurugan and Incharoensakdi 2018). During thermal hydrolysis, the lignin has been reported to release various phenolics and organic acids.

2.6 Detoxification of Hemicellulose Hydrolysate

2.6.1 *Physical Methods*

2.6.1.1 Vacuum Evaporation

Vacuum evaporation is a physical treatment which simultaneously removes volatile compounds and concentrate the sugars. The disadvantage of this method is the increase of non-volatile compounds (lignin derivatives) along with the sugar syrup (Parajo et al. 1997). Larsson et al. (2001) achieved complete removal of furfural and 4% reduction in hydrolymethyl furfural content from wood hydrolysate by decreasing 10% of total volume. Similarly, Rodrigues et al. (2001) obtained the furfural removal of about 98% by vacuum evaporation from sugarcane bagasse hydrolysate. On the other hand, Parajo et al. (1997) observed increase in concentration of lignin derivatives in wood hydrolysate when the volume of hydrolysate was reduced to about 1/3. Silva and Roberto (2001) converted xylose from detoxified rice straw hydrolysate to xylitol through microbial fermentation, with the reduction of conversion in the presence of non-volatile compounds.

2.6.1.2 Solvent Extraction

Solvents are potential chemical to extract various biomolecules from plant sources. Similar to the extraction, the hemicellulose hydrolysate has been detoxified with the help of solvents like di-ethyl ether. The phenolics removal efficiency of 65% has been achieved when the hard wood hemicellulosic hydrolysate was treated with diethyl ether (Converti et al. 1999). On the other hand, the diethyl ether was reported to remove volatile compounds from detoxified Eucalyptus wood hydrolysates (Parajo et al. 1997). Persson et al. (2002) developed supercritical fluid to extract the furans, phenolics and aliphatic acids from lignocellulosic hydrolysate and reported increase of the fermentation by *Saccharomyces cerevisiae*. Wang and feng (2010) used aqueous two-phase system with two polymers or one polymer with salt for the removal of inhibitory compounds from lignocellulosic hydrolysate. The advantage of this method is the requirement of mild temperature, which does not affect sugar concentration and is safe to fermenting microorganisms. However, the cost of polymer and poor selection are important concern. Griffin and Shu (2004) used various boronic acid based extraction for detoxification of sugarcane bagasse hydrolysate. Among the boronic acids, naphthlene-boronic acid increased 90% reduction in acid soluble lignin from acid hydrolysate and increased sevenfold in xylose concentration.

2.6.1.3 Separation Through Molecular Sieves

Molecular size-based separation of biomolecules through sieves is a successful method at industrial level. Tran and Chambers (1986) used molecular sieves for the removal of inhibitors from oak hydrolysates in which the acetic acid and furfural removal was 40% and 82%, respectively. The application of two-stage membrane filtration (200 and 100 Da molecular weight membranes) in detoxification of sugar maple hemicellulose hydrolysate was found to remove the acetic acid, furfural and HMF (Stoutenburg et al. 2008), which eventually increased the fermentation by *Pichia stipitis* NRRL Y-7124. The commercial nanofiltration membranes (NF 90 and NF 270) was found to be suitable for removal of furfural from lignocellulosic hydrolysate (Qi et al. 2011). Wickramasinghe and Grzenia (2008) used anion exchange membranes (Sartobind Q) for the removal of acetic acid from biomass hydrolysates and observed better results than ion exchange resin (Amberlyst A21).

2.6.2 *Chemical Methods*

The chemical methods were applied for detoxification based on charges present in biomolecules/chemicals. Generally, the alkaline chemicals such as sodium hydroxide, calcium hydroxide, ammonium hydroxide, sodium sulfite, and sodium di-thionite have been applied to remove the inhibitors through simple precipitation. Besides, the adsorbing materials such as activated charcoal or ion-exchange resins have been reported to efficiently remove the toxic compounds.

2.6.2.1 Neutralization

Neutralization is the necessary process for microbial fermentation as the hydrolysate obtained from chemical hydrolysis mostly be acidic characteristics. The common alkaline chemical used for neutralization are hydroxides of calcium and sodium. It is reported that the neutralization process itself has the ability to remove certain inhibitory products of hydrolysate. However, inhibitor specific detoxification is vital when the concentration reaches the inhibitory level.

2.6.2.2 Overliming

The overliming is used for detoxification that utilizes calcium hydroxide to obtain alkaline condition (pH 10–11). As a result of overliming, most of the inhibitory components were removed through precipitate formation. The overliming is reported to remove the volatile compounds such as furfural, hydroxymethyl furfural and phenolics.

According to Roberto et al. (1991), pH adjusting with a combination of bases and acids is a low-cost treatment that gives good results. By adjusting the pH of sugarcane bagasse hemicellulosic hydrolysate first to 10 with $Ca(OH)_2$, and then to 6.5 with H_2SO_4, these authors obtained a partial removal of phenolic and other compounds, and a xylitol yield of 0.48 g/g. In contrast, the pH adjustment first with $Ca(OH)_2$ and then with NaOH enhanced the hydrolysate fermentability (Van Zyl et al. 1998). Similarly, Martinez et al. (2001) proved that using $Ca(OH)_2$ to adjust the pH of sugarcane bagasse hemicellulose hydrolysate to 9.0 improve the detoxification of acid hydrolysate.

2.6.2.3 Chemical Adsorption

The chemical adsorption removes the inhibitory compounds as most of the inhibitory compounds are charged molecule. The adsorption has been observed to remove furfural, hydroxymethyl furfural and phenolics, which are the most predominant inhibitors present in acid hydrolysates (Dominguez et al. 1996; Silva et al. 1998; Lee et al. 1999; Mussatto and Roberto 2001). Among the adsorbant, the charcoal is proved as detoxifying agent for hydrolysate from corn cob (Gupta et al. 2017), cornfibre (Buhner and Agblevor 2004), oak wood (Converti et al. 1999), rice straw (Mussatto et al. 2004) and sago trunk (Mustapa-Kamal et al. 2011). Toxic compounds may also be adsorbed on diatomaceous earth (Ribeiro et al. 2001), and on ion-exchange resins (Van Zyl et al. 1998; Lee et al. 1999; Larsson et al. 2001; Nilvebrant et al., 2001). Recently, Gupta et al. (2017) used charcoal for detoxification of corncob acid hydrolysate, in which the un-detoxified hydrolysate showed no xylitol production whereas the detoxified hydrolysate produced 4.8 g/L of xylitol.

2.6.3 Biological Adsorption

The use of microbial biomass as biosorbant for detoxification of hydrolysate received much attention in recent past. Biological adsorption method has various advantages over chemical adsorption such as cheaper, easily recyclable and biodegradable. The biosorption has been applied and the reports indicate that the yeast biomass has the ability to adsorb phenolic compounds, acetic acid and furans (Kim et al. 2015; Rossetto et al. 2020; Wang and Chen 2006). Recently, Jofre et al. (2021) used dry yeast for detoxification of hemicellulosic hydrolysates for xylitol production. The maximum phenolic compounds, acetic acid, furfural and 5-hydroxymethyl furfural removal of 27.0%, 27.5% and 25.8% was reported, respectively. In case of hemicellulose hydrolysate detoxification, the in-situ preparation of biosorbants from fermentation process can be a potential detoxifying agent.

2.6.4 Combined Detoxification

The combined detoxification is the process of using various techniques to remove most of the inhibitory components based on chemical properties. In this method, the detoxification of most of the inhibitory compound is possible as it undergoes product specific separation. Kumar et al. (2019) used combination of activated charcoal, membrane separation and ion exchange resin for detoxification of corn cob hydrolysate and obtained 62% of xylitol yield. Vallejos et al. (2016) used precipitation ($Ca(OH)_2$), anionic resins (IR-120 and IRA-120) and adsorption (activated charcoal) for detoxification. The adsorption process was reported for the removal of furfural, hydroxymethyl furfural and phenolic compounds, whereas the anionic exchange resin was effective for acetic acid and formic acids removal. The combination of detoxification techniques and their influence on xylitol production is represented in Table 2.2.

Table 2.2 Comparison of detoxification strategies applied for xylitol production

Feedstock	Detoxification	Yeast	Remarks	References
Sugarcane bagasse	Neutralization and activated charcoal	Candida guiliemondii FTI 20,037	Lowers furans; increase xylitol production	Carvalho et al. (2004)
Sugarcane bagasse	$Ca(OH)_2$, ion exchange resin and activated charcoal	*Candida tropicalis*	Removed all inhibitors; yield was 0.46 g/g	Vallejos et al. (2016)
Poplar wood	Vacuum evaporation and solvent extraction	*Candida guilliermondii*	Acetic acid and furfural removal was 80 and 98%; yield was 0.59 g/g	Dalli et al. (2017)
Rapeseed straw	Activated charcoal and ion-exchange resin	*Debaryomyces hansenii* NRRL Y-7426	Xylitol yield was 0.55 g/g	Lopez-Linares et al. (2018)
Corncob	Activated charcoal and membrane filtration	*Candida tropicalis* MTCC 6192	Xylitol yield was 62%	Kumar et al. (2019)

2.7 Further Prospective and Conclusions

The xylitol production from renewable feedstock through biological method is the potential route for eco-friendly process development. However, the generation of inhibitory components during acid hydrolysis is un-avoidable, which results in reduced xylitol yield. To obtain a successful xylitol production strategy, it is essential to detoxify the acid hydrolysate before the microbial fermentation. As discussed in this chapter, each detoxification method is strict to the compound and each method has advantages and disadvantages. In context, the combined methods are mostly accepted technique which is used in detoxification of acid hydrolysate from various feedstocks. In order to develop an efficient detoxification, the detoxifying agent should be optimized in terms of stability, reactivity and recyclability before commercialization. In addition, the metabolic engineering of xylitol producing strain to tolerate or utilize the inhibitors for their growth can be more advantageous.

Acknowledgements R.V. is thankful to the Graduate School and Faculty of Science, Chulalongkorn University (CU), for post-doctoral fellowship from Rachadaphiseksomphot Endowment Fund. A.I. acknowledges the research grants from CU on the Frontier Research Energy Cluster (CU-59-048-EN) and from Thailand Research Fund (IRG 5780008).

References

Badger PC (2002) Ethanol from cellulose: a general review. In: Janick J, Whipkey A (eds) Trends in new crops and new uses. V.A., ASHS Press, Alexandria, pp 17–21

Buhner J, Agblevor FA (2004) Effect of detoxification of dilute-acid corn fiber hydrolysate on xylitol production. Appl Biochem Biotechnol 119:13–30

Carvalheiro F, Duarte LC, Girio FM (2008) Hemicellulose biorefineries: a review on biomass pretreatments. J Sci Ind Res 67:849–864

Carvalho W, Canilha L, Mussatto SI, Dragone G, Morales ML, Solenzal AIN (2004) Detoxification of sugarcane bagasse hemicellulosic hydrolysate with ion-exchange resins for xylitol production by calcium alginate-entrapped cells. J Chem Technol Biotechnol 79:863–868

Chandel AK, Es C, Rudravaram R, Narasu ML, Rao LV, Ravindra P (2007) Economics and environmental impact of bioethanol production technologies: an appraisal. Biotechnol Mol Biol Rev 2:14–32

Chandel AK, Silva SS, Singh OV (2011) Detoxification of lignocellulosic hydrolysates for improved bioconversion of bioethanol. In: Bernardes MAS (ed) Biofuel production: recent developments and prospects. InTech, Rijeka

Converti A, Dominguez JM, Perego P, Silva SS, Zilli M (2000) Wood hydrolysis and hydrolysate detoxification of subsequent xylitol production. Chem Eng Technol 23:1013–1019

Converti A, Perego P, Dominguez JM (1999) Xylitol production from hardwood hemicellulose hydrolysates by *P. tannophilus*, *D. hansenii* and *C. guilliermondii*. Appl Biotechnol Biochem 82:141–151

Dalli SS, Patel M, Rakshit SK (2017) Development and evaluation of poplar hemicellulose prehydrolysate upstream processes for the enhanced fermentative production of xylitol. Biomass Bioenergy 105:402–410

Demirbas A (2005) Bioethanol from cellulosic material: a renewable motor fuel from ethanol. Energy Sour 27:327–337

Dominguez JM, Gong CS, Tsao GT (1996) Pretreatment of sugar cane bagasse hemicellulose hydrolyzate for xylitol production by yeast. Appl Biochem Biotechnol 57–58:49–56

Filho EXF, Tuohy MG, Puls J, Coughlan MP (1991) The xylan-degrading enzyme systems of *Penicillium capsulatum* and *Talaromyces emersonii*. Biochem Soc Trans 19:25

Griffin GJ, Shu L (2004) Solvent extraction and purification of sugars from hemicelluloses hydrolysates using boronic acid carriers. J Chem Technol Biotechnol 79:505–511

Gusakov AV, Salanovich TN, Antonov AI, Ustinov BB, Okunev ON, Burlingame R (2007) Design of highly efficient cellulase mixtures for enzymatic hydrolysis of cellulose. Biotechnol Bioeng 97:1028–1038

Gupta R, Hemansi, Gautama S, Shukla R, Kuhad RC (2017) Study of charcoal detoxification of acid hydrolysate from corncob and its fermentation to xylitol. J Environ Chem Eng 5:4573–4582

Haltrich D, Nidetzky B, Kulbe KD, Steiner W, Zupancic S (1996) Production of fungal xylanases. Bioresour Technol 58:137–161

Howard RL, Abotsi E, Jansen van Rensburg EL, Howard S (2003) Lignocellulose biotechnology: issues of bioconversion and enzyme production. Afr J Biotechnol 2:602–619

Jeon WY, Shim WY, Lee SH, Choi JH, Kim JH (2013) Effect of heterologous xylose transporter expression in Candida tropicalis on xylitol production rate. Bioprocess Biosyst Eng 36:809–817

Jofre FM, Hernandez-Perez AF, Santos JCD, Felipe MDGD (2021) Use of dry yeast biomass as a new approach for detoxification of hemicellulosic hydrolysates aiming to xylitol production. Ind Crop Product 170:113812

Jorgensen H, Eriksson T, Borjesson J, Tjerneld F, Olsson L (2003) Purification and characterization of five cellulases and one xylanase from *Penicillium brasilianum* IBT 20888. Enzyme Microb Technol 32:851–861

Katahira S, Mizuike A, Fukuda H, Kondo A (2006) Ethanol fermentation from lignocellulosic hydrolysate by a recombinant xylose- and cellooligosaccharide-assimilating yeast strain. Appl Microbiol Biotechnol 72:1136–1143

Kim TY, Lee JW, Cho SY (2015) Application of residual brewery yeast for adsorption removal of reactive orange 16 from aqueous solution. Adv Powder Technol 26:267–274
Kirk TK, Connors WJ, Zeikus JG (1977) Advances in understanding the microbiological degradation of lignin. In: Loewus FA, Runeckles VC (eds) The structure, biosynthesis and degradation of wood. Plenun, New York, pp 369–394
Kumar V, Sandhu PP, Ahluwalia V, Mishra BB, Yadav SK (2019) Improved upstream processing for detoxification and recovery of xylitol produced from corncob. Bioresour Technol 291:121931
Larsson S, Cassland P, Jönsson LJ (2001) Development of a Saccharomyces cerevisiae strain with enhanced resistance to phenolic fermentation inhibitors in lignocellulose hydrolysates by heterologous expression of laccase. Appl Environ Microbiol 67:1163–1170
Lee WG, Lee JS, Shin CS, Park SC, Chang HN, Chang YK (1999) Ethanol production using concentrated oak wood hydrolysates and methods to detoxify. Appl Biochem Biotechnol 77–79:547–559
Lopez-Linares JC, Romero I, Cara C, Castro E, Mussatto SI (2018) Xylitol production by *Debaryomyces hansenii* and *Candida guilliermondii* from rapeseed straw hemicellulosic hydrolysate. Bioresour Technol 247:736–743
Lynd LR, Weimer PJ, Zyl WH, Pretorius IS (2002) Microbial cellulose utilization: fundamentals and biotechnology. Microbiol Mol Biol Rev 66:506–577
Lynd LR (1996) Overview and evaluation of fuel ethanol from cellulosic biomass: technology, economics, the environment, and policy. Annu Rev Energy Environ 21:403–465
Martinez A, Rodriguez ME, Wells ML, York SW, Preston JF, Ingram LO (2001) Detoxification of dilute acid hydrolysates of lignocellulose with lime. Biotechnol Progr 17:287–293
Mussatto SI, Santos JC, Roberto IC (2004) Effect of pH and activated charcoal adsorption on hemicellulosic hydrolysate detoxification for xylitol production. J Chem Technol Biotechnol 79:590–596
Mustapa-Kamal SM, Mohamad NL, Abdullah AGL, Abdullah N (2011) Detoxification of sago trunk hydrolysate using activated charcoal for xylitol production. Procedia Food Sci 1:908–913
Mussatto SI, Roberto IC (2001) Hydrolysate detoxification with activated charcoal for xylitol production by Candida guilliermondii. Biotechnol Lett 23:1681–1684
Mussatto SI, Roberto IC (2004) Alternatives for detoxification of diluted-acid lignocellulosic hydrolyzates for use in fermentative processes: a review. Bioresour Technol 93:1–10
Nair NU, Zhao H (2010) Selective reduction of xylose to xylitol from a mixture of hemicellulosic sugars. Metab Eng 12:462–468
Nilvebrant NO, Reimann A, Larsson S, Jönsson LJ (2001) Detoxification of lignocellulose hydrolysates with ion exchange resins. Appl Biochem Biotechnol 91–93:35–49
Oh EJ, Ha SJ, Rin Kim S, Lee WH, Galazka JM, Cate JH, Jin YS (2013) Enhanced xylitol production through simultaneous co-utilization of cellobiose and xylose by engineered Saccharomyces cerevisiae. Metab Eng 15:226–234
Parajo JC, Dominguez H, Dominguez JM (1997) Improved xylitol production with Debaryomyces hansenii Y-7426 from raw or detoxified wood hydrolyzates. Enzyme Microb Technol 21:18–24
Persson P, Larsson, S, Jonsson LJ, Nilvebrant NO, Sivik B, Munteanu F, Thoerneby L, Gorton L (2002) Supercritical fluid extraction of a lignocellulosic hydrolysate of spruce for Detoxification and to facilitate analysis of inhibitors. Biotechnol Bioeng. 79:694–700
Qi B, Luo J, Chen X, Hang X, Wan Y (2011) Separation of furfural from monosaccharides by nanofiltration. Bioresour Technol 14:7111–7118
Rahman SHA, Choudhury JP, Ahmad AL, Kamaruddin AH (2007) Optimization studies on acid hydrolysis of oil palm empty fruit bunch fiber for production of xylose. Bioresour Technol 98:554–559
Rao RS, Jyothi CP, Prakasham RS, Sarma PN, Rao LV (2006) Xylitol production from corn fiber and sugarcane bagasse hydrolysates by *Candida tropicalis*. Bioresour Technol 97:1974–1978
Reis SD, Costa MAF, Peralta RM (2003) Xylanase production by a wild strain of *Aspergillus nidulans*. Acta Sci Biol Sci 1:221–225

Ribeiro MHL, Lourenco PAS, Monteiro JP, Ferreira-Dias S (2001) Kinetics of selective adsorption of impurities from a crude vegetable oil in hexane to activated earths and carbons. Eur Food Res Technol 213:132–138

Roberto IC, Felipe MGA, Lacis LC, Silva SS, Mancilha IM (1991) Utilization of sugar cane bagasse hemicellulosic hydrolysate by *Candida uilliermondii* for xylitol production. Bioresour Technol 36:271–275

Rodrigues RCLB, Felipe MGA, Silva JBA, Vitolo M, Villa PV (2001) The influence of pH, temperature and hydrolysate concentration on the removal of volatile and non-volatile compounds from sugarcane bagasse hemicellulosic hydrolysate treated with activated charcoal before or after vacuum evaporation. Braz J Chem Eng 18:299–311

Rossetto R, Maciel GM, Bortolini DG, Ribeiro VR, Haminiuk CWI (2020) Acai pulp and seeds as emerging sources of phenolic compounds for enrichment of residual yeasts (Saccharomyces cerevisiae) through biosorption process. Lwt 128

Saha BC (2003) Hemicellulose bioconversion. J Ind Microbiol Biotechnol 30:279–291

Sasaki M, Jojima T, Inui M, Yukawa H (2010) Xylitol production by recombinant Corynebacterium glutamicum under oxygen deprivation. Appl Microbiol Biotechnol 86:1057–1066

Sheehan J, Himmel M (1999) Enzymes, energy, and the environment: a strategic perspective on the U.S. Department of Energy's research and development activities for bioethanol. Biotechnol Prog 15:817–827

Silva CJSM, Roberto IC (2001) Improvement of xylitol production by *Candida guilliermondii* FTI 20037 previously adapted to rice straw hemicellulosic hydrolysate. Appl Microbiol 32:248–252

Silva SS, Felipe MGA, Vitolo M (1998) Xylitol production by *Candida guilliermondii* FTI 20037 grown in pretreated sugar cane bagasse hydrolysate. Sustain Agr Food Energ Ind 1116–1119

Stoutenburg RM, Perrotta JA, Amidon TE, Nakas JP (2008) Ethanol production from a membrane purified hemicellulosic hydrolysate derived from sugar maple by *Pichia stipitis* NRRL Y-7124. BioResources 3:4

Sun R, Tomkinson J (2003) Characterization of hemicelluloses isolated with tetraacetylethylenediamine activated peroxide from ultrasound irradiated and alkali pre-treated wheat straw. Eur Polym J 39:751–759

Tran AV, Chambers RP (1986) Ethanol fermentation of red oak prehydrolysate by the yeast *Pichia stipitis* CBS 5776. Enzyme Microb Technol 8:439–444

Tuohy MG, Puls J, Claeyssens M, Vrsanska M, Coughlan MP (1993) The xylandegrading enzyme system of *Talaromyces emersonii*: novel enzymes with activity against aryl beta-D-xylosides and unsubstituted xylans. Biochem J 290:515–523

Vallejos ME, Chade M, Mereles EB, Bengoechea DI, Brizuela G, Felissia E, Area MC (2016) Strategies of detoxification and fermentation for biotechnological production of xylitol from sugarcane bagasse. Ind Crop Product 91:161–169

Van Zyl C, Prior BA, du Preez JC (1998) Production of ethanol from sugar cane bagasse hemicellulose hydrolysate by Pichia stipitis. Appl Biochem Biotechnol 17:357–369

Velmurugan R, Incharoensakdi A (2018) Nanoparticles and organic matter: process and Impact. In: Ahmad P (ed) Nanomaterials in plants, algae and micro-organisms, vol 1. Academic press, Elsevier, pp 407–428

Virupakshi S, Babu KG, Gaikwad SR, Naik GR (2005) Production of a xylanolytic enzyme by a thermoalkaliphilic *Bacillus* sp. JB-99 in solid state fermentation. Process Biochem 40:431–435

Wang B, Feng H (2010) Detoxification of lignocellulosic hydrolysates. In: Blaschek HP, Ezeji TC, Scheffran J (eds) Biofuels from agricultural wastes and by-products. Wiley-Blackwell, Oxford. https://doi.org/10.1002/9780813822716.ch11

Wang J, Chen C (2006) Biosorption of heavy metals by Saccharomyces cerevisiae: a review. Biotechnol Adv 24:427–451

Wickramasinghe SR, Grzenia D (2008) Adsorptive membranes and resins for acetic acid removal from biomass hydrolysates. Desalination 234:144–151

Wiselogel A, Tyson S, Johnson D (1996) Biomass feedstock resources and composition. In: Wyman CE (ed) Handbook on bioethanol, production and utilization, vol 6, pp 105–118

Wong KKY, Saddler JN (1992) *Trichoderma* xylanases, their properties and applications. In: Visser J, Beldman G, Kusters-van Someren M, Voragen AGJ (eds) Xylans and xylanases. Amsterdam, Elsevier, pp 171–186
Wyman CE (1996) Ethanol production from lignocellulosic biomass: overview. In: Wyman CE (ed) Handbook on ethanol production: production and utilization. Taylor and Francis, Bristol, Pa, pp 1–18
Zhang YHP, Himmel ME, Mielenz JR (2006) Outlook for cellulase improvement: screening and selection strategies. Biotechnol Adv 24:452–481
Zhuang J, Li L, Pang C (2012) Detoxification of wheat straw formic acid hydrolysis and xylitol production. Adv Mat Res 383–390:5453–5457

Chapter 3
Fermentative Production of Xylitol from Various Lignocellulosic Hydrolysates

Sarah Souza de Queiroz, Fanny Machado Jofre, Italo Andrade de Bianchini, Fernanda Weber Bordini, Tatiane Silva da Boaes, Anuj Kumar Chandel, and Maria das Graças de Almeida Felipe

Abstract The increasing scarcity and price inflation of fossil fuels in association with environmental pollution and climate change have led researchers, policy analysts, and organizations to pursue alternative production ways to minimize the impact of petroleum-derived products utilization. Due to this, there is a growing interest in the implementation of a bio-based economy, in which production systems employ renewable biological resources for the generation of the broad spectrum of products required by society, such as fuels, chemicals, materials, and energy/power. In this context, several studies have been carried out over the last years on the microbial production of xylitol from hemicellulosic hydrolysates. The continuous interest in the biotechnological production of xylitol is due to several factors, among them the fact that it mainly uses lignocellulosic biomass (LCB) as raw material. There is a great diversity of LCB cultivated around the world, such as corn, sugarcane, rapeseed, sugar beet, rice, among others. However, due to the recalcitrant nature of cell wall structure, different biomass pretreatment methods must be employed to facilitate its conversion into products. The literature reports that among the various physical, chemical, and biological pretreatments, dilute acid hydrolysis is the most employed method due to its high efficiency and low cost compared to other pretreatments. Thus, the use of agro-industrial waste as feedstock to obtain xylose-rich hemicellulosic hydrolysates in the production of xylitol and other high-added value products in a context of biorefineries contributes to the growth of a bio-based economy. This chapter presents the LCB distribution in the world, methods to obtain xylose coupled with xylose fermentation for xylitol production from a variety of feedstock.

Keywords Lignocellulosic biomass · Hemicellulose hydrolysis · Fermentation · Xylitol

S. S. de Queiroz · F. M. Jofre (✉) · I. A. de Bianchini · F. W. Bordini · T. S. da Boaes · A. K. Chandel · M. das G. de Almeida Felipe
Departamento de Biotecnologia, Escola de Engenharia de Lorena, Universidade de São Paulo, Lorena CEP 12602-810, SP, Brasil
e-mail: jofrefanny@usp.br

© The Author(s), under exclusive license to Springer Nature Switzerland AG 2022

M. das G. de Almeida Felipe and A. K. Chandel (eds.), *Current Advances in Biotechnological Production of Xylitol*,
https://doi.org/10.1007/978-3-031-04942-2_3

3.1 Introduction

The constant interest in lignocellulosic biomass (LCB) as raw materials for chemicals and energy production is attributable to its chemical composition, which is abundant in carbohydrates and other compounds which can be converted into several products. Among the three main fractions of lignocellulose, lignin, cellulose and hemicellulose, the last is the target fraction regarding the biotechnological production of xylitol (Zoghlami and Paës 2019; Chandel et al. 2021).

Xylitol, a functional sweetener, is a five-carbon sugar alcohol which occurs naturally in small amounts in plums, strawberries, cauliflower, and pumpkin, among others. However, commercially it is synthesized from xylose sugar from hemicellulosic hydrolysate of lignocellulosic biomass. It is a white crystalline solid, soluble in water and low-calorific sweetener compared to sugar. Taking into consideration the growing demand of xylitol, US Department of Energy (USDOE) has included xylitol in top 12 biochemicals made by lignocellulosic biomass (Chandel and Segato 2021). Xylitol has many health benefits such as an alternative sweetener for diabetic patients, ability to prevent demineralization of teeth and bones, otitis media infection, respiratory tract infections, among others (Silva and Chandel 2012).

Several methodologies for deconstruction of hemicellulose can be used, which are divided into physical, chemical, physicochemical, or biological techniques. Among the existing techniques, dilute-acid hydrolysis is commonly employed, and consists of using an acid (sulfuric or phosphoric acid) as a catalyst for the hydrolysis reaction. This technique features high efficiency, short reaction time and is cost effective compared to other methods (Hernández-Pérez et al. 2019a). Some of the most used agro-industrial and forestry residues in this process are sugarcane bagasse and straw, switchgrass, eucalyptus, corn cob, sorghum bagasse, among others. Liquid hot water, dilute acid hydrolysis and steam explosion with or without acid catalyst are the preferred method to obtain hemicellulose hydrolysate, which is predominantly made of xylose, which is a 5-carbon sugar along with glucose, galactose, mannose, arabinose, weak acids, and lignin derived phenolics (Hernández-Pérez et al. 2020a, b, c, d).

The hemicellulosic hydrolysate is submitted to the fermentation process by the selected microorganism employing the desired fermentation conditions. Yeasts particularly *Candida guilliermondii* and *C. utilis* are one of the most widely used microorganisms for the fermentative production of xylitol from a variety of lignocellulose hydrolysates employing a battery of mode of fermentative cultivation techniques (Hernández-Pérez et al. 2019b).

This chapter aims to provide the cutting-edge information on geographical distribution of LCB in the world, cell wall composition of different LCB substrates, methods of hemicellulose hydrolysis and finally the fermentative production of xylitol.

3.2 Lignocellulosic Biomass Distribution Worldwide

LCB as a renewable feedstock has raised great interest in several industrial sectors. Technology development is the key to enhance its use as feedstock to produce fuels and value-added bioproducts through valorization (Chandel et al. 2020). The world's population is approaching 8 billion people (Ali et al. 2020); as well as this rapid growth is also the demand for food production (Rockström et al. 2020). In this context, the scale-up of agriculture based intensification approaches to obtain high yield/productivity is favored rather than the expansion of the planting area (Crist et al. 2017). These intensification strategies are part of a set of agricultural policies to promote carbon neutral bio-economy and environmental protection.

In 2019, the estimated global planted area was about 1.61 billion hectares, and agricultural activities were responsible for the largest occupation of Earth's surface (FAOSTAT 2019). Figure 3.1 shows the production of lignocellulosic biomass in 20 principal agrarian countries, considering all the countries reported to the FAO dataset. A determining factor in the type of commodity exploited in each of these regions is the climate conditions. The climate and temperature directly influence the variety of biomass and their distribution across the world (Antar et al. 2021).

The large-scale produced agricultural commodities are sugarcane, corn, rice, potato, soybeans, cassava, sugar beets, sorghum and wheat. These crops are mainly produced for food but can also be used as raw materials to diversify production chains (FAO 2021). Corn, sugarcane, wheat, soybean, rapeseed, and sunflower are the main crops used in the production of biofuels worldwide, however, their availability for

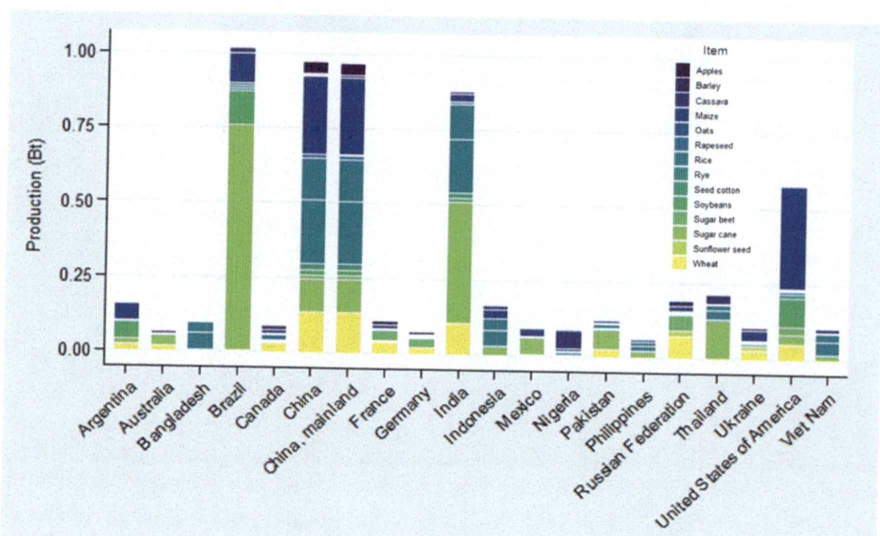

Fig. 3.1 Top 20 countries agricultural production based on lignocellulosic biomass crops worldwide (Adapted from FAOSTAT 2019)

the energy sector is directly or indirectly related to the supply of the food industry. This dispute for raw material contributes to increase the demand and to elevate the prices. In addition, raw material availability is also dependent on crop seasonality and adverse weather conditions (Antar et al. 2021).

LCB is geographically abundant worldwide (Nanda et al. 2015). The top 5 producers among the evaluated commodities were highlighted (Figs. 3.1 and 3.2). In North America, the most important crops are corn, soybean, and wheat. In South America, there is intensive sugarcane, soybeans, and maize (corn) production. Together, these crops occupy more than 80% of the continent's agricultural production (Magalhães et al. 2019). In this context, Brazil occupies a leading position in the global production of crops (Fig. 3.1). Sugarcane takes the first position in the ranking as the principal commodity produced by the country, estimated at 592,031.3 thousand tons in the 2021/22 harvest (CONAB 2021). On the other hand, the largest agricultural production currently occurs in the Asian continent, mainly concentrated in China and India (FAOSTAT 2019).

Most of the crops produced on a large scale already have a defined reuse destination. On the other hand, the waste obtained from smaller scale agro-industrial crops remains undefined. One example is rapeseed straw, whose generation is increasing due to the growth in production and sale of rapeseed oil, resulting in poorly utilizing its high sugar content. (Pińkowska et al. 2013; López-Linares et al. 2018).

Fig. 3.2 Global distribution of major crop producers and crop-specific producing countries of corn, soybeans, sugarcane, sugar beet, rice, and wheatn (Adapted from FAOSTAT 2019)

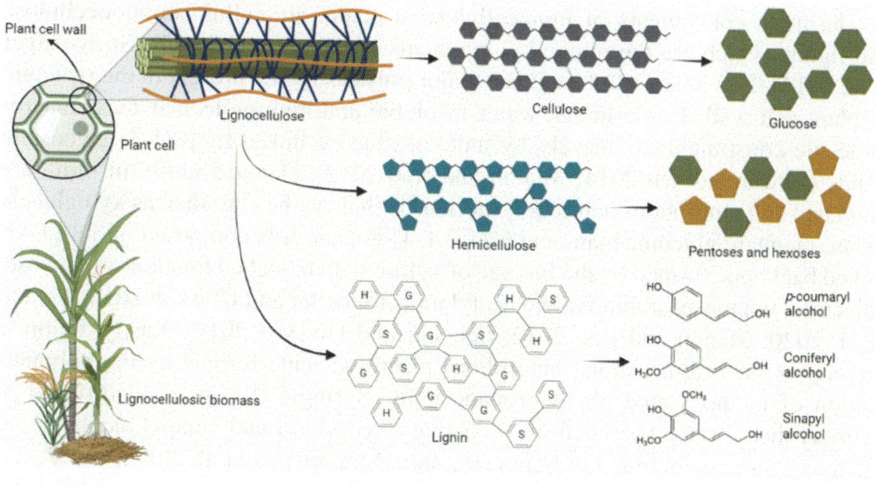

Fig. 3.3 Molecular architecture of plant cell wall

As well as the intensification of agricultural production, the generation of residues from biomass processing is increasing among geographical areas and crop groups. Waste residues from agriculture and forestry have a great potential to be integrated into processes aiming at sustainable development. In this context, their valorization is one of the most prospective resources for the growth of a bio-based economy. Green technologies directed at the production of fuels and bioproducts can contribute to the economy and collaborate to the reduction of greenhouse gas emissions and global warming (Nanda et al. 2015; Antar et al. 2021).

The relationship between agriculture and industry is a key point in the development of a more sustainable society. In this context, science and technological innovation are expanding production capacity to meet the increased global demand for food, biofuels, and plant-based bioproducts. To summarize, "Modern Agriculture" looks for the full use of biomass waste. These strategies can favor the uplifting of a pragmatic bioeconomy and amplification of integrated production chains.

3.3 LCB as Sustainable Feedstock for the Production of High-Added Value Products

LCB is considered as one of the most promising feedstock for the production of bio-based products, mainly because of their foreseeable amount i.e. approximately production of 200×10^9 tons per year (Romaní et al. 2020). These materials encompass several types of feedstocks, which are widely available worldwide, such as agro-industrial by-products, forestry wastes, energy crops, and municipal solid wastes (Antunes et al. 2017). In addition, the diversified composition of LCB also makes them suitable to be converted into several bio-based products.

The main constituents of lignocellulose structure are cellulose, hemicellulose and lignin, which are the integral components of the plant cell-wall (Strassberger et al. 2014) (Fig. 3.3). Cellulose is the major polysaccharide present in the structure of plant cell-wall. It is a linear, water insoluble and high molecular mass macromolecule constituted exclusively by units of glucose linked by β-(1-4) glycosidic bonds (Juturu and Wu 2014; Menon and Rao 2012). Hemicellulose, in turn, is a branched and amorphous heteropolysaccharide that can be classified as xyloglucan, xylan, mannan, glucomannan, and β-(1-3,1-4)-glucan. It is composed of a β-(1-4)-linked backbone formed by the linkage of xylose with branched ligations of glucose, galactose, mannose, arabinose and acetyl units (Scheller and Ulvskov 2010; Arcaño et al. 2020; Menon and Rao 2012; Scheller and Ulvskov 2010). Lastly, lignin is a complex three-dimensional amorphous macromolecule formed by the polymerization of methoxylated phenylpropane units. Syringyl (S), guaiacyl (G) and *p*-hydroxyphenyl (H), derived from *p*-coumaryl, coniferyl and sinapyl alcohols, are the major monomers found in lignin structure (Strassberger et al. 2014).

The lignocellulose structure can be fully converted into several high-added value products. The first step to achieve this goal is pretreat the biomass to isolate the targeted lignocellulosic fraction. Depending on the pretreatment method used, each fraction can be solubilized into its respective constituent monomers. Then, the fractions themselves or their monomers can be converted into a wide range of bioproducts (Fig. 3.3) (Menon and Rao 2012; Chandel et al. 2019). Fermentative processes are frequently employed to convert the monomers released from lignocellulose into different bio-based products. Through employing microorganisms with distinct characteristics, several bioproducts can be obtained from cellulosic and hemicellulosic sugars, such as ethanol, sugar alcohols (xylitol, sorbitol, mannitol, erythritol, and arabitol), organic acids (citric acid, lactic acid, acetic acid, propionic acid, and burytic acid), biosurfactants, biopigments, biogas, single cell protein, among countless others (Hernández-Pérez et al. 2020a, b, c, d; Antunes et al. 2017; Chandel et al. 2018). Some studies have also shown the capacity of some microorganisms, mainly bacteria, to convert the monomers derived from lignin into different high-added products. After lignin depolymerization, the monomers can be converted into bioproducts, such as lipids, aromatics, polyhydroxyalkanoates, pyruvate, lactate, pyrogallol, and others (Lago et al. 2012; Strassberger et al. 2014; Xu et al. 2019) (Fig. 3.4).

Besides the fermentative production of biomolecules, physical, chemical, physicochemical, and/or enzymatic processes can also be employed in the generation of high-added value products (Fig. 3.2) (Werpy and Petersen 2004). Xylooligosaccharides, used as prebiotics due their health benefits, are often produced from the hemicellulosic fraction of lignocellulosic biomass by employing thermal or enzymatic treatments (Santibáñez et al. 2021). Nanocellulose is also a high-added value product that can be obtained from lignocellulosic biomass. To this, hemicellulose, lignin, and other compounds are first removed by a pretreatment method, which can be acid or alkaline treatments. After that, nanocellulose can be extracted from cellulose fibrils by different extraction methods, as acid and enzymatic hydrolysis or by the application of mechanical processes, such as homogenization at high pressure, ultrasonication, and ball milling (Phanthong et al. 2018). The application of lignin

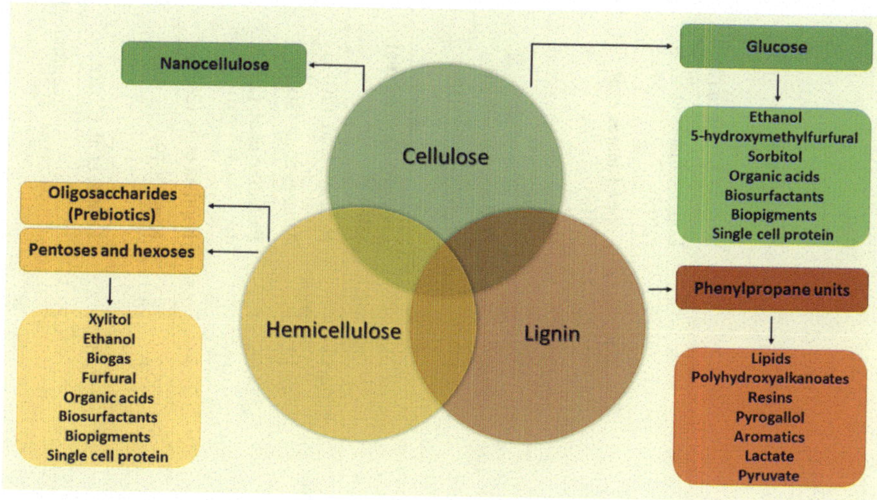

Fig. 3.4 Examples of bio-based products that can be obtained from lignocellulosic biomass

itself in the production of bio-based products is also described by some works, as in the generation of bio-based phenolic resins. In this field, lignin acts as a macromolecular substitute for phenol, being cross-linked by an aldehyde cross-linker to produce the resins (Nieuwenhove et al. 2020).

3.4 Lignocellulosic Biomass Used for the Biotechnological Production of Xylitol

The use of lignocellulosic biomass, the most abundant and renewable raw material available on the planet, for value-added molecules products is steadily growing. Xylitol is a biomolecule of great commercial interest and its production from the biotechnological route is promising compared to the chemical conversion (Ghaffar et al. 2017).

LCB from agriculture represents more than 60% of the biomass produced worldwide (Ghaffar et al. 2017; Espinoza-Acosta 2020; Cortivo et al. 2018). Different biomass sources are used as raw materials for xylitol production. Among the by-products obtained from LCB, the straw from corn, and corn bran originating from grain milling, are rich in pentosans and can be used as an alternative feedstock for xylitol production (Irmak et al. 2017). A study by Du et al. (2020) reached a production of 24.2 gL^{-1} of xylitol using corn cob slurry from biomass hydrolysate containing approximately 30 gL^{-1} of initial xylose concentration. Table 3.1 summarizes the xylitol production from a variety of lignocellulosic sources employing fermentation process.

Table 3.1 Various lignocellulosic biomass used in the biotechnological production of xylitol

Biomass	Microorganism	Initial xylose (gL^{-1})	Xylitol			Ref.
			$Y_{P/S}$ (gg^{-1})	Qp (gL^{-1} $\mathbf{h^{-1}}$)	Titer (gL^{-1})	
Corn cob slurry	*Kluyveromyces marxianus* CICC 1727–5	~30.0	0.82	–	24.2	Du et al. (2020)
Sugarcane bagasse	*Candida tropicalis* MTCC 184	42.6	0.83	0.86	34.5	Raj and Krishnan (2020)
Rapeseed straw	*C. guilliermondii* FTI 20,037	41.51	0.55	0.42	–	López-Linares et al. (2018)
Rapeseed straw	*Debaryomyces hansenii* NRRL Y-7426	41.51	0.45	0.41	–	López-Linares et al. (2018)
Sugarcane bagasse	*C. tropicalis*	–	0.86	0.55	38.96	Hernández-Pérez et al. (2020a)
Sugarcane bagasse	*Pichia fermentans* E015	150	0.54	–	79.0	Prabhu et al. (2020)
Oat and soybean hull	*Saccharomyces cerevisiae* YRH 396	28.3	0.33	–	8.17	Cortivo et al. (2018)
White sorghum straw	*D. hansenii*	30.0	1.16	0.61	29.23	Ledezma-Orozco et al. (2018)
Banana peel	*C. tropicalis* DSM 7524	67.8	–	–	24.57	Rehman et al. (2013)
Barley straw	*C. guilliermondii* FTI 20,037	60.0	–	0.65	47.12	Moraes et al. (2020)
Sugarcane bagasse	*Cyberlindnera xylosilytica* UFMG-CM-Y-309	40.0	0.63	0.20	14.06	Palladino et al. (2021)
Corn cob	*S. cerevisiae* PE-2-GRE3	36.8	–	0.83	71.7	Baptista et al. (2020)
Palm fruit bunch fiber	*C. tropicalis*	82.3	0.44	–	35.2	Kim (2019)
Apple pomace	*C. guilliermondii* FTI 20037	11.10	0.38	0.10	9.35	Leonel et al. (2020)

(continued)

Table 3.1 (continued)

Biomass	Microorganism	Initial xylose (gL^{-1})	Xylitol			Ref.
			$Y_{P/S}$ (gg^{-1})	Qp (gL^{-1} $\mathbf{h^{-1}}$)	Titer (gL^{-1})	
Olive pits	*Pichia fermentans*	50	0.74	–	71.9	Narisett et al. (2021)
Sugarcane bagasse, straw and syrup	*C. guilliermondii FTI 20037*	60.0	0.86	0.51	38.96	Hernández-Pérez et al. (2020b)
Rice straw	*C. tropicalis* MTCC 6192	–	0.60	0.26	25.8	Singh et al. (2021)

López-Linares et al. (2018) used rapeseed straw, an agro-industrial waste with high sugar content. This study found a xylitol production of 0.55 gg^{-1}, using the microorganism *Candida guilliermondii* and rapeseed straw with an initial xylose content of 41.41 gL^{-1} (Table 3.1). Sugarcane is a commodity crop widely cultivated in tropical countries. This fractionation generates high amounts of agro-industrial by-products, such as bagasse and straw, which can be used as feedstock for xylitol production (Hernández-Pérez et al. 2020a). Raj and Krishnan (2020) showed the potential of wastes in the xylitol titer, reaching 34.5 gL^{-1} by *C. tropicalis* MTCC 184 using sugarcane bagasse with an initial xylose content of 42.6 gL^{-1}.

The mixture of different by-products as feedstock for xylitol production can also be employed. Cortivo et al. (2018) studied the combination of soy hulls and oats, two agro-wastes with very limited applications in animal feed and low-heat combustion ovens, respectively. When integrated into biorefineries, it reached 8.17 gL^{-1} for xylitol titer, showing a positive balance of sugars for the sweetener production. Hernández-Pérez et al. (2020b) also use a combination of sugarcane straw, bagasse, and syrup for xylitol fermentation by *Candida guilliermondii* FTI 20037 following the integrated biorefinery approach. In this research study, it was reported a titer of approximately 26.19 gL^{-1} with 60 gL^{-1} of initial xylose concentration.

Alternative by-products obtained from smaller-scale agriculture are also promising in the production of high added-value products, such as xylitol. Some examples are the use of banana peel (Rehman et al. 2013), olive pits (Narisett et al. 2021), apple pomace (Leonel et al. 2020), and palm fruit bunch fiber (Kim 2019) (Table 3.1).

As for the microorganisms used for the production of xylitol, the preference for the *Candida* genus is common, due to its high yield even under limited oxygen conditions (Ledezma-Orozco et al. 2018). However, studies with engineered or wild strains are mentioned in the literature in order to expand the range of microorganisms and search for even better production yields (Baptista et al. 2020; Cunha et al. 2019). Study by Prabhu et al. (2020) used a wild *P. fermentans* of food origin and found high levels of xylitol accumulation (79 gL^{-1}) from a non-detoxified sugarcane bagasse pre-hydrolysate rich in xylose (150 gL^{-1}) (Table 3.1). The authors highlight the fact that *P. fermentans* accumulates xylitol in a non-detoxified medium, that is, with the presence of toxic compounds that could have been overcome by the initial inoculum concentration together with the cultivation parameters and/or chemical mutagenesis performed with methanesulfonate of ethyl. In contrast, the study performed by Baptista et al. (2020) found that removing acetic acid and increasing inoculum concentration in corn cob hydrolysate increased xylitol productivity.

The volume of xylitol production by biotechnological route is influenced by several factors, separately or combined, such as the chosen microorganism, the biomass feedstock, parameters of the processes, among others. As well as the study of the best combination conditions, other relevant factors are the associated costs, etc. The biotechnological route integrated into a biorefinery is promising, but bottlenecks still exist and need to be overcome (Narissett et al. 2021).

3.5 Process for Obtaining Hydrolysate from Different Biomass

As presented in the previous topics, lignocellulosic biomass is rich in cellulose, hemicellulose and lignin, which can be used as raw material to produce energy and high-added value products, such as xylitol. However, the utilization of monomeric compounds present in biomass requires first the breakdown of the rigid structure of lignocellulose (Albuquerque et al. 2014). Different techniques of biomass pretreatment focus on the deconstruction of the hemicellulosic fraction from lignocellulose to obtain a xylose-rich solution for xylitol production.

There are several pretreatments that can be used to obtain hemicellulosic hydrolysates, which are physical, chemical and/or biological methods (Rao et al. 2016). Physical pretreatment disrupts the structure of biomass to facilitate the access of chemical or biological catalysts. Some examples of physical disruption methods are grinding, milling, electron beam, microwave among others (Taherzadeh and Karimi 2008). On the other hand, chemical pretreatment uses acid, alkali, organic solvents and other chemicals as catalysts to deconstruct the plant cell wall structure (Rao et al. 2016). Biological pretreatment is mediated by living microorganisms or specific enzymes, which degrades lignin, a complex macromolecule responsible for the recalcitrance of lignocellulose (Vasco-Correa et al. 2016). In addition, other hydrolysis techniques reported in the literature uses inorganic salts, active solvents, solids, supercritical fluids, ionic liquids, steam explosion, lime and ammonia processing and hot water (Arcaño et al. 2020).

It was reported by Mirfakhar et al. (2020) a yield of 18.12 gL^{-1} of xylose from sweet sorghum bagasse by employing the auto-hydrolysis pretreatment of biomass, which uses hot water in a solid:liquid ratio of 8% (w/v), at 210 °C, during 60 min. In the case of rice straw pretreatment for xylitol production, Roberto et al. (1994) performed the acid hydrolysis with 70 mg of concentrated sulfuric acid per gram (dw), 10:1 of liquid to solid ratio, at 145 °C for 20 min and obtained a yield of 16.2 gL^{-1} of xylose.

The most traditional pretreatment method employed is the chemical hydrolysis with dilute acid solution since it has advantages such as greater efficiency, low cost, easy operation and short reaction times (Arcaño et al. 2020). The dilute acid hydrolysis of sugarcane straw was performed by Hernández-Pérez et al. (2016) under the conditions of 1% (w/v) of H_2SO_4, temperature at 121 °C for 20 min, with 1:10 of solid/liquid ratio, resulting in 18.6 gL^{-1} of xylose. Under these same operational conditions, Sene et al. (2011) obtained a high xylose content (17.69 gL^{-1}) from sorghum straw, and, Leonel et al. (2020) obtained 11.1 gL^{-1} from apple pomace.

Recent strategies have employed the combination of dilute acid hydrolysis with different pretreatment to obtain a high-xylose content hydrolysate. Bonfiglio et al. (2021) studied a two-step pretreatment of switchgrass (*Panicum virgatum*) and Eucalyptus (*Eucalyptus globulus*) through steam explosion (200 °C, for 10 min) followed by acid hydrolysis (4% of H_2SO_4, at 121 °C, for 60 min). The authors reported a xylose yield of 39.99 gL^{-1} in switchgrass hydrolysate and 58.06 gL^{-1} of xylose in eucalyptus hydrolysate. In another study, the pretreatment of corn cob was performed through the extraction of hemicellulose with 55% of tetrabutylammonium hydroxide (TBAH) under conditions of 90 min, at 60 °C, liquid solid ratio of 12 (v/w), followed by the dilute acid hydrolysis resulted in a yield of 80.07% hemicellulose extraction (Jia et al. 2016).

Therefore, within this context, it can be concluded that an efficient hydrolysis process must release the maximum amount of sugars, with also minimum formation of inhibitory products (Kumar et al. 2020). The proper choice of the hydrolysis technique is a challenging step that will depend on the type of biomass used, type of equipment, costs and sustainability of the process.

3.6 Conclusions and Future Perspectives

LCB produced on a large scale in agriculture are widely distributed across the globe and the diversity of cultivated species depends on some regional features, such as climate, soil, economy, among other factors. The increase in worldwide demand over the years has caused an intensification in agricultural activities. Although these crops are cultivated with the main objective of food production, the waste generated presents great potential for use as feedstock to produce energy and high added-value chemicals, such as xylitol. In addition, the forestry waste generated also has the potential to be integrated into biotechnological processes. Steam explosion, hot water hydrolysis and dilute acid hydrolysis are the major pretreatment method to recover the xylose from biomass which is a principle carbon source for the xylitol production employing various fermentation approaches. Commercial xylitol production seems techno-economically viable in integrated corn and sugarcane processing mills eventually strengthening the value chain in low carbon economy.

In this context, the development of sustainable technologies for the biotechnological production of xylitol can be achieved with the cooperation and association between the agriculture and forestry sectors together with the industry. Thus, the integral use of vegetal biomass including their waste and by-products in a biorefinery context allows the diversification of the production chain, without competing with the food industry, in order to consolidate a bio-based economy.

Acknowledgements Anuj K. Chandel acknowledges the CNPq-Brazil for the scientific productivity program (Process number: 309214/2021-1).

References

Albuquerque TL, Silva JrIJ, Macedo GR et al (2014) Biotechnological production of xylitol from lignocellulosic wastes: a review. Process Biochem 49:1779–1789

Ali G, Ghulam M, Said SZ, et al (2020). Population clock: World. Int J Curr Microbiol Appl Sci V: 12–23. https://www.census.gov/popclock/world. Zugegriffen: 29. Sept 2021

Antar M et al (2021) Biomass for a sustainable bioeconomy: an overview of world biomass production and utilization. Renew Sustain Energy Rev 139:110691. https://doi.org/10.1016/j.rser.2020.110691

Antunes FAF, Santos JC, Cunha MAA et al (2017) Biotechnological production of xylitol from biomass. In: Fang Z, Smith, JrR, Qi X (eds) Production of platform chemicals from sustainable resources. Springer, Singapore. https://doi.org/10.1007/978-981-10-4172-3_10

Arcaño YD, García ODV, Mandelli D et al (2020) Xylitol: a review on the progress and challenges of its production by chemical route. Catal Today 344:2–14. https://doi.org/10.1016/j.cattod.2018.07.060

Baptista SL, Carvalho LC, Romaní A et al (2020) Development of a sustainable bioprocess based on green technologies for xylitol production from corn cob. Ind Crops Prod 156:112867. https://doi.org/10.1016/j.indcrop.2020.112867

Bonfiglio F, Cagno M, Yamakawa CK et al (2021) Production of xylitol and carotenoids from switchgrass and *Eucalyptus globulus* hydrolysates obtained by intensified steam explosion pretreatment. Ind Crops Prod 170:113800. https://doi.org/10.1016/j.indcrop.2021.113800

Chandel AK, Garlapati VK, Singh AK, Antunes FAF, Silva SS (2018) The path forward for lignocellulose biorefineries: bottlenecks, solutions, and perspective on commercialization. Bioresour Technol 264:370–381. https://doi.org/10.1016/J.BIORTECH.2018.06.004

Chandel AK, Albarelli JQ, Santos DT, Chundawat SP, Puri M, Meireles MAA (2019) Comparative analysis of key technologies for cellulosic ethanol production from Brazilian sugarcane bagasse at a commercial scale. Biofuels, Bioprod Biorefin 13:994–1014. https://doi.org/10.1002/BBB.1990

Chandel AK, Garlapati VK, Kumar SPJ et al (2020) The role of renewable chemicals and biofuels in building a bio-economy. Biofuels Bioprod Biorefin 14:830–844. https://doi.org/10.1002/bbb.2104

Chandel AK, Forte MBS, Gonçalves IS et al (2021) Brazilian biorefineries from second generation biomass: critical insights from industry and future perspectives. Biofuels, Bioprod Biorefin. https://doi.org/10.1002/bbb.2234

Chandel AK, Segato F (2021) Production of Top 12 biochemicals selected by USDOE from renewable resources. Elsevier, Amsterdam, The Netherland. ISBN: 9780128235317

CONAB. Companhia Nacional De Abastecimento (2021) Acompanhamento da safra brasileira: cana-de-açúcar. Safra 2020/21. Primeiro levantamento. Brasília

Cortivo PRD, Hickert LR, Hector R et al (2018) Fermentation of oat and soybean hull hydrolysates into ethanol and xylitol by recombinant industrial strains of *Saccharomyces cerevisiae* under diverse oxygen environments. Ind Crops Prod 113:10–18. https://doi.org/10.1016/j.indcrop.2018.01.010

Crist E, Mora C, Engelman R (2017) The interaction of human population, food production, and biodiversity protection. Science 356:260–264

Cunha JT, Soares PO, Romaní A et al (2019) Xylose fermentation efficiency of industrial *Saccharomyces cerevisiae* yeast with separate or combined xylose reductase/xylitol dehydrogenase and xylose isomerase pathways. Biotechnol Biofuels 12:1–14. https://doi.org/10.1186/s13068-019-1360-8

Du C, Li Y, Zong H et al (2020) Production of bioethanol and xylitol from non-detoxified corn cob via a two-stage fermentation strategy. Bioresour Technol 310:123427. https://doi.org/10.1016/j.biortech.2020.123427

Espinoza-Acosta JL (2020) Biotechnological production of xylitol from agricultural waste. Biotecnia 22:126–134. https://doi.org/10.18633/BIOTECNIA.V22I1.1160

FAO (2021) World Food and Agriculture – Statistical Yearbook 2021. Rome. https://doi.org/10.4060/cb4477en

FAOSTAT. Database collection of the Food and Agriculture Organization of the United Nations. http://www.fao.org/faostat/ (2019). (Accessed 22 Sept 21)

Ghaffar A, Yameen M, Aslam N et al (2017) Acidic and enzymatic saccharification of waste agricultural biomass for biotechnological production of xylitol. Chem Cent J 11:1–6. https://doi.org/10.1186/s13065-017-0331-z

Hernández-Pérez AF, Arruda PV, Sene L et al (2019a) Xylitol bioproduction: State-of-the-art, industrial paradigm shift and opportunities for integrated biorefineries. Crit Rev Biotechnol 39:924–943. https://doi.org/10.1080/07388551.2019.1640658

Hernández-Pérez AF, Antunes FAF, Santos JC et al (2020a) Valorization of the sugarcane bagasse and straw hemicellulosic hydrolysate through xylitol bioproduction: effect of oxygen availability and sucrose supplementation as key factors. Biomass Convers Biorefinery. https://doi.org/10.1007/s13399-020-00993-7

Hernández-Pérez AF, Chaves-Villamil AC, Arruda PV et al (2020b) Sugarcane syrup improves xylitol bioproduction from sugarcane bagasse and straw hemicellulosic hydrolysate. Waste and Biomass Valorization 11:4215–4224. https://doi.org/10.1007/s12649-019-00742-6

Hernández-Pérez AF, Jofre FM, Queiroz SS et al (2020c) Biotechnological production of sweeteners. Biotechnol Prod Bioact Compd 261–292. https://doi.org/10.1016/B978-0-444-64323-0.00009-6

Hernández-Pérez AF, Jofre FM, Queiroz SS, et al (2019b) Biotechnological production of sweeteners. In: Verma, ML, Chandel AK (eds) Biotechnological production of bioactive compounds. Elsevier Press, Amsterdam, The Netherlands

Hernández-Pérez AF, Valadares F, Queiroz SS, et al (2020d) Traditional economy versus modern technology based bioeconomy. In: Kataki R, Pandey A, Khanal SK, Pant D (eds) Current developments in biotechnology and bioengineering-sustainable bioresources for emerging bio-economy, pp. 495–505. Elsevier Press, Amsterdam, The Netherland. ISBN: 978-0-444-64309-4

Irmak S, Canisag H, Vokoun C et al (2017) Xylitol production from lignocellulosics: are corn biomass residues good candidates? Biocatal Agric Biotechnol 11:220–223. https://doi.org/10.1016/j.bcab.2017.07.010

Jia H, Shao T, Zhong C et al (2016) Evaluation of xylitol production using corncob hemicellulosic hydrolysate by combining tetrabutylammonium hydroxide extraction with dilute acid hydrolysis. Carbohyd Polym 151:676–683. https://doi.org/10.1016/j.carbpol.2016.06.013

Juturu V, Wu JC (2014) Microbial cellulases: engineering, production and applications. Renew Sustain Energy Rev. https://doi.org/10.1016/j.rser.2014.01.077

Kim S (2019) Xylitol production from byproducts generated during sequential acid-/alkali-pretreatment of empty palm fruit bunch fiber by an adapted *Candida tropicalis*. Front Energy Res 7:1–9. https://doi.org/10.3389/fenrg.2019.00072

Kumar V, Yadav SK, Kumar J, Ahluwalia V (2020) A critical review on current strategies and trends employed for removal of inhibitors and toxic materials generated during biomass pretreatment. Bioresour Technol 299:122633. https://doi.org/10.1016/j.biortech.2019.122633

Lago AC, Bonomi A, Cavalett O et al (2012) Sugarcane as a carbon source: the Brazilian case. Biomass Bioenerg 46:5–12. https://doi.org/10.1016/j.biombioe.2012.09.007

Ledezma-Orozco E, Ruíz-Salazar R, Bustos-Vázquez G et al (2018) Producción de xilitol a partir de hidrolizados ácidos no detoxificados de bagazo de sorgo por *Debaryomyces hansenii*. Agrociencia 52:1095–1106

Leonel LV, Sene L, Cunha MAA et al (2020) Valorization of apple pomace using bio-based technology for the production of xylitol and 2G ethanol. Bioprocess Biosyst Eng 43:2153–2163. https://doi.org/10.1007/s00449-020-02401-w

López-Linares JC, Romero I, Cara C et al (2018) Xylitol production by *Debaryomyces hansenii* and *Candida guilliermondii* from rapeseed straw hemicellulosic hydrolysate. Bioresour Technol 247:736–743. https://doi.org/10.1016/j.biortech.2017.09.139

Magalhães AI et al (2019) Lignocellulosic biomass from agro-industrial residues in South America: current developments and perspectives. Biofuels Bioprod Biorefin 13:1505–1519. https://doi.org/10.1002/bbb.2048

Menon V, Rao M (2012) Trends in bioconversion of lignocellulose: biofuels, platform chemicals & biorefinery concept. Prog Energy Combust Sci 38:522–550. https://doi.org/10.1016/j.pecs.2012.02.002

Moraes EJC, Silva DDV, Dussán KJ et al (2020) Xylitol-sweetener production from barley straw: optimization of acid hydrolysis condition with the energy consumption simulation. Waste Biomass Valoriz 11:1837–1849. https://doi.org/10.1007/s12649-018-0501-9

Nanda S, Ramin A, Ajay KD et al (2015) An assessment on the sustainability of lignocellulosic biomass for biorefining. Renew Sustain Energy Rev 50:925–941. https://doi.org/10.1016/j.rser.2015.05.058

Narisett V, Castro E, Durgapal S et al (2021) High level xylitol production by Pichia fermentans using non-detoxified xylose-rich sugarcane bagasse and olive pits hydrolysates. Bioresour Technol 342. https://doi.org/10.1016/j.biortech.2021.126005

Nieuwenhove IV, Renders T, Lauwaert J et al (2020) Biobased resins using lignin and glyoxal. ACS Sustain Chem Eng 8:18789–18809. https://doi.org/10.1021/ACSSUSCHEMENG.0C07227

Palladino F, Rodrigues RCLB, Cadete RM et al (2021) Novel potential yeast strains for the biotechnological production of xylitol from sugarcane bagasse. Biofuels Bioprod Biorefining 15:690–702. https://doi.org/10.1002/bbb.2196

Phanthong P, Reubroycharoen P, Hao X et al (2018) Nanocellulose: extraction and application. Carbon Resour Convers 1:32–43. https://doi.org/10.1016/J.CRCON.2018.05.004

Pińkowska H, Wolak P, Oliveros E (2013) Production of xylose and glucose from rapeseed straw in subcritical water—Use of Doehlert design for optimizing the reaction conditions. Biomass Bioenerg 58:188–197. https://doi.org/10.1016/j.biombioe.2013.09.005

Prabhu AA, Bosakornranut E, Amraoui Y et al (2020) Enhanced xylitol production using non-detoxified xylose rich pre-hydrolysate from sugarcane bagasse by newly isolated *Pichia fermentans*. Biotechnol Biofuels 13:1–15. https://doi.org/10.1186/s13068-020-01845-2

Raj K, Krishnan C (2020) Improved co-production of ethanol and xylitol from low-temperature aqueous ammonia pretreated sugarcane bagasse using two-stage high solids enzymatic hydrolysis and *Candida tropicalis*. Renew Energy 153:392–403. https://doi.org/10.1016/j.renene.2020.02.042

Rao LV, Goli JK, Gentela J et al (2016) Bioconversion of lignocellulosic biomass to xylitol: an overview. Biores Technol 213:299–310

Rehman S, Nadeem M, Ahmad F et al (2013) Biotechnological production of xylitol from banana peel and its impact on physicochemical properties of rusks. J Agric Sci Technol 15:747–756

Rockström J, Ottmar E, Juliana G, et al (2020) Planet-proofing the global food system. Nature Food 2020 1(1 1):3–5

Romaní A, Morais ES, Soares PO, Freire MG, Freire CSR, Silvestre AJD, Domingues L (2020) Aqueous solutions of deep eutectic systems as reaction media for the saccharification and fermentation of hardwood xylan into xylitol. Bioresour Technol. https://doi.org/10.1016/j.biortech.2020.123524

Santibáñez L, Henríquez C, Corro-Tejeda R et al (2021) Xylooligosaccharides from lignocellulosic biomass: a comprehensive review. Carbohydr Polym 251:117118. https://doi.org/10.1016/J.CARBPOL.2020.117118

Scheller HV, Ulvskov P (2010) Hemicelluloses. Annu Rev Plant Biol 61:263–289. https://doi.org/10.1146/annurev-arplant-042809-112315

Sene L, Arruda PV, Oliveira SMM et al (2011) Evaluation of sorghum straw hemicellulosic hydrolysate for biotechnological production of xylitol by *Candida guilliermondii*. Braz J Microbiol 42:1141–1146. https://doi.org/10.1590/S1517-83822011000300036

Silva SS, Chandel AK (2012) D-xylitol: fermentative production, application and commercialisation (2012). Springer Verlag, Heidelberg, Germany. ISBN: 978-3-642-31886-3

Singh S, Kaur D, Yadav SK, et al (2021) Process scale-up of an efficient acid-catalyzed steam pretreatment of rice straw for xylitol production by *C. Tropicalis* MTCC 6192. Bioresour Technol 320:124422. https://doi.org/10.1016/j.biortech.2020.124422

Strassberger Z, Tanase S, Rothenberg G (2014) The pros and cons of lignin valorisation in an integrated biorefinery. RSC Adv 4:25310–25318. https://doi.org/10.1039/C4RA04747H

Taherzadeh MJ, Karimi K (2008) Pretreatment of lignocellulosic wastes to improve ethanol and biogas production: A review. Int J Mol Sci 9:1621–1651. https://doi.org/10.3390/ijms9091621

Vasco-Correa J, Ge X, Li Y (2016) Biological pretreatment of lignocellulosic biomass. In: Biomass fractionation technologies for a lignocellulosic feedstock based biorefinery, pp. 561–585. Elsevier

Werpy T, Petersen G (2004) Top value-added chemicals from biomass. In: Volume I: Results of screening for potential candidates from sugars and synthesis gas. URL https://www.osti.gov/biblio/15008859-top-value-added-chemicals-from-biomass-volume-results-screening-potential-candidates-from-sugars-synthesis-gas (Accessed 26 Sept 21).

Xu Z, Lei P, Zhai R et al (2019) Biotechnology for Biofuels Recent advances in lignin valorization with bacterial cultures: microorganisms, metabolic pathways, and bio-products. Biotechnol Biofuels. https://doi.org/10.1186/s13068-019-1376-0

Zoghlami A, Paës G (2020) Lignocellulosic biomass: understanding recalcitrance and predicting hydrolysis. Front Chem. https://doi.org/10.3389/fchem.2019.00874

Chapter 4
Strain Improvement Methods for Enhanced Xylitol Production

Rishi Gupta, Lethiwe D. Mthembu, and Nirmala Deenadayalu

Abstract Depleting fossil fuels and recent advances in biorefineries development have led an urge for the development for various renewable value-added products such as xylitol. Xylitol is a diabetic sweetener that helps in preventing tooth decay and ear infections. Current production strategies are based on chemical processes, which are environmentally hazardous. Researchers are now shifting their interest towards the biological production of xylitol but low xylitol productivities of natural microorganisms are one of the major limiting factors. The current chapter briefly reviews various strain improvement/development strategies for the cost-effective production of xylitol.

Keywords Xylitol · Xylose · Biomass · Genetic engineering · Strain development

4.1 Introduction

Increasing trends in developing biorefineries has shown immense potential towards production of xylitol from biomass. It has been listed in the top 12 renewable value-added commodity chemicals by US Department of Energy (USDOE). Xylitol is a sugar alcohol with five carbons that is found in small amounts in nature. It has gained worldwide interest due to its sweetening ability, which is comparable to sucrose but with far less calories. Since xylitol is digested in the human body via insulin-independent pathways, it can be used as a sugar replacement by diabetics. Besides xylitol also poses anticariogenic effects of xylitol and hence it can also aid in the promotion of dental health and the prevention of caries (Huang et al. 2011). Furthermore, xylitol is also used to treat ear infections in children and as a parenteral nutrition

---

R. Gupta (✉)
Anton Paar India Pvt Ltd, Udyog Vihar, Gurugram, Haryana, India
e-mail: rishi.gupta@anton-paar.com

L. D. Mthembu · N. Deenadayalu
Department of Chemistry, Durban University of Technology, Steve Biko Road, Berea, Durban 4001, South Africa

© The Author(s), under exclusive license to Springer Nature Switzerland AG 2022

M. das G. de Almeida Felipe and A. K. Chandel (eds.), *Current Advances in Biotechnological Production of Xylitol*,
https://doi.org/10.1007/978-3-031-04942-2_4

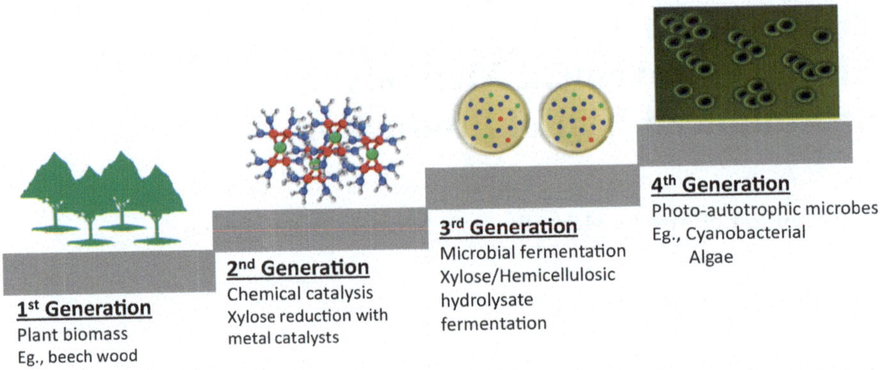

Fig. 4.1 Different generations for xylitol production

supplement in infusion treatment. Though xylitol has several pharmacological benefits, however, from a commercial point of view, it is primarily used as a sweetener or coating agent for pharmaceutical products in the pharmaceutical industry, as well as in certain sweetened products such as confectionery, personal health products such as mouthwash and toothpaste, and in the pharmaceutical industry as a sweetener or coating agent for pharmaceutical products (Rafiqul and Sakinah 2013).

Similar to biofuels, xylitol production can also be categorised among various generations (Fig. 4.1). It was first discovered in 1890 from beech wood syrup by Emil Fischer (Zacharis 2012), however, industrially it is manufactured by chemical hydrogenation of xylose-containing hemicellulosic hydrolysate in the presence of a metal catalyst at high temperatures and pressures. These chemical methods are expensive and energy-intensive, and also necessitate a complicated purification and separation steps.

Besides chemical synthesis, xylitol can also be produced through the biological routes using enzymes and/or microbes. Different mechanisms for xylose assimilation in different organisms are shown in Fig. 4.2. D-xylose is transformed to xylitol in a single step by xylose reductase (XR; EC 1.1.1.21). The resulting xylitol is either secreted or oxidized further into xylulose by xylitol dehydrogenase (XDH; EC 1.1.1.9). Both the enzymes require pyridine nucleotide cofactors in varied forms among different yeasts. XR is a NADH/NADPH-dependent enzyme, whereas XDH needs NAD+ (Parajó et al. 1998). After phosphorylation, D-xylulose is further metabolized via the pentose phosphate pathway. The production of xylitol is favoured by oxygen-limited conditions. Under oxygen-limited conditions an imbalance in redox occurs due to non-regeneration of cofactors, which favours the synthesis of xylitol by downregulating the activity of XDH (Prakasham et al. 2009). While, bacterial xylose metabolism begins with the xylose isomerase (XI; EC 5.3.1.5)-mediated conversion of xylose to xylulose and continues through the pentose phosphate pathway for cell maintenance and development, skipping xylitol production route due to absence of XR and XDH (Parajó et al. 1998).

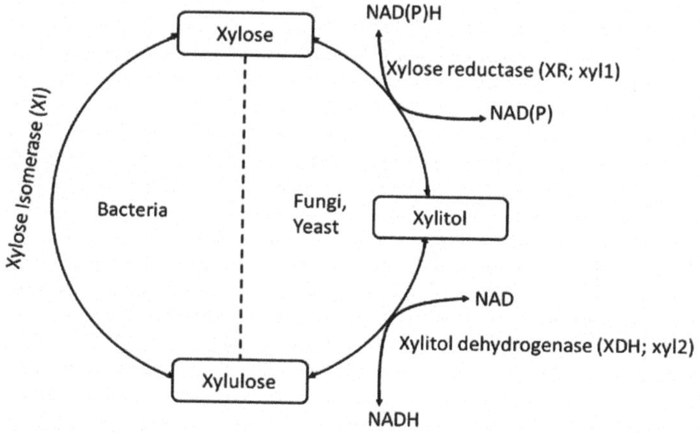

Fig. 4.2 Xylose assimilation mechanisms in different microbes

4.2 Current Status of Xylitol Production from Xylose and Hemicellulose Hydrolysate

4.2.1 From Xylose

Biotechnological xylitol production from xylose is a potentially attractive alternative to chemical xylitol production because it occurs under much milder process conditions and can be based on sugar mixtures such as lignocellulosic hydrolysates, which saves energy and costs associated with substrate purification (Huang et al. 2011). Whole cell biocatalysts incorporating bacteria, fungus, yeast, and/or recombinant strains, or cell free/immobilized enzyme systems, are used to convert biotechnological materials. Due to its high yield on laboratory scale, cell free enzymatic reduction or immobilized enzyme catalysis may be an alternative choice. However, due to the high cofactor required, large-scale synthesis may not be cost-effective (Dasgupta et al. 2017). Filamentous fungi have also been shown to synthesize xylitol, however their yields are substantially lower. *Enterobacter liquefaciens* and *Corynebacterium* sp. are two bacteria that ferment xylose into xylitol. However, the strains' poor production, which is a result of their long incubation times, limits their applicability in a broader sense. Yeasts are recommended for xylitol fermentation because of their high pentose absorption rates and xylitol productivity (Dasgupta et al. 2017). Table 4.1 illustrate the xylitol production from various strain.

Table 4.1 List of xylitol producing microorganisms

Microorganism	Xylitol (g/L)	Productivity (g/L/h)	Reference
Aspergillus niger	0.36	0.003	Sampaio et al. (2003)
Candida amazonenses	25.2	0.520	Cadete et al. (2012)
C. boidinii	59.3	0.32–0.46	Vandeska et al. (1996)
C. guilliermindii	54.6	0.58	Mussatto and Roberto (2008)
C. mogii NRRL Y-17032	30	0.4	Mayerhoff et al. (1997)
C. parapsilosis	210	3.18	Kim et al. (1997)
C. tropicalis ATCC 13803	237	2.0	Park et al. (2014)
Corneybacterium sp. B 4247	75	3.1	Rangaswamy and Agblevor (2002)
Cyberlindnera galapagoensis f.a., sp. nov	24	0.33	Guamán-Burneo et al. (2015)
C. xylosilytica	33.02	0.459	Cadete et al. (2015)
Debaryomyces hansenii	56.23	2.34	Pal et al. (2013)
Enterobacter liquefaciens	33.3	0.35	Yoshitake et al. (1973)
Fusarium oxysporum	1	0.02	Suihko et al. (1983)
Hansenula polymorpha	58	0.6	Suryadi et al. (2000)
Petromyces albertensis	39.8	0.16	Dahiya (1991)
Penicillium crustosum; P. citrinum; P. expansum; P. griseoroseum; P. italicum; P. janthinellum; P. perperogenum; P. roqueforti	0.14–0.52	0.003–0.006	Sampaio et al. (2003)

4.2.2 *From C-5 Hydrolysate*

Lignocellulosic biomass can serve as an excellent feedstock for the production of xylose from one of its structural polymers, hemicellulose. Hemicellulose, by virtue of its nature is comprised of various type of monomeric sugars including xylose, glucose, arabinose etc. and acid hydrolysis of lignocellulosic biomass is one of the most preferred approach to extract the xylose. However, these hydrolysates along with the sugars also contains few fermentation inhibitory compounds, which need to be detoxified prior to the xylitol fermentation. There are several methods to remove these inhibitory compounds from the hydrolysate and these are also associated with significant amount of sugar loss. The most feasible way is to develop inhibitory tolerant microbes, which can produce xylitol directly from the lignocellulosic hydrolysate (Ledezma-Orozco et al. 2018). Besides, use of co-culture technique, such as combination of *Saccharomyces cerevisiae* and *Candida tropicalis*, have also been attempted to directly use the lignocellulosic hydrolysates (Sehnem et al. 2017). In co-culture, *S. cerevisiae* can ferment C-6 sugars, while xylose is converted to

Table 4.2 List of few studies comprising xylitol production from lignocellulosic biomass

Microorganisms	Biomass	Xylitol yield (g/g)	References
C. guilliermondii BL 13	Soja husk	0.46	Cunha-Pereira et al. (2017)
C. guilliermondii FTI 20037	Sugarcane bagasse	0.69	Vaz de Arruda et al. (2017)
C. guilliermondii FTI 20037	Sugarcane straw	0.47	Vaz de Arruda et al. (2017)
C. mogii NRRL Y-17032	Rice straw	0.65	Mayerhoff et al. (1997)
C. shehatae HM 52.2	Rice husk	0.11	Hickert et al. (2013)
C. tropicalis	Corn cobs	0.58	Misra et al. (2013)
C. tropicalis	Corn straw	0.71	Wang et al. (2015)
C. tropicalis CCTCC M2012462	Corn cobs	0.71	Ping et al. (2013)
C. tropicalis CICC1779	Corn cobs	0.77	Jia et al. (2016)
D. hansenii and *C. guilliermondii*	Canola straw	0.45–0.55	López-Linares et al. (2018)
H. polymorpha ATCC 34438	Sunflower stalks	0.023	Martínez et al. (2012)
Pichia stipitis FPL YS30	Corn stover	0.61	Rodrigues et al. (2011)
Pachysolen tannophilus	Olive stones	0.44	Mateo et al. (2015)
S. cerevisiae YRH 396	Soja husk	0.45	Cortivo et al. (2018)
W. anomalus WA-HF5.5	Rice and soja husk	0.86	Sehnem et al. (2017)

xylitol by *C. tropicalis* (Cheng et al. 2014; Huang et al. 2011; Mateo et al. 2015). In another study Yuan et al. (2019) used hemicellulose hydrolysate to produce xylitol. After a 63-h fed-batch fermentation in a 15 L bioreactor, *Escherichia coli* strain WZ51 generated 131.6 g/L xylitol from detoxified hemicellulosic hydrolysate. The xylitol yield was 0.95 g g1 (grams of xylitol generated per gram of xylose ingested), and the xylitol productivity was 2.09 g/L/h. These results show that *E. coli* has the maximum xylitol output and productivity when using hemicellulosic hydrolysate as the raw material. Furthermore, nearly little residual xylose, arabinose, or arabitol was discovered in the culture, which favours xylitol purification by crystallization. Table 4.2 summarizes few studies on xylitol production from lignocellulosic biomass.

4.3 Strain Improvement Strategies

The industrial manufacturing of chemicals via microbial fermentation focuses on genetic stability, regulatory and safety issues, growth dynamics, recombinant protein expression levels, high productivity, and finally ease of product recovery. Existing pentose fermenting yeasts have limitations, mostly due to poor xylitol titre, among other factors. Following the growing richness of the genetic database and greater understanding about metabolic mapping of several non-*Saccharomyces* yeasts, selective gene modification to boost titre and productivity has been targeted. The section has briefly reviewed various strategies used to improve the microbial strains for enhanced cell growth vis-à-vis xylitol production.

4.3.1 Genetic Engineering

Genetic engineering is an excellent tool for developing novel microbial strains to generate desired products from various feed stocks materials. Various genetic engineering approaches including the optimization of selective xylose transporter system to increase xylose uptake rate, over-expression of XR to maximize reduction of xylose into xylitol, increase in intracellular hexose mono phosphate (HMP) flux by over-expression of NADP dependent glucose-6-phosphate dehydrogenase (G6PD) and 6-phosphogluconate dehydrogenase (6-PGD) to enhance supply of NADPH, and deletion of XDH to stop oxidation of xylitol into xylulose have been targeted for improved xylitol production (Dasgupta et al. 2017). A list of few genetically engineered microbes for xylitol production are shown in Table 4.3.

S. cerevisiae was known to be a non-xylose fermenting yeast due to its lack of the D-xylose metabolic pathway, but its GRAS status (generally regarded as safe) and great tolerance to inhibitors present in lignocellulose hydrolysates attracted the interest of researchers. As a result, scientists created the recombinant *S. cerevisiae* strain by inserting xylitol metabolism genes into it. Chung et al. (2002) reported on the integration of a xylose reductase gene into the chromosomes of *S. cerevisiae* using two distinct vectors. The recombinant strains were able to generate 0.90 g of D-xylitol for every gram of D-xylose. While in another report, the xylose reductase (XR) gene (xyl1), which was cloned from *P. stipitis* CBS 6054, was converted into *S. cerevisiae* in 1991 and was controlled by the phosphoglycerate kinase (PGK) promoter. The conversion ratio of D-xylose to D-xylitol in this recombinant strain was greater than 95% (Chen et al. 2010).

Because of their inherent D-xylose uptake mechanisms and ability to maintain the reduction–oxidation balance during D-xylitol accumulation, Candida yeasts were deemed better prospective candidates than genetically modified *S. cerevisiae*. However, due to the opportunistic pathogenic character of several *Candida spp.*, their applicability in the food business was restricted. Therefore, reports on metabolic engineering strategies in *Candida* strains are in scarce. Ko et al. (2006) attempted to develop a recombinant *C. tropicalis* strain by disrupting the xyl2 gene, which encodes the xylitol dehydrogenase (XDH) using the Ura-blasting approach. However, in absence of xyl2 gene the yeast was not able to grow and further to support its growth and NADH regeneration, glycerol was used as a co-substrate. The system was found capable to have a D-xylose to D-xylitol conversion yield of approximately 98% with a volumetric productivity of 3.23 g/L/h (Ko et al. 2006). In another study, when *C. tropicalis* was given the *C. parapsilosis* xyl1 gene under the regulation of

Table 4.3 Summary of few genetically modified organisms for xylitol production

Organism	Genetic modifications	Productivity (g/L/h)	Reference
C. tropicalis	Expressed At5g17010	1.14	Jeon et al. (2013)
C. tropicalis	Expressed xyl1	1.44	Jeon et al. (2012)
C. tropicalis	Overexpressed zwf and gnd	1.25	Ahmad et al. (2012)
C. tropicalis	Expressed NADH-preferring xyl1	5.09	Lee et al. (2003)
S. cerevisiae	Expressed xyl1, cdt-1 and gh1-	0.55	Oh et al. (2013)
K. marxianus YZJ74	Expressed CiGXF1	4.14	Zhang et al. (2015)
E. coli	Expressed xyl1 replaced crp with crp*, deleted xyl2	0.79	Cirino et al. (2006)
E. coli	Expressed xyl1 and Xyl3, deleted xyl2	0.27	Akinterinwa and Cirino (2009)
C. glutamicum	Expressedxyl1 and araE, deleted ldhA, ptsF, and 3xyl2	7.9	Sasaki et al. (2010)
C. glutamicum	Expressed NAD(P)H-dependent xyl1 + araA + dpe + xyl1	0.28 ± 0.05	Dhar et al. (2016)
Gluconobacter oxydans	Disrupted adhB	0.98	Suzuki et al. (2002)
Trichoderma reesei	Partial mutation of xdh1	0.02	Wang et al. (2005)
C. tropicalis	Mutation of xyl2 gene	3.23	Ko et al. (2006)
N. crassa	Multicopy integration of xyl1	2.09	Yuan et al. (2019)
S. cerevisae BJ3505	Chromosomal xyl1 gene integration	2.34	Bae et al. (2004)
S. cerevisae D452-2	Chromosomal xyl1 gene integration	1.1	Oh et al. (2013)
C. tropicalis LNG2	Chromosomal gene integration of Codon optimized XR and deletion of xyl2	1.44	Jeon et al. (2012)
K. marxianus YZJ017	Episomal expression of xyl1	0.89	Zhang et al. (2014)
C. tropicalis ARSdR-16	Knockout of xyl2 by site directed mutagenesis	0.62	Ko et al. (2011)
D. hansenii CBS767	Xyl2 disruption using His4	0.33	Pal et al. (2013)
P. stipitis	Knockout of xyl2	0.31	Kim et al. (2001)

an alcohol dehydrogenase promoter, the recombinant yeast produced more D-xylitol than the wild type strain (Chung et al. 2002). The oxygen availability was the essential parameter in optimizing the D-xylitol-fermenting conditions, such as aeration, temperature, and pH, for D-xylitol synthesis from D-xylose. An excess of NADH reduced the action of xylitol dehydrogenase under a temporary oxygen-limited environment, resulting in D-xylitol accumulation. The optimum temperature and pH for D-xylitol-yielding yeasts were found to be 30–37 °C and 4–6, respectively (Chen et al. 2010; Chung et al. 2002; Silva and Afschar et al. 1994).

The development of the xylose synthesis pathway from glucose resulted in a green and pollution-free xylose synthesis technique. Owing to its rapid growth under commonly used culture conditions and metabolic plasticity, as well as the availability of significant biochemical and physiological information and genetic engineering techniques, *E. coli* is one of the best host organisms for metabolic engineering and synthetic biology research. The genetic engineering of *E. coli* strains require insertion of XR genes along with NADPH cofactor to generate ethanol from hemicellulosic hydrolysates, which is also the limiting step (Yuan et al. 2019). Cirino et al. (2006) reported xylitol synthesis utilizing *E. coli* W3110, a genetically engineered strain, and achieved xylitol production of up to 38 g/L (30 °C, 250 rpm for 80 h). Further a series of *E. coli* strains with multiple copies of XR-encoding genes were created using a CRISPR/Cas9 markerless gene-editing approach. The generated plasmid-free strains had higher stability and a lower metabolic load than plasmid-harbored bacteria. Moreover, in a fed-batch fermentation of hemicellulosic hydrolysate in 15L bioreactor, multi-copy integration of xyl1 gene from *Neurospora crassa* in engineered *E. coli* resulted in an improved xylitol titre of 131.6 g/L with productivity of 2.09 g/L/h (Yuan et al. 2019).

Interestingly use of photoautotroph organisms for the production of xylitol is also attracting significant interests. In general, cyanobacteria do not assimilate xylose and require xylose transporters to facilitate xylose transport. Attempts have been made to overexpress the xylose symporter Ec-XylE and ATP-dependent transporter Ec-XylFGH in *Synechococcus elongatus* PCC 7942 (McEwen et al. 2013) and *Synechocystis* PCC 6803 (Lee et al. 2015), respectively. Recently, xylose transporter from *E. coli* and the xylose reductase gene from *C. boidinii* (Cb-XR) were overexpressed in *S. elongatus* (Fan et al. 2020). While in another study, a gene xylose reductase from *N. crassa* was expressed in *Chlamydomonas reinhardtii*, which resulted in 0.38 g/L xylitol with a yield of 0.05 g/g (Pourmir et al. 2013).

4.3.2 *Metabolic Engineering*

The fermentation of xylose is required for the bioconversion of lignocellulose to fuels and chemicals, but because wild-type strains of *S cerevisiae* cannot metabolize xylose, scientists have engineered xylose metabolism in this yeast (Jeffries and Jin 2004). Toivari et al. (2007) constructed a strains of *S. cerevisiae* by metabolic engineering, that was able to produce sugar alcohol xylitol, from d-glucose in a single fermentation step. A *S. cerevisiae* strain lacking transketolase accumulated d-xylulose 5-phosphate intracellularly and released ribitol and pentose sugars (d-ribose, d-ribulose, and d-xylulose) into the growth medium. The expression of *P. stipitis* xylitol dehydrogenase-encoding gene xyl2 in a transketolase-deficient strain resulted in an 8.5-fold increase in the total quantity of excreted sugar alcohols xylitol. Finally, inactivation of the endogenous xylulokinase-encoding gene xks1 was required to elevate xylitol excretion to 50% of the 5-carbon sugar alcohols excreted. Further to improve the xylose consumption, several sugar transpoter (ST) genes have been isolated from different yeast species. Among the ST gene families, first studies started from the identification of the hexose transporter genes (Hxt) in *S. cerevisiae* (Kruckeberg 1996). Within this gene family, transporter proteins expressed under conditions of low glucose concentrations can also transport the xylose molecule. The high-affinity and moderate hexose transporters proteins, Hxt7 and Hxt4, for example, have been widely explored for the construction of a xylose consumption recombinant *S. cerevisiae* strain (Farwick et al. 2014; Sedlak and Ho 2004; Saloheimo et al. 2007). However, the glucose transporter HXT7 shows about 100-fold lower affinity for xylose compared to glucose (Saloheimo et al. 2007) The identification of glucose/xylose facilitator 1 (GXF1) and glucose/xylose symport 1 (GXS1) proteins in *C. intermedia* revealed the importance of substrate consumption on xylose metabolism (Leandro et al. 2006). Other gene families related to xylose uptake have also been studied such as *P. stipitis* sucrose transporter proteins (SUT's) and xylose trans-porter proteins (XUT's). CiGxf1 and CiGxs1, *P. stipitis* Sut1 and Xut1, 4, 6 and, 7 are some examples of characterized yeast xylose transporters reported in the literature (Fonseca et al. 2011; Moon et al. 2013; Young et al. 2011). Although promising, the development of recombinant strains with genes that encode transporter proteins to xylose consumption towards the xylitol production still comprises only a small part of the studies on this topic.

4.3.3 Cell Surface Display

Cell-surface engineering approach has emerged as an advantageous way to obviate the addition of commercial enzymes. This technology consists in anchoring enzymes (e.g., those capable of degrading biomass) to the yeast cell wall. Cell surface engineering constitutes a cost-effective solution eliminating the need for the addition of exogeneous enzyme; additionally, the displayed enzymes can be re-used in repeated batch process. Guirimand et al. (2016) developed a xylose reductase co-displaying *S. cerevisiae* strain, which showed 5.8 g/L xylitol production with 79.5% theoretical yield from xylose contained in rice straw hydrolysate. The process was further improved to produce 37.9 g/L xylitol under consolidate bioprocessing (CBP) following the nanofiltration approach to remove the fermentation inhibitors in the hydrolysate. Later the same group attempted to co-display xylose reductase with other enzymes on the cell surface of recombinant strain of *S. cerevisiae*. The recombinant strain *S. cerevisiae* (YPH499-XR-BGL-XYL-XYN) was observed not only to improve the xylitol production under consolidated bioprocessing but also was found to reduce the enzyme requirement for pretreatment of lignocellulosic biomass (Guirimand et al. 2019).

4.3.4 Adaptive Evolution

Adaptive evolution of microbes is a promising strategy to enhance their performance by promoting their adaptation for some specific stress condition. The strategy poses to be a very important tool to improve the xylitol production from lignocellulosic hydrolysate by acclimatizing the cells against fermentation inhibitors such as hydroxy-methyl furfural, phenolic compounds, acetic acid etc., under low pH conditions (Koppram et al. 2012; Yamakawa et al. 2018). Here the concentration of these inhibitors created a selection pressure and the strains with improved inhibitor tolerance to these compounds outgrow with respect to other strains (Huang et al. 2016). The strategy of adaptive evolution not only can improve the stress tolerance of strains but will also minimize the xylitol production cost by avoiding the lignocellulosic hydrolysate's detoxification step. Also, the strategy found to be efficient in improving the cell growth and xylose uptake rate (Kim 2019; Sharma et al. 2016). The adaptive evolution strategy is imposed by transferring the microbes into medium with increased concentration of inhibitory compounds in a sequential fashion. This can be done in batch as well as in a continuous fermentation mode by using a chemostat system. Koppram and colleagues (2015), observed 11% increase in the xylitol yield following the adaptive evaluation approach over a recombinant *S. cerevisiae* strain cultivated in spruce hydrolysate medium in a chemostat system. Similar reports are also been reported to produce xylitol from *C. tropicalis* over lignocellulose hydrolysate (Misra et al. 2013). Though till date the reports of applying the approach of adaptive evolution to improve the strains for xylitol production are scarce but the technique holds potential to becoming an emerging area of research.

4.4 Conclusion and Future Prospects

Xylitol holds immense potential as sweetener in different industrial sectors but its production processes are suffering from various technological bottlenecks. On one hand chemical synthesis of xylitol is environmentally foe, on the other hand the microbial processes are lacking cost-effectiveness. In this aspect the role of bioprospecting of new microbes and improving existing xylitol producing strains have become the need of the hour. Genetic modifications and genome editing may help in developing new strains for xylitol production from lignocellulosic biomass with improved inhibitor- and osmo-tolerance. Also, the biomass collection, pretreatment, fermentation and down-streaming processes need to be optimized in line with the lifecycle analysis of xylitol production from the lignocellulosic biomass.

References

Ahmad I, Shim WY, Jeon WY, Yoon BH, Kim JH (2012) Enhancement of xylitol production in *Candida tropicalis* by co-expression of two genes involved in pentose phosphate pathway. Bioprocess Biosyst Eng 35:199–204

Akinterinwa O, Cirino PC (2009) Heterologous expression of D-xylulokinase from *Pichia stipitis* enables high levels of xylitol production by engineered *Escherichia coli* growing on xylose. Metab Eng 11:48–55

Bae SM, Park YC, Lee TH, Kweon DH, Choi JH, Kim SK, Ryu YW, Seo JH (2004) Production of xylitol by recombinant *Saccharomyces cerevisiae* containing xylose reductase gene in repeated fed-batch and cell-recycle fermentations. Enzyme Microb Biotechnol 35:545–549

Cadete RM, Melo MA, Dussán KJ, Rodrigues RC, Silva SS, Zilli JE, Vital MJ, Gomes FC, Lachance MA, Rosa CA (2012) Diversity and physiological characterization of D-xylose-fermenting yeasts isolated from the Brazilian amazonian forest. PLoS ONE 7:431–435

Cadete RM, Cheab MA, Santos RO, Safar SV, Zilli JE, Basso LC, Lee CF, Kurtzman CZ (2015) *Cyberlindnera xylosilytica* sp. nov., a xylitol-producing yeast species isolated from lignocellulosic materials. Int J Syst Evol Microbiol 65:2968–2974

Chen X, Jiang Z-H, Chen S, Qin W (2010) Microbial and bioconversion production of D-xylitol and its detection and application. Int J Biol Sci 6:834

Cheng KK, Wu J, Lin ZN, Zhang JA (2014) Aerobic and sequential anaerobic fermentation to produce xylitol and ethanol using non-detoxified acid pretreated corncob. Biotechnol Biofuels 7(1):1–9

Chung Y-S, Kim M-D, Lee W-J, Ryu Y-W, Kim J-H, Seo J-H (2002) Stable expression of xylose reductase gene enhances xylitol production in recombinant *Saccharomyces cerevisiae*. Enzyme Microb Technol 30:809–816

Cirino PC, Chin JW, Ingram LO (2006) Engineering *Escherichia coli* for xylitol production from glucose–xylose mixtures. Biotechnol Bioeng 95:1167–1176

Cortivo PRD, Hickert LR, Hector R, Ayub MAZ (2018) Fermentation of oat and soybean hull hydrolysates into ethanol and xylitol by recombinant industrial strains of *Saccharomyces cerevisiae* under diverse oxygen environments. Ind Crops Prod 113:10–18

Da Silva S, Afschar A (1994) Microbial production of xylitol from D-xylose using *Candida tropicalis*. Biopro Eng. 11:129–134

Dahiya JS (1991) Xylitol production by *Petromyces albertensis* grown on medium containing D-xylose. Can J Microbiol 37:14–31

Dasgupta D, Bandhu S, Adhikari DK, Ghosh D (2017) Challenges and prospects of xylitol production with whole cell bio-catalysis: a review. Microbiological Res 197:9–21

Dhar KS, Wendisch VF, Nampoothiri KM (2016) Engineering of *Corynebacterium glutamicum* for xylitol production from lignocellulosic pentose sugars. J Biotechnol 230:63–71

Fan ES, Lu KW, Wen RC, Shen CR (2020) Photosynthetic reduction of xylose to xylitol using Cyanobacteria. J Biotechnol 15:1900354

Farwick A, Bruder S, Schadeweg V, Oreb M, Boles E (2014) Engineering of yeast hexose transporters to transport D-xylose without inhibition by D-glucose. Proc Natl Acad Sci 111:5159–5164

Fda C-P, Hickert LR, Rech R, Dillon AP, Ayub MAZ (2017) Fermentation of hexoses and pentoses from hydrolyzed soybean hull into ethanol and xylitol by *Candida guilliermondii* BL 13. Braz J Chem Eng 34:927–936

Fonseca C, Olofsson K, Ferreira C, Runquist D, Fonseca LL, Hahn-H¨agerdal et al (2011) The glucose/xylose facilitator Gxf1 from *Candida intermedia* expressed in a xylose-fermenting industrial strain of Saccharomyces cerevisiae increases xylose uptake in SSCF of wheat straw. Enzym Microb Technol 48:518–525

Guamán-Burneo MC, Dussán KJ, Cadete RM, Cheab MAM, Portero P, Carvajal- Barriga EJ, da Silva SS, Rosa CA (2015) Xylitol production by yeasts isolated from rotting wood in the Galápagos Islands, Ecuador, and description of *Cyberlindnera galapagoensis* f.a., sp. nov. Antonie Van Leeuwenhoek 108:919–931

Guirimand G, Sasaki K, Inokuma K, Bamba T, Hasunnuma T, Kondo A, (2016) Cell surface engineering of *Saccharomyces cerevisiae* combined with membrane separation technology for xylitol production from rice straw hydrolysate. Appl Microbiol Biotechnol 100:3477–3487

Guirimand G et al (2019) Cell-surface display technology and metabolic engineering of *Saccharomyces cerevisiae* for enhancing xylitol production from woody biomass. Green Chem 21:1795–1808

Hickert LR, da Cunha-Pereira F, de Souza-Cruz PB, Rosa CA, Ayub MAZ (2013) Ethanogenic fermentation of cocultures of *Candida shehatae* HM 522 and *Saccharomyces cerevisiae* ICV D254 in synthetic medium and rice hull hydrolysate. Biores Technol 131:508–514

Huang C-F, Jiang Y-F, Guo G-L, Hwang W-S (2011) Development of a yeast strain for xylitol production without hydrolysate detoxification as part of the integration of co-product generation within the lignocellulosic ethanol process. Biores Technol 102:3322–3329

Huang M, Peabody G, Kao KC (2016) Tolerance of microbial biocatalysts to feedstocks, products, and environmental conditions. Metab Eng Bioprocess Commer 2016:73–100. Springer International Publishing

Jeffries T, Jin Y-S (2004) Metabolic engineering for improved fermentation of pentoses by yeasts. Appl Microbiol Biotechnol 63:495–509

Jeon WY, Yoon BH, Ko BS, Shim WY, Kim JH (2012) Xylitol production is increased by expression of codon-optimized *Neurospora crassa* xylose reductase gene in *Candida tropicalis*. Bioproc Biosyst Eng 35:191–198

Jeon WY, Shim WY, Lee SH, Choi JH, Kim JH (2013) Effect of heterologous xylose transporter expression in *Candida tropicalis* on xylitol production rate. Bioproc Biosyst Eng 36:809–817

Jia H, Shao T, Zhong C, Li H, Jiang M, Zhou H, Wei P (2016) Evaluation of xylitol production using corncob hemicellulosic hydrolysate by combining tetrabutylammonium hydroxide extraction with dilute acid hydrolysis. Carb Polym 151:676–683

Kim S (2019) Xylitol production from byproducts generated during sequential acid-/alkali-pretreatment of empty palm fruit bunch fiber by an adapted *Candida tropicalis*. Front Energy Res 7:1–9

Kim SY, Kim JH, Oh DK (1997) Improvement of xylitol production by controlling oxygen supply in *Candida parapsilosis*. J Ferment Bioeng 83:267–270

Kim MS, Chung YS, Seo JH, Jo DH, Park YH, Ryu YW (2001) High-yield production of xylitol from xylose by a xylitol dehydrogenase defective mutant of *Pichia stipitis*. J Microbiol Biotechnol 11:564–569

Ko BS, Kim J, Kim JH (2006) Production of xylitol from D-xylose by a xylitol dehydrogenase gene-disrupted mutant of Candida tropicalis. Appl Environ Micobiol 72:4207–4213

Ko BS, Kim DM, Yoon BH, Bai S, Lee HY, Kim JH, Kim IC (2011) Enhancement of xylitol production by attenuation of intracellular xylitol dehydrogenase activity in *Candida tropicalis*. Biotechnol Lett 33:1209–1213

Koppram R, Albers E, Olsson L (2012) Evolutionary engineering strategies to enhance tolerance of xylose utilizing recombinant yeast to inhibitors derived from spruce biomass. Biotechnol Biofuels 5:1–12

Kruckeberg AL (1996) The hexose transporter family of *Saccharomyces cerevisiae*. Arch Microbiol 166:283–292

Leandro MJ, Gonçalves P, Spencer-Martins I (2006) Two glucose/xylose transporter genes from the yeast *Candida intermedia*: first molecular characterization of a yeast xylose–H +symporter. Biochem J 395:543–549

Ledezma-Orozco E, Ruíz-Salazar R, Bustos-Vázquez G, Montes-García N, Roa-Cordero V, Rodríguez-Castillejos G (2018) Producción de xilitol a partir de hidrolizados ácidos no detoxificados de bagazo de sorgo por Debaryomyces hansenii. Agrociencia 52:1095–1106

Lee JK, Koo BS, Kim SY (2003) Cloning and characterization of the xyl1 gene, encoding an NADH-preferring xylose reductase from *Candida parapsilosis*, and its functional expression in *Candida tropicalis*. Appl Environ Microb 69:6179–6188

Lee T-C, Xiong W, Paddock T, Carrieri D, Chang I-F, Chiu H-F, Ungerer J, Hank Juo S-H, ,Maness P-C, Yu J (2015) Engineered xylose utilization enhances bio-products productivity in the cyanobacterium *Synechocystis* sp PCC 6803. Metab Eng 30:179–189

López-Linares JC, Romero I, Cara C, Castro E, Mussatto SI (2018) Xylitol production by Debaryomyces hansenii and Candida guilliermondii from rapeseed straw hemicellulosic hydrolysate. Bioresour Technol 247:736–743

Martínez ML, Sánchez S, Bravo V (2012) Production of xylitol and ethanol by *Hansenula polymorpha* from hydrolysates of sunflower stalks with phosphoric acid. Ind Crops Prod 40:160–166

Mateo S, Puentes JG, Moya AJ, Sánchez S (2015) Ethanol and xylitol production by fermentation of acid hydrolysate from olive pruning with *Candida tropicalis* NBRC 0618. Biores Technol 190:1–6

Mayerhoff ZDVL, Roberto IC, da Silva SS (1997) Xylitol production from rice straw hemicellulose hydrolysate using different yeast strains. Biotechnol Lett 19:407–409

McEwen JT, Machado IMP, Connor MR, Atsumi S (2013) Engineering *Synechococcus elongatus* PCC 7942 for continuous growth under diurnal conditions. Appl Environ Microbiol 79:1668

Misra S, Raghuwanshi S, Saxena RK (2013) Evaluation of corncob hemicellulosic hydrolysate for xylitol production by adapted strain of *Candida tropicalis*. Carb Polym 92:1596–1601

Moon J, Lewis Liu Z, Ma M, Slininger PJ (2013) New genotypes of industrial yeast *Saccharomyces cerevisiae* engineered with YXI and heterologous xylose transporters improve xylose utilization and ethanol production. Biocatal Agric Biotechnol 2:247–254

Mussatto SI, Roberto IC (2008) Establishment of the optimum initial xylose concentration and nutritional supplementation of brewer's spent grain hydrolysate for xylitol production by *Candida guilliermondii*. Process Biochem 43:540–546

Oh EJ, Ha SJ, Rin Kim S, Lee WH, Galazka JM, Cate JH, Jin YS (2013) Enhanced xylitol production through simultaneous co-utilization of cellobiose and xylose by engineered *Saccharomyces cerevisiae*. Metab Eng 15:226–234

Pal S, Choudhary V, Kumar A, Biswas D, Mondal AK, Sahoo DK (2013) Studies on xylitol production by metabolic pathway engineered *Debaromyces hansenii*. Bioresour Technol 147:449–455

Parajó JC, Domínguez H, Domínguez JM (1998) Biotechnological production of xylitol Part 1: Interest of xylitol and fundamentals of its biosynthesis. Bioresour Technol 65:191–201

Park YC, Kim SK, and Seo JH (2014) Recent advances for microbial production of xylitol. Bioproc Ren Res Comm Bioprod 497–518

Ping Y, Ling H-Z, Song G, Ge J-P (2013) Xylitol production from non-detoxified corncob hemicellulose acid hydrolysate by *Candida tropicalis*. Biochem Eng J 75:86–91

Pourmir A, Noor-Mohammadi S, Johannes TW (2013) Production of xylitol by recombinant microalgae. J Biotechnol 165:178–183

Prakasham RS, Rao RS, Hobbs PJ (2009) Current trends in biotechnological production of xylitol and future prospects. Curr Trends Biotechnol Pharm 3:8–36

Rafiqul SAM (2013) Processes for the production of xylitol—a review. Food Rev Int 29:127–156

Rangaswamy S, Agblevor FA (2002) Screening of facultative anaerobic bacteria utilizing D-xylose for xylitol production. Appl Microbiol Biotechnol 60:88–93

Rodrigues RCLB, Kenealy WR, Jeffries TW (2011) Xylitol production from DEO hydrolysate of corn stover by *Pichia stipitis* YS-30. J Ind Microbiol Biotechnol 38:1649–1655

Saloheimo A, Rauta J, Stasyk Sibirny AA, Penttil¨a M, Ruohonen L (2007) Xylose transport studies with xylose-utilizing *Saccharomyces cerevisiae* strains expressing heterologous and homologous permeases. Appl Microbiol Biotechnol 74:1041–52

Sampaio FC, da Silveira WB, Chaves-Alves VM, Passos FML, Coelho JLC (2003) Screening of filamentous fungi for production of xylitol from d-xylose. Braz J Microbiol 34:325–328

Sasaki M, Jojima T, Inui M, Yukawa H (2010) Xylitol production by recombinant *Corynebacterium glutamicum* under oxygen deprivation. Appl Microbiol Biotechnol 86:1057–1066

Sedlak M, Ho NWY (2004) Characterization of the effectiveness of hexose transporters for transporting xylose during glucose and xylose co-fermentation by a recombinant Saccharomyces yeast. Yeast 1:671–684

Sehnem NT, Hickert LR, da Cunha-Pereira F, de Morais MA, Ayub MAZ (2017) Bioconversion of soybean and rice hull hydrolysates into ethanol and xylitol by furaldehyde-tolerant strains of *Saccharomyces cerevisiae*, *Wickerhamomyces anomalus* and their cofermentations. Biomass Conver Bioref 7:199–206

Sharma NK, Behera S, Arora R, Kumar S (2016) Enhancement in xylose utilization using *Kluyveromyces marxianus* NIRE-K1 through evolutionary adaptation approach. Bioproc Biosyst Eng 39:835–843

Suihko ML, Suomalainen I, Enari TM (1983) D-xylose catabolism in *Fusarium oxysporum*. Biotechnol Lett 5:525–530

Suryadi H, Katsuragi T, Yoshida N, Suzuki S, Tani Y (2000) Polyol production by culture of methanol-utilizing yeast. J Biosci Bioeng 89:236–240

Suzuki S, Sugiyama M, Mihara Y, Hashiguchi K, Yokozeki K (2002) Novel enzymatic method for the production of xylitol from d-arabitol by *Gluconobacter oxydans*. Biosci Biotechnol Biochem 66:2614–2620

Toivari MH, Ruohonen L, Miasnikov AN, Richard P, Penttilä M (2007) Metabolic engineering of *Saccharomyces cerevisiae* for conversion of D-glucose to xylitol and other five-carbon sugars and sugar alcohols. Appl Environ Microbiol 73:5471–5476

Vandeska E, Amartey S, Kuzmanova S, Jeffries TW (1996) Fed-batch culture for xylitol production by *Candida boidinii*. Process Biochem 31:265–270

Vaz de Arruda P, dos Santos JC, de Cássia Lacerda Brambilla Rodrigues R, da Silva DDV, Yamakawa CK, de Moraes Rocha GJ, das Graças de Almeida Felipe M (2017) Scale up of xylitol production from sugarcane bagasse hemicellulosic hydrolysate by Candida guilliermondii FTI 20037. J Ind Eng Chem 47:297–302

Wang TH, Zhong YH, Huang W, Liu T, You YW (2005) Antisense inhibition of xylitol dehydrogenase gene, xdh1 from *Trichoderma reesei*. Lett Appl Microbiol 40:424–429

Wang W, Ling H, Zhao H (2015) Steam explosion pretreatment of corn straw on xylose recovery and xylitol production using hydrolysate without detoxification. Proc Biochem 50:1623–1628

Yamakawa CK, Qin F, Mussatto SI (2018) Advances and opportunities in biomass conversion technologies and biorefineries for the development of a bio-based economy. Biomass Bioenergy 119:54–60

Yoshitake J, Ishizaki H, Shimamura M, Imai T (1973) Xylitol production by an Enterobacter species. Agric Biol Chem 37:2261–2267

Young E, Poucher A, Comer A, Bailey A, Alper H (2011) Functional survey for heterologous sugar transport proteins, using *Saccharomyces cerevisiae* as a host. Appl Environ Microbiol 77:3311–3319

Yuan X, Wang J, Lin J, Yang L, Wu M (2019) Efficient production of xylitol by the integration of multiple copies of xylose reductase gene and the deletion of Embden–Meyerhof–Parnas pathway-associated genes to enhance NADPH regeneration in *Escherichia coli*. J Ind Microbiol Biotechnol 46:1061–1069

Zacharis C (2012) Xylitol. In: K. O'Donnell, M. Kearsley (eds), Sweeteners and Sugar, 2nd edition. Alternatives in Food Technology, pp 347–382

Zhang J, Zhang B, Wang D, Gao X, Hong J (2014) Xylitol production at high temperature by engineered *Kluyveromyces marxianus*. Bioresour Technol 152:192–201

Zhang J, Zhang B, Wang D, Gao X, Hong J (2015) Improving xylitol production at elevated temperature with engineered *Kluyveromyces marxianus* through over-expressing transporters. Bioresour Technol 175:642–645

Chapter 5
An Overview of Different Approaches and Bioreactors for Xylitol Production by Fermentation

Carina A. Prado, Felipe A. F. Antunes, Ruly Terán-Hilares, Erick Díaz-Ruiz, Fanny M. Jofre, Gabriel L. Arruda, Monica M. Cruz-Santos, Yasmin C. S. Melo, and Júlio C. Santos

Abstract Due to the many beneficial characteristics and possibilities of industrial applications of xylitol, an adequate addressing of engineering strategies is fundamental, aiming profitable yield and productivity of fermentative processes at industrial scale. In the developing biotechnological route, the different kinds of bioreactors and operation modes must be considered and systematically evaluated, mainly taking into account the particularities of the bioprocess, e.g., control of aeration rate and the presence of microbial inhibitors in the medium. At bench and pilot scale, stirred tank reactors have been reported for xylitol production using yeasts; however, other alternatives as bubble columns and fluidized bed reactors were also evaluated. Additionally, regarding bioreactor operation modes, batch, fed-batch, and continuous processes have been studied, presenting each one their specific advantages and concerns. Commonly, xylitol batch fermentation allows the study of process variables with profit-making control of operational conditions, while fed-batch and continuous processes are interesting alternatives to enhance process productivity. In this chapter, the main bioreactors and operation modes reported for xylitol production are discussed, including authors' investigations presenting specific approaches. The focus was a review exploring an overview of different approaches for fermentative methods for xylitol production, besides current techniques, e.g., applied to metabolic engineering, also discussing advancements and future perspectives.

Keywords Biorefinery · Xylitol · Sugarcane bagasse · Biomass · Bioreactor

---

C. A. Prado · F. A. F. Antunes · E. Díaz-Ruiz · F. M. Jofre · G. L. Arruda · M. M. Cruz-Santos · Y. C. S. Melo · J. C. Santos (✉)
Departamento de Biotecnologia, Escola de Engenharia de Lorena, Universidade de São Paulo, Lorena 12602-810, SP, Brasil
e-mail: jsant200@usp.br

R. Terán-Hilares
Laboratório de Materiales, Universidad Católica de Santa María, Urb. San José S/N, Umacollo, Arequipa, Peru

© The Author(s), under exclusive license to Springer Nature Switzerland AG 2022

M. das G. de Almeida Felipe and A. K. Chandel (eds.), *Current Advances in Biotechnological Production of Xylitol*,
https://doi.org/10.1007/978-3-031-04942-2_5

5.1 Introduction

Xylitol is one of the top 12 renewable added-value chemicals that can be obtained from biomass, according to the US Department of Energy (USDOE). It has application in food, pharmaceutical, and nutraceutical industries, being a functional sweetener which can to reduce glucose, triglyceride, and cholesterol levels in the blood (Ur-Rehman et al. 2015). According to Grand View Research (2021), the xylitol market is expected to reach USD 738.10 million by 2028.

This polyalcohol can be produced by chemical or biochemical route. The chemical route requires use of high energy and extensive purification steps; while biotechnological methods by using microorganisms and/or enzymes allows the production process in mild conditions as well as the possibility of using agricultural and forestry wastes and by-products as carbon source without requirements of extensive xylose purification steps (Arcaño et al. 2020; Felipe Hernández-Perez et al. 2019).

However, the fermentative production of xylitol has already presented certain bottlenecks that need to be overcome in order to be economically competitive with the chemical production route. Mainly, strategies to obtain higher results of xylitol productivity have been evaluated, such as the selection of yeast species or genetic modification of strains, selection of suitable culture medium conditions, such as temperature, pH, substrate concentration, aeration, as well as the bioreactor and operation mode (Antunes et al. 2021; Zhang et al. 2021). These strategies have allowed the development of new technologies, which could potentially solve some process concerns. Additionally, scaling-up studies have been performed in order to promote biotechnological production of xylitol at industrial scale, aiming to obtain this high value-added product in a sustainable way as well as with reduced costs of production in a biorefinery context (Santos et al. 2003; Mussatto 2012).

The use of bioreactors for the fermentative production of xylitol has a significant importance in the scaling-up of this process (Moraes et al. 2020). There is a wide range of bioreactor designs (stirred tank, fluidized and membrane bioreactors) which are suitable and can be operated in different operation modes. For xylitol production, the fermentation has been performed through three main modes of operation: batch (simple and repeated batches), fed-batch or continuous fermentation. These operating configurations have their particularities that provide specific characteristics for each process. Indeed, the choice of fermentation operation mode, together with the determination of the appropriate parameters (air flow rate, pH, initial substrate concentrate, and others), can increase the xylitol productivity and yield values via microbial conversion (Rao et al. 2016; Felipe Hernández-Pérez et al. 2019).

This chapter provides a survey on the circumscribed literature, as well as in recent advances for xylitol biotechnological production, focusing on their general aspects and current bottlenecks related to process conditions and the use of different options of bioreactors and operation modes, as well as future perspectives for this polyalcohol production via fermentation.

5.2 Production of Xylitol by Fermentation: General Aspects and Process Conditions

Given the important applications of xylitol in several industrial segments such as food, pharmaceutical and odontological, there is a great interest for possible feasible and economical bioproduction routes.

Nowadays, xylitol is commercially produced through chemical synthesis from xylose-rich hemicellulosic hydrolysates. However, this process is considered costly, non-environmentally friendly, high energy demanding and polluting (Dasgupta et al. 2017; Arcaño et al. 2020). Thus, microbiological routes from biomass hydrolysates instead of chemical processes are of increased interest (Carneiro and Almeida 2019). Figure 5.1 shows a briefly schematic diagram regarding the technological routes for chemical and biotechnological xylitol production.

The xylitol production via fermentation uses microorganisms as biocatalysts in the conversion of xylose into xylitol. These microorganisms can be wild or genetically modified strains of bacteria, filamentous fungi, and yeasts, capable of assimilating pentose sugars, mainly xylose (Dasgupta et al. 2017). Although it is reported in the literature that some species of bacteria, such as *Pseudomonas putida* and *Escherichia coli*, as well as filamentous fungi, such as *Aspergillus*, *Penicillium* sp., are applied in the biotechnological production of xylitol, yeasts are the most used microorganisms, since their use results in high xylitol yield and productivity rates (Lugani and Sooch 2020; Baptista et al. 2020). Among the yeasts naturally capable of bio-converting

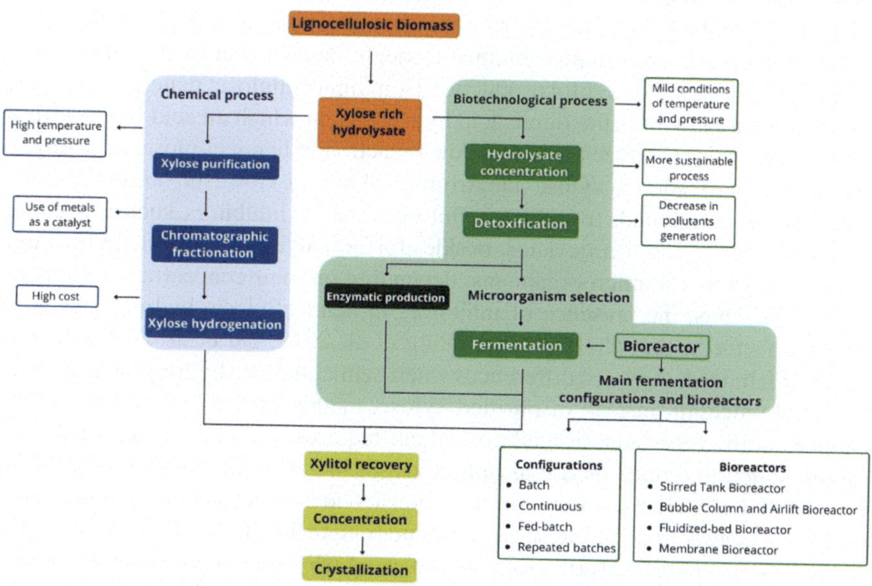

Fig. 5.1 Overall possibilities and remarks related to the main steps involved in chemical and biotechnological processes for the production of xylitol from lignocellulosic biomass

xylose to xylitol, *Candida* sp. genus stands out, especially the species *C. tropicalis*, *C. guilliermondii* and *C. magnoliae* (Raj and Krishnan 2020; Arruda et al. 2017; Wannawilai et al. 2017; Wannawilai et al. 2017). Furthermore, species belonging to the genus *Debaryomyces*, *Kluyveromyces* and *Pichia* have also been reported in the literature as good producers of xylitol (Narisetty et al. 2021; Zhang et al. 2021; Pappu and Gummadi 2018).

Extensive research has been carried out to determine optimized conditions for the fermentative production of xylitol to achieve an economically competitive process compared to the chemical production route. In this way, studies have evaluated the influence of some important parameters and alternatives regarding to the fermentation process, such as initial concentration of substrate and/or toxic compounds, nutritional supplementation, oxygen availability, mode of operation, type of bioreactor, use of free or immobilized cells, among others (Agblevor and Waleed 2005; Cortivo et al. 2020; Jofre et al. 2021).

An important parameter that influences the xylitol bioproduction is the initial concentration of substrate used in the process (Hernández-Pérez et al. 2016; Bonfiglio et al. 2021). Indeed, the initial substrate concentration is directly related to cell growth and, hence, to the amount of biomass. Zhang et al. (2018a, b) studied the variation in xylose concentration between 100 and 300 g/L (medium based on commercial xylose) and found that the xylitol yield was directly proportional to the increase in xylose concentration, up to the initial concentration of 250 g/L, when the yield reached 0.82 g/g in 192 h. However, when hydrolysate-based medium is used, lower levels of initial xylose concentration is employed. For example, Buhner and Agblevor (2004) observed yield value of 0.4 g of xylitol/g of xylose when using a medium based on a detoxified corn fiber hydrolysate containing about 60 g/L of xylose.

The hydrolysates are usually obtained through the chemical treatment of lignocellulosic biomasses. For xylitol production from lignocellulosic materials, the most used biomass treatment is the dilute acid hydrolysis, in which an acid (e.g., sulfuric or phosphoric acid) catalyzes the breakdown reaction of hemicellulose, releasing its monomeric sugars, such as xylose (De Arruda et al. 2011). However, this process also results in other compounds from lignocellulose, so called inhibitors, such as phenolic compounds, acetic and formic acids, besides furfural and 5-hydroxymethylfurfural, which are toxic to the microorganisms, depending on their concentration (Jönsson et al. 2013). Thus, the presence of inhibitors in hemicellulosic hydrolysates is an important factor in the whole process (Ping et al. 2013; Du et al. 2020a, b; Rao et al. 2021). Indeed, there are differences when hemicellulosic hydrolysates are used as substrate medium instead of purified xylose, mainly due to the presence of the inhibitors in the fermentation medium, which requires a following step of detoxification, aimed to reduce their concentrations. In this way, Carvalho et al. (2002a) worked with CaO, H_3PO_4, ammonium sulfate, and activated charcoal for hemicellulosic hydrolysate detoxification before fermentation. In this study, the authors verified that the bioconversion performance was enhanced when the hydrolysate was previously detoxified as following: CaO was used to increase pH to 7.0, followed by the use of H_3PO_4 to decrease the pH to 5.5, and, then, 2.5% w/v of active charcoal was used for removal of phenolic compounds.

Actually, in different studies, strains demonstrated higher results of xylitol production in fermentation medium previously treated with a detoxification technique, aimed to reduce the concentrations of toxic compounds (Vallejos et al. 2016; Silva-Fernandes et al. 2017; Kumar et al. 2018). For example, Villareal et al. (2006) investigated the effect of inhibitors from eucalyptus hemicellulosic hydrolysate, using *Candida guilliermondii*, and some detoxification methods, such as treatments with activated charcoal and cationic and anionic resins. In that work, authors observed that the best detoxification technique for the overall xylitol production process was using cationic resins, resulting in 32.7 $g{\cdot}L^{-1}$ of xylitol ($Y_{p/s}$ 0.57 $g{\cdot}g^{-1}$ and Qp 0.68 $g{\cdot}L^{-1}{\cdot}h^{-1}$) after 72 h of fermentation. In another study, Rodrigues et al. (2003) used a vacuum evaporation process which was considered a hydrolysate detoxification method prior to fermentation.

There are also some investigations that show that the nutritional supplementation in culture medium can improve the yeast fermentative performance for xylitol production. According to Arruda et al. (2017), and Pérez-Bibbins et al. (2016), the culture medium supplementation with yeast extract, NH_4SO_4, and $CaCl_2.2H_2O$ can significantly improve the xylitol production. In other study, Mussatto et al. (2008) performed xylitol production with *Candida guilliermondii* in a medium based on barley grain residue hydrolysate, and tested different concentrations of xylose (55, 75, and 95 $g.L^{-1}$), with and without supplementation (sulfate ammonium, calcium chloride, and rice bran extract). Thus, the authors verified that the optimal initial concentration of xylose in the culture medium was 60 $g.L^{-1}$ with supplementation of 2.0 g/L of yeast extract, 0.1 $g{\cdot}L^{-1}$ of $(NH_4)_2SO_4$, and 0.1 $g{\cdot}L^{-1}$ of $CaCl_2{\cdot}2H_2O$. Furthermore, Hernández-Pérez et al. (2016) evaluated the substrate supplementation composed by maltose, sucrose, cellobiose and glycerol in sugarcane straw hemicellulosic hydrolysate as a strategy to improve xylitol production by *Candida guilliermondii* FTI 20,037. In that work, authors observed a significant increase in xylitol concentration and volumetric productivity when the hydrolysate was supplemented with 10 $g{\cdot}L^{-1}$ of sucrose. Regarding to supplementation with different nitrogen sources, Hovnanyan et al. (2019) studied the production of xylitol by using the yeast *Candida guilliermondii*, verifying that the optimized nitrogen medium (3 $g{\cdot}L^{-1}$ of NH_4OH) contributed to an increase in xylitol production of 44% if compared with the results observed using others fonts (NH_4Cl, $(NH_4)_2$ SO_4).

One of the major parameters in the biotechnological production of xylitol is the oxygen availability during the fermentation, driving direct influence in the process yield ($Y_{p/s}$) and productivity (Qp). Commonly, high rates of $Y_{p/s}$ can be achieved by controlling the aeration at lower levels, that drives the metabolic pathway for xylitol production (Granström et al. 2007). For example, Branco et al. (2007) evaluated the aeration parameter in the production of xylitol, observing in the studied ranges that lowest value of aeration (1.33 vvm) produced the highest concentration of xylitol. Other authors, such as Mareczky et al. (2016), analyzed the effect of aeration of the medium during the xylitol production phase using *Candida parapsilosis* yeast. In that work, the aeration conditions were varied by changing flasks stirring (750 mL) from 125 to 220 rpm, and the fermentation was carried out for 72 h. The authors confirmed that less aeration produced greater amount of xylitol and, when there was no aeration, the yeast was unable to assimilate xylose, resulting in the non-formation of xylitol. The authors aimed to study these parameters for better understanding of how the aeration contributes for production of xylitol with *Candida guilliermondii*. Additionally, Felipe Hernández-Pérez et al. (2019) observed that oxygen transfer coefficient (K_La) of 6.5 h^{-1} resulted in maximum values of $Y_{p/s}$ (0.86 g/g) and Qp (0.51 $g{\cdot}L^{-1}$) during fermentation with *Candida tropicalis* in sugarcane bagasse and straw hemicellulosic hydrolysate. In other work, Bianchi et al. (2021) observed that the oxygen transfer coefficient (Kla) of 17 h^{-1} resulted in maximum values of Yp/s (0.59 g/g) and Qp (0.39 $g/L{\cdot}h^{-1}$) during fermentation with *Candida tropicalis* in sugarcane bagasse hemicellulosic hydrolysate.

Cell immobilization for xylitol fermentation is another strategy used in some approaches to optimize xylitol production to obtain higher productivity rates. This technique consists of using a support in which the cells remain attached/immobilized. This procedure presents the advantage of facilitating cell handling and recovery, favoring reuse of cells and continuous fermentation (Soleimani and Tabil 2013). There are various supports for the process of xylitol bioproduction, such as multi-walled carbon nanotubes, biocomposite-based carriers, oxidized carbon fiber, porous glass beads, calcium alginate, among others (Wang et al. 2021; Soleimani and Tabil 2013).

Within this concept, e.g., Santos et al. (2003) operated in batch mode the xylitol production from sugarcane bagasse hemicellulosic hydrolysate in an immobilized cell fluidized bed reactor, verifying that the xylitol production decreased according to the increase of aeration rate to 0.093 min^{-1}. In this work, the highest production occurred with of aeration rate of 0.080 min^{-1}. Besides, the authors reported lower values of xylitol yield when increasing carrier concentration (125 $g{\cdot}L^{-1}$), due to the break of air bubbles, which increases the oxygen availability in the medium.

Aiming to scale up the xylitol production, studies in bioreactors (in bench or pilot scale) in different operation modes are fundamental. Although simple batch is the most frequent applied operation mode, fed-batch and continuous batch fermentation are also reported in the literature. Each mode operation for xylitol production has advantages and disadvantages, as well as specific characteristics, as shown in more details in Sect. 5.4. Furthermore, the choose of a profitable bioreactor for the bioprocess, e.g., stirred tank reactor (STR), column reactor, and others, can also improve the xylitol production, as more deeply discussed in Sect. 5.3 (Carvalho et al. 2003; Hickert et al. 2013; Hernandez-Escoto et al. 2014; Antunes et al. 2021).

5.3 Bioreactors Used for Xylitol Production by Fermentation

The adequate choice of bioreactors is fundamental, allowing a controlled production, providing optimal cultivation conditions and favoring industrial-scale production. Each bioreactor is designed and used to meet the environmental required conditions by microorganisms or enzymes involved in the bioproducts production. Similarly, in xylitol production, the different bioreactor configurations can result in very different bioprocess performance.

Xylitol production by fermentation presents some particularities that can explore the use of different configurations of existing bioreactors. Among the main parameters, aeration deserves to be highlighted, since it is directly related to oxygen transfer rate. The importance of this parameter in the equipment configuration allows the control and balance of production of cells or xylitol, especially when working with yeasts (Abbott et al. 2016).

The main types of bioreactors commonly used in the production of xylitol are briefly presented below.

5.3.1 Stirred Tank Bioreactor (STR)

Stirred tank bioreactors (STR), illustrated in Fig. 5.2, are the most commonly used in xylitol production, mainly in fermentations using free cells (Felipe Hernández-Pérez et al. 2019). It allows an easy temperature control with the use of heat exchangers, besides pH control, monitored with adequate probes and the addition of acidic and basic solutions (Abbott et al. 2016).

They are cylindrical in configuration, with a common diameter-to-height ratio of 1:2 or 1:3. The working volume is filled with a liquid medium that can correspond to 70–80% of the total volume of the bioreactor. It is a good alternative in scale-up of processes, with bench configurations similar to those ones used in industrial scale (Rocha et al. 2012). Moreover, the main advantage of the STR is its great capacity for homogenizing the culture medium, as it makes use of mechanical stirrers, the most

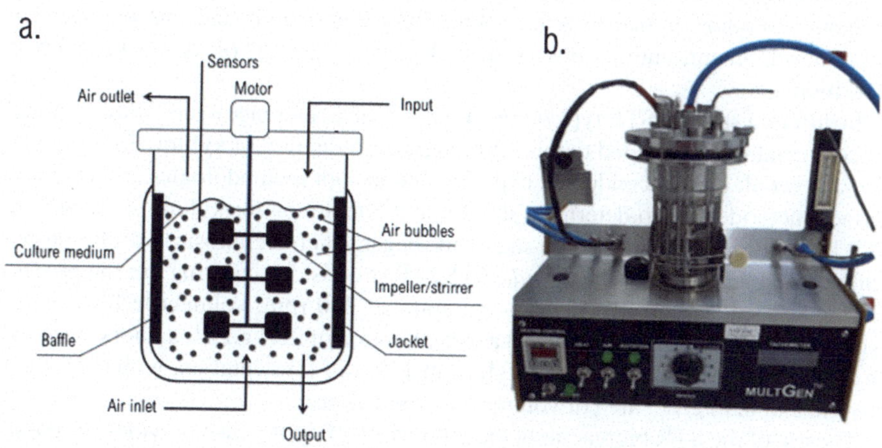

Fig. 5.2 **a** Representation of a stirred tank bioreactor (STR) **b** Photography of a MultGen bioreactor (STR) Tachometer, New Brunswick Scientific CO. Model F-1000, New Jersey, USA

common being a set of a six flat-blade disc turbines (Rushton). This agitation system aids in the transfer of oxygen from the gas bubble to the medium (Fig. 5.2) (Visscher et al. 2013). However, agitation can result in shear stresses to the cells, a not very decisive factor when using free cells for xylitol production, though important when immobilized cells are used, due to the shear stress imposed to the immobilization matrix (Kourkoutas et al. 2004).

The production of xylitol in STR bioreactor can be used in scale-up studies when evaluating the reproducibility of optimized conditions in flasks. Thus, Arruda et al. (2017) cultivated *Candida guilliermondii* for xylitol production from sugarcane bagasse detoxified hemicellulose hydrolyzate in bench scale and pilot scale, all in STR-type bioreactors, with 2.4 L, 18 L, and 125 L capacity, with the volumetric oxygen mass transfer coefficient (kL) of 16 h^{-1} as the key to scaling up. The authors observed an efficiency of 60% in conversion of xylose to xylitol, with 0.55 $g.g^{-1}$ of xylitol yield, and 0.31 $g.L^{-1}.h^{-1}$ of volumetric productivity.

Furthermore, there are some works in the literature that report xylitol production by using immobilized cells in STR, even considering the support can be damaged by stirring system. For example, Carvalho et al. (2003) investigated a medium based on sugarcane bagasse hemicellulosic hydrolysate, comparing three different systems for the production of xylitol with Ca-alginate entrapped cells of *Candida guilliermondii* yeast: 125 mL Erlenmeyer flasks, STR bioreactor, and basket-type stirred-tank reactor (BSTR). In this work, the best fermentation system corresponded to the STR type bioreactor, with 2.4 L of capacity and 1.3 L of the medium, operated with the agitation of 500 rpm and airflow of 1.7 $L.min^{-1}$, resulting in a production of 23.5 $g.L^{-1}$ of xylitol after 60 h of cultivation.

In another work using the same bioreactor (Carvalho et al. 2004), medium volume, and immobilized cells, the STR-type bioreactor was operated with hydrolysate-based medium and low oxygenation rates (300 rpm agitation and 1.0 min^{-1} airflow). By using hemicellulosic hydrolysate detoxified with a combination of different ion exchange columns, aimed to remove inhibitory compounds and improve fermentation development, authors observed production of 29 g/L of xylitol in 120 h in STR bioreactor.

In different studies, STR type bioreactor has been a reference to determine optimal process conditions intended to biotechnological production of xylitol. For example, Carvalho et al. (2004) conducted experimental design methodologies and response surface methodology to determine the optimal conditions for xylitol production in STR by using a medium based on sugarcane bagasse hemicellulosic hydrolysate and immobilized *Candida guilliermondii* FTI 20,037 cells. By carrying out the process with 5 × concentrated hydrolysate, agitation of 300 rpm, volumetric flow of 1.3 L.min-1, initial pH of 6.0 and cells at a concentration of 31.4 $g.L^{-1}$, authors observed production of up to 47.5 $g.L^{-1}$ of xylitol in 120 h of fermentation, with 0.81 g.g-1 of yield and 0.40 $g.L^{-1}.h^{-1}$ of volumetric productivity.

The use of the STR bioreactor in the production of xylitol also provides the possibility to carry out the process in different modes, as in repeated batches operation. Santos et al. (2003) cultivated *Candida guilliermondii* FTI 20,037 cells immobilized on Ca-alginate beads, performing five successive batches of xylose-containing culture medium in a 2.4 L STR reactor, with a total of 600 h of fermentation. In this system, at the end of each batch, the fermented medium was removed and the immobilized cells kept in the bioreactor for the next batch, with insertion of new substrate. Authors concluded that the average production values were satisfactory with an average production of 51.6 g.L^{-1} xylitol, observing productivity of 0.43 g.L^{-1}.h^{-1} and a yield of 0.71 g.g^{-1}.

Although STR has been largely used, the limitations are mainly linked to the high investment costs, operation costs, need for periodically ongoing maintenance, and estimated energy costs that can reach up to 58% of total costs in the case of industrial-scale bioreactors (Zhuang et al. 2021).

5.3.2 *Column Reactors: Bubble, Airlift and Fluidized Bed Bioreactors*

Column reactors are usually cylindrical devices that promote both oxygenation and agitation for the culture medium, dispensing the use of mechanical stirrers. They allow a great diversity of sizes and configurations, e.g., Bubble column and Airlift (Hong et al. 2014). The bubble column reactor uses the column filled with liquid fluid, and gas is inserted at the bottom, producing bubbles that homogenize the system. Airlift reactors have a draft tube (intern or extern), which drives the flow of medium, as shown in Fig. 5.3.

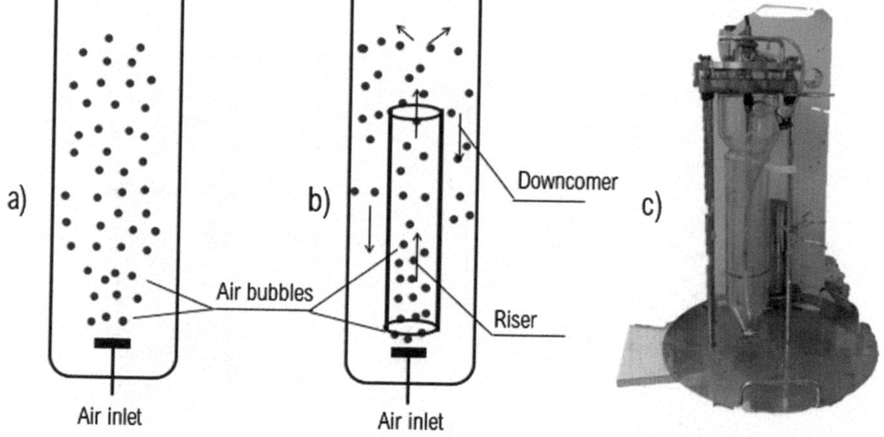

Fig. 5.3 Representation of usual column bioreactors. **a** Bubble column, **b** Air lift reactor and **c** Photography of a commercially available bench reactor—Bioengineering PID Fermenter AWS

The great advantage of those column bioreactors is the possibility of promote high oxygen transfer rate with high homogeneity of the medium. Airlift bioreactor was reported by Pérez-Bibbins et al. (2013) as an adequate system for xylitol production by immobilized *Debaryomyces hansenii*, since the minimum air flow required for homogenization is lower than that leading to the microaerobic conditions that trigger xylitol accumulation. In the same work, 0.43 $g.L^{-1}.h^{-1}$ of productivity and 0.71 $g.g^{-1}$ of yield was achieved.

Branco et al. (2007) evaluated the xylitol production from hemicellulosic hydrolysate by *Candida guilliermondii* immobilized in calcium-alginate using bubble column reactor. In that work, the authors reported a volumetric production of 0.21 $g.L.^{-1}.h^{-1}$ of xylitol at 1.33 vvm aeration rate and using 40% of immobilized cells.

In bubble columns and airlift bioreactors for xylitol production, studies should be continued targeting to evaluate other parameters, such as the bubble size distribution, gas hold-up, volumetric mass transfer coefficient, superficial gas velocity, operating conditions, and column dimensions (Matrawy et al. 2021; Zhuang et al. 2021).

Additionally, other configuration of column reactors is the fluidized bed. In this type of system, the flow of a fluid (gas or liquid) provides fluidization by moving a solid bed kept in the reactor, fluidizing it and homogenizing the medium (Fig. 5.4).

Fluidized bed reactors are commonly used in culture with immobilized cells or enzymes, and can have either ascending or descending supply and outlets (Zhuang et al.2021). The great advantage is its low shear injury promoted to the cells, with preservation of immobilization matrix, and good homogenization of medium promoted by the moving solids of the bed (Hong et al. 2014). Also, the same column reactor can be operated in a packed or fluidized mode, depending on the quantity of solids and fluid flow rate.

In the production of xylitol, this type of bioreactor was used in the study of Sarrouh et al. (2007), in which in a fluidized bed bioreactor using *Candida guilliermondii*

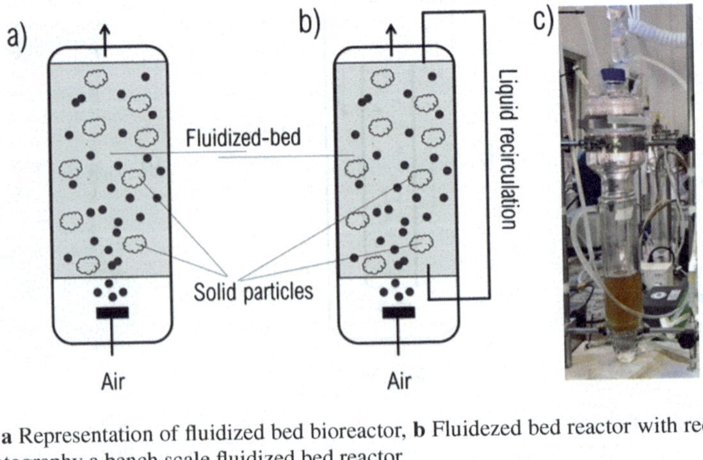

Fig. 5.4 **a** Representation of fluidized bed bioreactor, **b** Fluidezed bed reactor with recirculation, and **c** photography a bench scale fluidized bed reactor

cells immobilized on calcium alginate, resulted in a yield of 0.4 $g.g^{-1}$ of xylitol and an efficiency of 65% when using medium based on sugarcane bagasse hemicellulosic hydrolysate. In this study, the bioreactor consisted of a column measuring 540 mm × 55 mm, with an internal diameter of 9 mm and loaded with 1.5 L of culture medium, with 20% of this volume composed of the immobilized cells.

One of the most important operating parameters of a fluidized bed bioreactor is the aeration rate. For example, Santos et al. (2003) evaluated this parameter applied to xylitol production, in a column reactor (2 L capacity, 1.6 L working volume, dimensions of 540 mm × 55 mm, and central tube with an internal diameter of 9 mm), operated in fluidized bed reactor, with a medium based on sugarcane bagasse hemicellulosic hydrolysate and cells of *Candida guilliermondii* immobilized in porous glass spheres. In that work, authors found that the highest production of xylitol (17 $g.L^{-1}$) occurred at an aeration rate of 70 $mL.min^{-1}$, while 25 $mL.min^{-1}$ promoted a decrease in the yield of xylose consumption and biomass production (0.14 $g.g^{-1}$), despite bringing stability to immobilization and high yield of xylitol production (0.54 $g.g^{-1}$). When the aeration rate was increased to 140 mL/min, in addition to promoting a decrease in the fraction of immobilized cells and in the yield of xylitol production (0.36 $g.g^{-1}$), the consumption of xylose was stimulated.

More recently, fluidized bed bioreactor was evaluated with different approaches for xylitol production. For example, Antunes et al. (2017) used a cylindrical column of 600 mL with internal diameter of 5 cm, operated as a fluidized bed reactor supplied with 200 mL of culture medium and fed with air at 0.2 min^{-1} through a porous plate at the lower base. In this work, by using a wild Brazilian yeast *Candida tropicalis* UFMGX12-a, authors verified in 72 h of fermentation a production of 9.16 $g.L^{-1}$ of xylitol, yield of 0.39 $g.g^{-1}$ and productivity of 0.25 $g.L^{-1}.h^{-1}$.

5.3.3 *Membrane Bioreactor*

These systems receive in their structure separation units composed of membranes. This membrane structure allows separation, retention, and selectivity to the liquid and solid constituents present, helping in the removal of products, separation of cells and nutrient movement (Hong et al. 2014).

Membrane bioreactors can be further classified as external and internal types. In the external configuration, the membranes form a second system and are connected to the bioreactor. In this configuration, the liquid is recirculated between the bioreactor and the membrane system, a way widely used in processes that continuously recover the product from the fermentation. In the internal type of configuration, the membranes are inside the bioreactor, forming a single set (Zhuang et al. 2021).

Presenting a great potential on a laboratory scale, membrane bioreactors still shown many challenges when scale-up is proposed, especially in the production of high value-added liquid chemicals (Akkoyunlu et al. 2021).

Faria et al. (2002) evaluated the use of membrane bioreactor for the conversion of D-xylose to xylitol using *Candida guilliermondii* in a continuous process, with the membrane system used to separate the cells from the fermented broth. Authors verified that the best results were obtained using membranes with a maximum pore diameter of 0.2 μm and permeability of 42.3 $L \cdot m^2 \cdot h^{-1}.bar^{-1}$, obtaining a conversion of 86% and productivity of 1.14 $g \cdot L^{-1}$ of xylitol. In other work, Zahed et al. (2016) reported the xylitol production from rice straw hydrolysate by Candida tropicalis NCIM 3119 in a bioreactor coupled to crossflow microfiltration unit (pore size 0.45 μm, 0.1 m^2 surface area), also responsible for the separation of cells. In that work, in batch process, 26.5 $g.L^{-1}$ of xylitol (0.58 $g \cdot g^{-1}$ yield) were obtained, while in continuous process at dilution rate of 0.03 L/h, 31 $g.L^{-1}$ of xylitol were produced.

The possibility of membrane bioreactors to reuse cells was a strategy explored to increase the productivity of xylitol produced by *Candida tropicalis*, when using an inner membrane bioreactor. The reuse of cells in the bioreactor allowed the production of 8.5 $g.L^{-1}.h^{-1}$ of xylitol, an amount 3.4 times greater than batch fermentation with the same configuration (Kwon et al. 2006). However, the cost of membrane and fouling are the main challenges to be considered (Xu et al. 2019).

5.4 Operation Mode of the Bioprocess: Batch, Fed-Batch and Continuous

Each operation mode for xylitol biotechnological production has advantages and disadvantages, as well as specific characteristics (Pappu and Gummadi 2016). Table 5.1 shows some examples of xylitol production in different operation configurations and conditions, presenting main characteristics for each process. Three main operation modes have been evaluated, presenting advantages and disadvantages: batch, fed-batch, and continuous processes (Table 5.1) (Pérez-Bibbins et al. 2014).

5.4.1 Batch operation

A batch reactor is a system with no feed or output streams. For fermentation, the only inputs/outputs in batch operation of the bioreactor include sampling, medium evaporation, addition of concentrated acid or alkaline solutions to control pH, and antifoams. However, all of those inputs/outputs must be kept at minimum values, not resulting in great variation in useful volume of the reactor (lower than 10%). Besides, considering low solubility of oxygen in liquid medium, aeration is also continuous in the process, even in batch operation.

Batch fermentations are industrially used for generation of products of interest and can provide high substrate conversion and product yield (Pérez-Bibbins et al. 2016; López-Linares et al. 2018). In batch processing, specific changes that favor the system can be made and determined, allowing a focused study of the limiting

Table 5.1 Main characteristics and process conditions of xylitol production systems using different operation modes

Operation mode	Substrate	Microorganism	Process conditions	P_F $g{\cdot}L^{-1}$	$Y_{P/S}$ (g/g)	Q_P $(g{\cdot}L^{-1}{\cdot}h^{-1})$	Reference
Batch	Wheat straw hydrolyzate	*Candida guilliermondii*	Stirred tank reactor; 300 rpm; AR: 0.6 vvm; X_0: 0.5 $g{\cdot}L^{-1}$; S_0: 30.5 $g{\cdot}L^{-1}$ xylose; Temperature: 30 °C; F_T: 54 h	27.5	0.9	0.5	Canilha et al. (2003)
Batch	Sugarcane bagasse hemicellulosic hydrolysate	*Candida guilliermondii*	Stirred tank reactor 300 rpm; pH: 5.5; K_La: 22.5 h^{-1}; X_0: 1 $g{\cdot}L^{-1}$ Temperature: 30ºC; F_T:50 h	36.29	0.64	0.76	Rodrigues et al. (2003)
Batch	Corn cob hemicellulosic hydrolysate	*Candida tropicalis*	Erlenmeyer flasks; 200 rpm; AR: 0.4 vvm; X_0: 5% (v/v); S_0: 80 $g{\cdot}L^{-1}$ xylose; Temperature:30ºC; F_T:120 h	58.3	0.74	0.61	Guo et al. (2013)
Batch	Sugarcane bagasse hemicellulosic hydrolysate	*Candida tropicalis*	Erlenmeyer flasks; 200 rpm; K_{La} 89 h^{-1}; X_0: 0.03 $g{\cdot}L^{-1}$; S_0: 160 $g{\cdot}L^{-1}$ xylose; Temperature: 30 °C; F_T: 48 h	4.9	0.36	0.19	Unrean and Ketsub (2018a, b)
Batch	Synthetic media (based on xylose and glucose)	*Saccharomyces cerevisiae*	Erlenmeyer flasks; 200 rpm; 30 $g{\cdot}L^{-1}$xylose and 20 $g{\cdot}L^{-1}$ glucose; FR:4.8 $mL{\cdot}h^{-1}$;Temperature: 30 ºC;	23.24	0.91	0.74	Baptista et al. (2018)

(continued)

Table 5.1 (continued)

Operation mode	Substrate	Microorganism	Process conditions	P_F $g{\cdot}L^{-1}$	$Y_{P/S}$ (g/g)	Q_P $(g{\cdot}L^{-1}{\cdot}h^{-1})$	Reference
Fed-batch	Corn cob hemicellulosic hydrolysate	*Candida tropicalis*	Erlenmeyer flasks; 200 rpm; AR: 0.4 vvm; X_0: 5% (v/v); S_0: 160 $g{\cdot}L^{-1}$ xylose; FR: 70 $mL{\cdot}h^{-1}$ Temperature: 30 °C; F_T: 120 h	96.5	0.83	1.01	Guo et al. (2013)
Fed-batch	Xylose and glucose	Genetically modified *Saccharomyces cerevisiae* CK17 Δ XKS1	Stirred tank reactor; 300 rpm; AR: 0.025 vvm; S_0: 54 $g{\cdot}L^{-1}$ xylose and 6 $g{\cdot}L^{-1}$ glucose; FR: 2 $mL.h^{-1}$; Temperature: 35 °C;	34.97	0.94	0.73	Yang et al. (2020)
Fed-batch	Corn cob hemicellulosic hydrolysate	Genetically modified *Saccharomyces cerevisiae* PE-2-GRE3	RALF reactor; 400 rpm; AR: 2 vvm; S_0: 27.2 $g{\cdot}L^{-1}$ xylose; FR: 4.8 mLh^{-1} Temperature: 30 °C; pH: 5	19.0	0.89	0.39	Baptista et al. (2018)
Continuous	Sugarcane bagasse hemicellulosic hydrolysate	*Candida guilliermondii*	Stirred tank reactor; 300 rpm; DR: 0.03 h^{-1}, K_La: 20 h^{-1} Temperature: 30 °C; pH 5.5	23.1	0.58	0.7	Acosta Martínez et al. (2000)
Continuous	Sugarcane bagasse hemicellulosic hydrolysate	*Candida guilliermondii* FTI 20,037	Stirred tank reactor; 300 rpm; DR: 0.01 h^{-1}, K_La: 20 h^{-1} Temperature: 30 °C; pH 7	11.1	0.54	0.05	Martínez et al. (2002)

(continued)

Table 5.1 (continued)

Operation mode	Substrate	Microorganism	Process conditions	P_F $g \cdot L^{-1}$	$Y_{P/S}$ (g/g)	Q_P $(g \cdot L^{-1} \cdot h^{-1})$	Reference
Continuous	Vineshoot trimmings hemicellulosic hydrolysate	*Debaryomyces hansenii* NRRL Y-7426	Stirred tank reactor; 200 rpm; DR: 0.043 h^{-1}; K_La: 20 h^{-1} Temperature: 31.5 °C; pH 6.2	5.1	0.55	0.22	Salgado et al. (2012)

physicochemical factors. In addition, it helps to control cell growth and process asepsis (Zhuang et al. 2021).

The study of batch reactors has shown different contributions to xylitol production, with outstanding interest mainly in process conditions, as the effect of oxygen supply on yield and productivity (Pronk et al. 1990; Rafiqul and Sakinah 2013; Arcaño et al. 2020) In addition, different reactor configurations have been evaluated in conjunction with other techniques, such as cell recycle and immobilization, which also can favor the increase in xylitol production (Sreenath and Jeffries 1987). For example, Santos et al. (2003), using Erlenmeyer flasks operated in batch regime, studied the use of *C. guilliermondii* cells immobilized in Ca-alginate beads for the xylitol production from sugarcane bagasse hydrolysate. In their work, a full factorial design was applied by studying the concentration of sodium alginate, concentration of calcium chloride and bead curing time as independent variables. With this analysis, results were obtained for each of the response variables, viz, xylitol volumetric productivity, yield and solubilization of Ca-alginate beads, with values of 0.43 $g{\cdot}L^{-1}{\cdot}h^{-1}$, 0.47 $g{\cdot}g^{-1}$ and 1.9%, respectively.

Xylitol production by batch fermentation has been a challenging strategy. One of the main concerns regarding this operation is the high concentration of inhibitors at the beginning of the process (Guo et al. 2013). A batch operation model enabled Kelly et al. (2008) to determine the inhibition rate of the main inhibitors from lignocellulosic biomass hydrolysate, such as furfural, syringaldehyde, and vanillin, delaying the time and reducing the rate of xylitol production in *C. guilliermondii* cultures.

As previously mentioned, the batch regime allows the in-depth study of different fermentation parameters. López-Linares et al. (2018) verified that the batch regime helped in feasibility of producing xylitol by *D. hansenii or C. guilliermondii* from rapeseed straw hemicellulosic hydrolysate, concluding that *C. guilliermondii* presented higher resistance to toxic products in the hydrolysate. In other work, Tochampa et al. (2005) investigated the performance of xylitol production with the yeast *C. mogii* ATCC 18,364 from xylose to evaluate the influence of glucose as a co-substrate on xylitol yield. The authors concluded that with glucose levels of about 10% in co-substrate with xylose at the beginning of fermentation in a batch reactor increased xylitol yield; otherwise, with increasing glucose, xylitol production was suppressed.

Besides simple bacth, repeated-batch systems were reported for xylitol production with immobilized cells, and consisted of sequential fermentation using the previously used immobilized cells as inoculum (Santos et al. 2003; Sarrouh and Da Silva 2013; Dorantes-Landa et al. 2020).

Sarrouh and Da Silva (2013) worked with the xylitol production in repeated batch fermentation system using immobilized cells of *Candida guilliermondii* FTI20037. The authors verified after six successive batches a final xylitol concentration of 8 $g{\cdot}L^{-1}$, process yield of 0.11 $g{\cdot}g^{-1}$ and productivity of 0.08 $g{\cdot}L^{-1}$ h^{-1}, with reproducible results in the repeated fermentation batches.

5.4.2 Fed-Batch Operation

In addition to the classical batch operation, there is the fed-batch process, which is based on the supply of a substrate or nutrient in different periods along the process time, achieving an increase in production yield or productivity (Moran et al. 2013). Fed-batch fermentation is frequently used due to the propitious control of the C/N ratio by regulating the amount of carbon and nitrogen in the fermenter during the growth and accumulation phases of the product (Lee, Vadlani and Min 2017). Particularly, fed-batch is adequate to overcome inhibition issues; however, as in continuous systems, the large operation time could raise concerns about microbial contamination.

Specifically for xylitol production, a fed-batch operation can help to control the effect of inhibitory compounds. Guo et al. (2013) studied a two-stage fed-batch fermentation process, reporting high biomass production and xylose consumption, with 96.5 $g{\cdot}L^{-1}$ of xylitol produced with yield of 0.83 $g{\cdot}g^{-1}$ and productivity of 1.01 $g{\cdot}L^{-1}{\cdot}h^{-1}$. In that work, authors concluded that fed-batch fermentation decreased the inhibitory effect of the by-products of the corn cob hydrolyzed medium, avoiding the process of detoxification. In agreement, similar advantage of the fed-batch reactor could be also observed by Ping et al. (2013), when they performed xylitol production from corn cob hydrolysate by *Candida tropicalis* CCTCC M2012462, achieving control of inhibition by compounds present in the hemicellulosic hydrolysate. The maximum xylitol concentration obtained was 38.8 $g.L^{-1}$, showing yield of 0.7 $g.g^{-1}$, and a productivity of 0.46 $g.L^{-1}$ h^{-1} after 84 h of process.

Other authors have also reported that fed-batch mode of operation can maximize xylitol production. For example, Ramirez and Escoto (2021), performed an optimization study of xylitol production in fed-batches using a genetic algorithm. The operation was carried out with constant and variable feeding of the xylose flow for consumption of high amount of xylose, directing it to xylitol production. The authors' research studied the process design and parameters such as time, sugar concentration in the reactor, and the ratio of glucose to xylitol. Finally, the research showed that a constant flow rate promotes a higher xylitol concentration.

The main disadvantage of this operation mode is the high probability of contamination of the process without adequate precautions, due to the extended operation time, compared to the single batch (Jeevahan et al. 2020). Therefore, different authors have thoroughly studied this regimen to overcome this particular limitation. Some studies are related to the use of strains with hyper-acidophilic behaviors that facilitate production at the industrial level with lower probability of microbial contamination (Tamburini et al. 2015). An example is the study by Tamburini et al. (2015), which demonstrated the viability of the production of xylitol by *Candida tropicalis* DSM 7524 at the following conditions: temperature 32 °C, 80 $g{\cdot}L^{-1}$ of initial concentration of xylose and pH 2.5. In a time of 81 h, a final xylitol concentration of 71.34 $g{\cdot}L^{-1}$ was reached, with productivity of 0.79 $g{\cdot}L^{-1}{\cdot}h^{-1}$. In the same sense, Zhang et al. (2015) tested the possibility of generating up to 312.05 $g{\cdot}L^{-1}$ of xylitol by modified *K. marxianus YZJ017* in fed-batch fermentation, performing a process at high temperatures, thus saving the sterilization of the substrate.

5.4.3 *Continuous Operation*

Unlike batch and fed-batch operation modes, in continuous processes inlet and outlet streams in bioreactor are continuously flowing, at the same flow rate. Usually started as a batch or fed-batch operation, in continuous process inlet and outlet flows begin at an adequate time and, after a transient phase, stationary phase is reached, when the properties at any point of the reactor remain constant as function of the time. The objective is that continuous fermentation systems allow the maintenance of high rates of product formation over long periods (Faria et al. 2002; Zhuang et al. 2021).

In this operation mode, the dilution rate is a fundamental parameter for both xylitol production and cell growth (Martínez et al. 2003). Among the advantages, continuous process allows the use of smaller volume reactors to reach a desirable productivity in xylitol, positively impacting capital investment (Hong et al. 2014). However, it also has several disadvantages that lead to process failures, most notably natural or induced mutations, contamination, and clogging of pipelines due to cell agglomeration (Carvalho et al. 2003; Pérez-Bibbins et al. 2014). An alternative is the use of strategies as cell recycling or cells immobilization, which are interesting to allow the use of dilution rates higher than maximum specific growth of microorganism without cell washout.

Different authors reported xylitol production in continuous regime and concluded that it is an effective alternative to obtain higher yields with faster production (Hernandez-Escoto et al. 2014). Faria et al. (2002) observed a conversion of 86% of D-xylose into xylitol by a *Candida guilliermondii* strain in continuous fermentation. In that work, the separation of xylitol was also studied using membrane technology in a continuous regime. The conversion of D-xylose resulted in productivity of 1.14 g xylitol/L/h, using a dilution rate of 0.03 h^{-1}. Moreover, the membrane with the highest efficiency had a pore diameter of 0.2 μm and a permeability of 42.9 L/(m^2.h.bar).

5.5 Techniques Applied to Metabolic Engineering

In the context of genetic engineering and synthetic biology, scientific efforts have promoted the modification of microorganism strains that use alternative metabolic routes and have efficient xylitol production rates (Table 5.2). Some examples can be highlighted by the studies of Jin et al. (2005), who used *Pichia stipitis* FPL-YS30, a xyl3-Δ1 mutant to metabolize xylose into xylitol; Atzmüller et al. (2020) engineered *Meyerozyma guilliermondii* to increase xylitol production, and Kumar et al. (2018) used recombinant *Saccharomyces cerevisiae* to improve the productivity of the processes. Table 5.2 shows the production of xylitol by different engineered microorganisms.

Iverson et al. (2016) modified strains of *Escherichia colli* by the inclusion or deletion of operons in the genetic system, to optimize the use of NADH molecules generated in the catabolism of glucose in xylose reduction in xylitol. The work

Table 5.2 Production of xylitol from xylose by engineered microorganisms

Microorganism	Strain descriptions	Volumetric productivity ($g.L^{-1}.h^{-1}$)	Reference
Saccharomyces cerevisiae	Increased expression of the XR enzyme	1.50	Oh et al. (2013)
Candida tropicalis	Reduction of XDH activity	0.62	Ko et al. (2011)
Candida tropicalis	Heterologous expression of xylose transporters	1.57	Jeon et al. (2013)
Kluyveromyces marxianus	Overexpression NcXR gene	4.43	Zhuang et al. (2021)
Candida tropicalis	Overexpressed *zwf* and *gnd*	1.25	Ahmad et al. (2012)
Saccharomyces cerevisiae	Expressed *P. stipitis* cdt-1 and gh1-1	0.55	Oh et al. (2013)
Kluyveromyces marxianus	Overexpression of transporter genes KmFPS1, CiGXF1, and CiGXS1	3.40	Zhang et al. (2015)
Escherichia coli	Co-expression of xylose reductase and glucose dehydrogenase	8.0	Jin et al. (2019)
Trichoderma reesei	Overexpressed *xyl1* gene Silenced *xyiH* gene	0.026	Hong et al. (2014)

reported a high rate of xylose to xylitol conversion by catabolized glucose. In other work, Guo et al. (2013) reported a modification in the pentose phosphate route to increase the regeneration efficiency of a coenzyme of *Gluconobacter oxydans*, observing an increase of more than 3 times in the xylitol yield in relation to that obtained from the wild microorganism.

Other strategies within genetic modifications are focused on the suppression of the XDH gene, expression of the enzyme XR and the ability of cofactors (Jain and Ghosh 2021). The suppression of the XDH gene can prevent xylitol from being oxidized in xylose and thus increase its accumulation and decrease the need to regulate the oxygen supply (Jain and Mulay 2014). In a study by Pal et al. (2013), they projected the mutant *Debaryomyces hansenii* interrupting the XDH gene. In the case of the mutant, the xylitol concentration was 2.5 times higher than that of the native strain.

Jeon et al. (2012) built a mutant of *Candida tropicalis* incorporating an XR codon of *Neurospora crassa* using the pGAPDH promoter. Xylitol yield and productivity increased by 62% and 73%, respectively, compared to the original strain. Other studies have reported that xylitol production can also be improved by increasing the availability of the NADPH cofactor that indirectly increases the activity of the XR enzyme. The enzymes 6PGDH and G6PDH are directed to genetic alterations, as they promote the production of NAPDH (Felipe Hernández-Pérez et al. 2019).

5.6 Concluding Remarks and Future Perspectives

Xylitol has been gaining increasing attention over the years, given its properties of interest and range of applications. There is a high industrial and consumer demand, which drives scientific studies and research that aim to both improve production efficiency and analyze prospects in financial terms. Regarding the market scenario, estimates point to an increasing search for the molecule, which may indicate significant movements in the economy of the sectors involved (De Albuquerque et al. 2014).

The integration of the bioproduction of xylitol into a biorefinery could contribute to the techno-economic viability of this bioprocess, eventually increasing the profitability of a biorefinery (Felipe Hernández-Pérez et al. 2019). In this sense, the consolidation and expansion of technological alternatives in the biotechnological route are necessary.

Felipe Hernández-Pérez et al. (2019) described some research reports that estimated the world xylitol market to be expected more than \$1.15 billion by 2023, corresponding to a compound annual growth rate in volume and value of approximately 5.7% between 2018 and 2023. Similarly, another report also estimated that the world xylitol market will reach US\$ 1.37 billion (\$4.5.kg^{-1}) by 2025.

In particular, the xylitol market in Brazil is expected to achieve substantial growth, reaching over USD 1.7 million sales by 2023, owing to growing health awareness among consumers along with an increase in per capita income (Moraes et al. 2020). U.S. xylitol market is expected to witness significant growth over the coming seven years, owing to the rise in diabetic and obese population, promoting the product use as a low-calorie natural sweetener. Finland's xylitol market revenue is projected to reach USD 8.5 million by 2023, owing to rising health issues due to excess sugar consumption as well as increase in health claims related to intake of xylitol among the Finnish population (Felipe Hernández-Pérez et al. 2019).

Xylitol producing microorganisms such as *C. tropicalis and C. guilhermodii* show increased yields and high productivity from various lignocellulosic and biomass biomasses under different cultivation processes for the production of xylitol (Hernandez-Escoto et al. 2014; Abd Rahman et al. 2020). On the other hand, interventions aimed at improving strains such as metabolic engineering create improvements with respect to metabolic transport of xylose that can improve xylitol production driven for industrial requirements. Besides, some genetic modifications were studied in yeast species such as *C guilhermondi FTI 20,037* and *Kluyveromyces marxianus IZ 1339* (Arruda et al. 2017; De Souza Queiroz et al. 2021). These techniques include increased expression of the XR enzyme, reduction of XDH expression, and availability of cofactor NADPH. Despite the important advances observed in metabolic engineering strategies, the development of metabolically stable robust recombinant strains with a high tolerance to inhibitor compounds from hemicellulosic hydrolysates remains a challenge.

Regarding the metabolic engineering, the main lines of xylitol production research is related to the regulation of metabolic routes, such as the tricarboxylic acid cycle

(TCA) and the pathway of pentose phosphate, and also more refined gene editing techniques, as: the expression and deletion of genes constituting the artificial route glucose-to-xylitol; Molecular Biology Techniques for the Development of microorganisms modified for high xylitol production titers in short-run fermentations and from high xylose concentrations (Hernández-Pérez et al. 2019; Xu et al. 2019); production of xylitol by enzymatic means (Su et al. 2013).

An interesting aspect to be considered is xylitol purification, the last step of the biotechnological route, which has consubstantial impact in the biorefinery, in addition to be an expensive step in the overall production and significantly important in the large-scale implementation of this bioprocess (Mareczky et al. 2016). Indeed, an important challenge is the downstream of the fermented medium, which has a large quantity of impurities as residual sugars, phenolic compounds, salts and proteins (De Souza Queiroz et al. 2021). When compared to the other steps, the downstream of biotechnological process has received less attention and significant efforts.

Besides studies of metabolic engineering and downstream, use of computational techniques for technoeconomic analysis of a biorefinery for xylitol production are fundamental for the viability of the bioprocess. In this way, Franceschin et al. (2011) presented a realistic model for a conversion of rye straw into fuel and xylitol by technical and economical assessment based on experimental data, with simulation aided by the software Aspen PlusTM for studies of biomass pretreatment and mass balances aimed to implement equipment size and capital investment calculations. A financial analysis was carried out to calculate costs and profitability; it was found out that, if a substantial portion of hemicellulose is transformed to xylitol, this product significantly contributes to the viability of the process. In other work, Unrean and Ketsub (2018a, b) perfomed economic analysis of integrated lignocellulosic bioprocess for co-production of ethanol and xylitol from sugarcane bagasse, considering the production process model simulated by using SuperPro Designer software (Intelligen Inc., USA).

It is worth to comment that all the aspects aforementioned in this chapter do not encompass the whole range of possibilities for innovation in xylitol production. Even so, it was demonstrated the great potential offered by biotechnological processes for the manufacture of the molecule. In this sense, it is necessary the constant search for improvement methodologies in the production chain, which embody the genetic improvement of strains to produce more and in a more sustainable way, added by the study of adequate bioreactors and operation modes, to improve the efficiency of unit operations and the use of residual biomasses. Stirred tank and column bioreactors were evaluated as interesting options, with fed-batch and continuous operation modes as alternatives to improve xylitol yield and productivity, helping to overcome drawbacks as the presence of inhibitors compounds in hemicellulosic hydrolysates. Finally, an important remark is the requirement of public and private sector investments to make the biotechnological production of xylitol a reality.

References

Abbott AP, D'Agostino C, Davis SJ, Gladden LF, Mantle MD (2016) Do group 1 metal salts form deep eutectic solvents?. Phys Chem Chem Phys 18(36):25528–25537. https://pubs.rsc.org/en/content/articlelanding/2016/cp/c6cp05880a/unauth

Abd Rahman NH, Md. Jahim J, Abdul Munaim MS, et al (2020) Immobilization of recombinant Escherichia coli on multi-walled carbon nanotubes for xylitol production. Enzyme Microb Technol 135:109495. https://doi.org/10.1016/j.enzmictec.2019.109495

Acosta Martínez E, Silva SS, Felipe MG (2000) Effect of the oxygen transfer coefficient on xylitol production from sugarcane bagasse hydrolysate by continuous stirred-tank reactor fermentation. In: Twenty-First Symposium on Biotechnology for Fuels and Chemicals (pp 633–641). Humana Press, Totowa, NJ. https://link.springer.com/chapter/10.1007/978-1-4612-1392-5_48

Agblevor F, Waleed K (2005) Water extraction of steam exploded Xylo-Oligosaccharides for Xylitol production. In: The 2005 annual meeting

Ahmad I, Shim WY, Jeon WY, Yoon BH, Kim JH (2012) Enhancement of xylitol production in Candida tropicalis by co-expression of two genes involved in pentose phosphate pathway. Bioprocess Biosyst Eng 35(1):199–204. https://doi.org/10.1007/s00449-011-0641-9

Antunes FAF, Santos JCD, Cunha MAAD, Brumano LP, Milessi TSDS, Terán-Hilares R et al (2017) Biotechnological production of xylitol from biomass. In: Production of Platform Chemicals from Sustainable Resources (pp 311–342). Springer, Singapore. https://link.springer.com/chapter/10.1007/978-981-10-4172-3_10

Antunes FAF, Thomé LC, Santos JC, Ingle AP, Costa CB, Dos Anjos V, Bell MJV, Rosa CA, Da Silva SS (2021) Multi-scale study of the integrated use of the carbohydrate fractions of sugarcane bagasse for ethanol and xylitol production. Renew Energy 163(2021):1343–1355

Arcaño YD, García ODV, Mandelli D, Carvalho WA, Pontes LAM (2020) Xylitol: A review on the progress and challenges of its production by chemical route. Catal Today 344:2–14. https://doi.org/10.1016/j.cattod.2018.07.060

Arruda PV et al (2017) Scale up of xylitol production fromsugarcane bagasse hemicellulosic hydrolysate by Candida guilliermondii FTI 20037. J Ind Eng Chem 47:297–302

Atzmüller D, Ullmann N, Zwirzitz A (2020) Identification of genes involved in xylose metabolism of Meyerozyma guilliermondii and their genetic engineering for increased xylitol production. Amb Express 10(1):1–11. https://link.springer.com/article/10.1186/s13568-020-01012-8

Baptista SL, Cunha JT, Romaní A, Domingues L (2018) Xylitol production from lignocellulosic whole slurry corn cob by engineered industrial Saccharomyces cerevisiae PE-2. Biores Technol 267:481–491. https://doi.org/10.1016/j.biortech.2018.07.068

Baptista SL, Carvalho LC, Romaní A, Domingues L (2020) Development of a sustainable bioprocess based on green technologies for xylitol production from corn cob. Ind Crops Prod 156. https://doi.org/10.1016/j.indcrop.2020.112867

Bianchini IDA, Sene L, da Cunha MAA, Felipe MDGDA (2021) Short-term adaptation strategy improved Xylitol production by Candida guilliermondii on Sugarcane Bagasse Hemicellulosic Hydrolysate. BioEnergy Res 1–13

Bonfiglio F, Cagno M, Yamakawa CK, Mussatto SI (2021) Fernando Production of xylitol and carotenoids from switchgrass and Eucalyptus globulus hydrolysates obtained by intensified steam explosion pretreatment. Ind Crop Prod 170:113800

Branco RF, Santos JC, Murakami LY, Mussatto SI, Dragone G, Silva SS (2007) Xylitol production in a bubble column bioreactor: influence of the aeration rate and immobilized system concentration. Process Biochem 42(2):258–262

Buhner J, Agblevor FA (2004) Effect of detoxification of dilute-acid corn fiber hydrolysate on xylitol production. Appl Biochem Biotechnol 119(1):13–30. https://doi.org/10.1385/ABAB:119:1:13

Canilha L, e Silva JBA, Felipe MG, Carvalho W (2003) Batch xylitol production from wheat straw hemicellulosic hydrolysate using Candida guilliermondii in a stirred tank reactor. Biotechnol Lett 25(21):1811-1814. https://doi.org/10.1023/A:1026288705215

Carneiro CVG, Almeida JR (2019) Xylitol production: identification and comparison of new producing yeasts. Microorganisms 7(11):484. https://doi.org/2076-2607/7/11/484

Carvalho W, Silva SS, Converti A, Vitolo M (2002a) Metabolic behavior of immobilized Candida guilliermondii cells during batch xylitol production from sugarcane bagasse acid hydrolyzate. Biotechnol Bioeng 79(2):165–169. https://doi.org/10.1002/bit.10319

Carvalho W, Silva SS, Santos JC, Converti A (2003) Xylitol production by Ca-alginate entrapped cells: comparison of different fermentation systems. Enzyme Microb Technol 32(5):553–559. https://www.sciencedirect.com/science/article/pii/S0141022903000073?casa_token=1hyNu23tgjAAAAAA:aqVsSwYSgd5DZBpu-10NLzknhaFPiT0u14hR9rYLsBzpvJdIVkx4REiF7uaZtMAgT6Nt-CzrlwF_

Carvalho W, Santos JC, Canilha L, e Silva JBA, Felipe MG, Mancilha IM, Silva SS (2004) A study on xylitol production from sugarcane bagasse hemicellulosic hydrolysate by Ca-alginate entrapped cells in a stirred tank reactor. Process Biochem 39(12):2135-2141. https://doi.org/10.1016/j.procbio.2003.11.021

Cortivo PRD, Hickert LR, Rosa CA, Ayub MAZ (2020) Conversion of fermentable sugars from hydrolysates of soybean and oat hulls into ethanol and xylitol by Spathaspora hagerdaliae UFMG-CM-Y303. Ind Crops Prod 146:112218

Cunha JT, Soares PO, Romaní A, Thevelein JM, Domingues L (2019) Xylose fermentation efficiency of industrial Saccharomyces cerevisiae yeast with separate or combined xylose reductase/xylitol dehydrogenase and xylose isomerase pathways. Biotechnol Biofuels 12(1):1–14. https://doi.org/10.1186/s13068-019-1360-8

Dasgupta D, Bandhu S, Adhikari DK, Ghosh D (2017) Challenges and prospects of xylitol production with whole cell bio-catalysis: a review. Microbiol Res 197:9–21. https://doi.org/10.1016/j.micres.2016.12.012

Dasgupta D, Kurmi AK, Adhikari DK, Ghosh D (2020) Xylitol production from lignocellulosic pentosans using Kluyveromyces marxianus: kinetic modelling of yeast growth and fermentation. Biofuels 11(3):309–319

De Albuquerque TL, da Silva Jr IJ, de Macedo GR, Rocha MVP (2014) Biotechnological production of xylitol from lignocellulosic wastes: a review. Process Biochem 49(11):1779–1789. https://doi.org/10.1016/j.procbio.2014.07.010

De Arruda PV, de Cássia Lacerda Brambilla Rodrigues R, da Silva DDV, de Almeida Felipe MDG (2011) Evaluation of hexose and pentose in pre-cultivation of Candida guilliermondii on the key enzymes for xylitol production in sugarcane hemicellulosic hydrolysate. Biodegradation 22(4):815–822. https://link.springer.com/article/10.1007/s10532-010-9397-1

De Faveri D, Perego P, Converti A, Del Borghi M (2002) Xylitol recovery by crystallization from synthetic solutions and fermented hemicellulose hydrolyzates. Chem Eng J 90(3):291–298

De Freitas Branco R, Dos Santos JC, Da Silva SS (2011) A novel use for sugarcane bagasse hemicellulosic fraction: xylitol enzymatic production. Biomass Bioenergy 35(7):3241-3246

De Mohamad NL, Mustapa Kamal SM, Mokhtar MN (2015) Xylitol biological production: a review of recent studies. Food Rev Intl 31(1):74–89

Dorantes-Landa DN, Cocotle-Ronzón Y, Morales-Cabrera MA, Hernández-Martínez E (2020) Modelling of the xylitol production from sugarcane bagasse by immobilized cells. J Chem Technol Biotechnol. https://doi.org/10.1002/jctb.6394

Du C, Li Y, Zong H et al (2020a) Bioresource Technology Production of bioethanol and xylitol from non-detoxified corn cob via a two- stage fermentation strategy. Bioresour Technol 310:123427. https://doi.org/10.1016/j.biortech.2020.123427

Du C, Li Y, Zong H, Yuan T, Yuan W, Jiang Y (2020b) Production of bioethanol and xylitol from non-detoxified corn cob via a two-stage fermentation strategy. Biores Technol 310:123427. https://doi.org/10.1016/j.biortech.2020.123427

Faria LFF, Pereira N Jr, Nobrega R (2002) Xylitol production from D-xylose in a membrane bioreactor. Desalination 149(1–3):231–236. https://doi.org/10.1016/S0011-9164(02)00766-X

Felipe Hernandez-Pérez AF et al (2019) Xylitol bioproduction: state-of-the-art, industrial paradigm shift, and opportunities for integrated biorefineries. Critical Rev Biotechnol 1549–1571. https://www.tandfonline.com/doi/abs/10.1080/07388551.2019.1640658

Franceschin G, Sudiro M, Ingram T, Smirnova I, Brunner G, Bertucco A (2011) Conversion of rye straw into fuel and xylitol: a technical and economical assessment based on experimental data. Chem Eng Res Des 89(6):631–640. https://doi.org/10.1016/j.cherd.2010.11.001

Granström TB, Izumori K, Leisola M (2007) A rare sugar xylitol. Part I: The biochemistry and biosynthesis of xylitol. Appl Microbiol Biotechnol 74:277–281. https://doi.org/10.1007/s00253-006-0761-3

Guo X, Zhang R, Li Z, Dai D, Li C, Zhou X (2013) A novel pathway construction in Candida tropicalis for direct xylitol conversion from corncob xylan. Biores Technol 128:547–552. https://www.sciencedirect.com/science/article/pii/S0960852412016641?casa_token=ib7ezjfBu_4AAAAA:Al--3Fs3jzVAdOkV5iJK05uyDaYHFyEFkVrZDpPIRLQ3DxbBi-y9OkaCg9SblcBGAka9QqPymIgT

Hernandez-Escoto H, Rodriguez-Gomez D, Morales-Rodriguez R (2014) Process design and control of a xylitol production reactor. Comput Aided Chem Eng 33:757–762. https://doi.org/10.1016/B978-0-444-63456-6.50127-7

Hernández-Pérez AF, Costa IAL, Silva DDV, Dussán KJ, Villela TR, Canettieri EV, Felipe MGA (2016) Biochemical conversion of sugarcane straw hemicellulosic hydrolyzate supplemented with co-substrates for xylitol production. Bioresour Technol 200:1085–1088

Hickert LR et al (2013) Simultaneous saccharification and co fermentation of un-detoxified rice hull hydrolysate by Saccharomyces cerevisiae ICV D254 and Spathaspora arborariae NRRL Y-48658 for the production of ethanol and xylitol. Biores Technol 143:112–116

Hong, Y., Dashtban, M., Kepka, G., Chen, S., & Qin, W. (2014). Overexpression of D-xylose reductase (xyl1) gene and antisense inhibition of D-xylulokinase (xyiH) gene increase xylitol production in Trichoderma reesei. *BioMed research international, 2014*. https://doi.org/10.1155/2014/169705

Hovnanyan KO, Gasparyan HV, Marutyan SV, Navasardyan LH, Trchounian AH (2019) Comparative structural analysis of yeasts *Candida Guilliermondii* NP-4 Cultivated with and without nitrogen source. Proceedings of the YSU b: Chemical and Biological Sciences 53(248):53–58

Iverson A, Garza E, Manow R, Wang J, Gao Y, Grayburn S, Zhou S (2016) Engineering a synthetic anaerobic respiration for reduction of xylose to xylitol using NADH output of glucose catabolism by Escherichia coli AI21. BMC Syst Biol 10(1):1–10. https://doi.org/10.1186/s12918-016-0276-1

Jain, V., & Ghosh, S. (2021). Biotransformation of lignocellulosic biomass to xylitol: an overview. *Biomass Conversion and Biorefinery*, 1–19.doi/ https://doi.org/10.1007/s13399-021-01904-0

Jain H, Mulay S (2014) A review on different modes and methods for yielding a pentose sugar: xylitol. Int J Food Sci Nutr 65(2):135–143. https://doi.org/10.3109/09637486.2013.845651

Jeevahan JJ, Chandrasekaran M, Venkatesan SP, Sriram V, Joseph GB, Mageshwaran G, Durairaj RB (2020) Scaling up difficulties and commercial aspects of edible films for food packaging: a review. Trends Food Sci Technol 100:210–222. https://doi.org/10.1016/j.tifs.2020.04.014

Jeon WY, Yoon BH, Ko BS, Shim WY, Kim JH (2012) Xylitol production is increased by expression of codon-optimized Neurospora crassa xylose reductase gene in Candida tropicalis. Bioprocess Biosyst Eng 35(1):191–198. https://doi.org/10.1007/s00449-011-0618-8

Jeon WY, Shim WY, Lee SH, Choi JH, Kim JH (2013) Effect of heterologous xylose transporter expression in Candida tropicalis on xylitol production rate. Bioprocess Biosyst Eng 36(6):809–817. https://doi.org/10.1007/s00449-013-0907-5

Jin YS, Cruz J, Jeffries TW (2005) Xylitol production by a Pichia stipitis D-xylulokinase mutant. Appl Microbiol Biotechnol 68(1):42–45. https://doi.org/10.1007/s00253-004-1854-5

Jin LQ, Yang B, Xu W, Chen XX, Jia DX, Liu ZQ, Zheng YG (2019) Immobilization of recombinant Escherichia coli whole cells harboring xylose reductase and glucose dehydrogenase for xylitol production from xylose mother liquor. Biores Technol 285:121344. https://doi.org/10.1016/j.biortech.2019.121344

Jofre FM, Hernández-Pérez AF, Santos JC dos, Felipe M das G de A (2021) Use of dry yeast biomass as a new approach for detoxification of hemicellulosic hydrolysates aiming to xylitol production. Ind Crops Prod 170. https://doi.org/10.1016/j.indcrop.2021.113812

Jönsson LJ, Alriksson B, Nilvebrant NO (2013) Bioconversion of lignocellulose: inhibitors and detoxification. Biotechnol Biofuels 6:1–10. https://doi.org/10.1186/1754-6834-6-16

Kamer DDA, Palabiyik I, Işık NO, Akyuz F, Demirci AS, Gumus T (2019) Effect of confectionery solutes on the rheological properties of fish (Oncorhynchus mykiss) gelatin. LWT 101:499–505

Kelly C, Jones O, Barnhart C, Lajoie C (2008) Effect of furfural, vanillin and syringaldehyde on Candida guilliermondii growth and xylitol biosynthesis. In: Biotechnology for fuels and chemicals. Humana Press, pp 615–626. https://doi.org/10.1007/978-1-60327-526-2_57

Kim JH, Han KC, Koh YH, Ryu YW, Seo JH (2002) Optimization of fed-batch fermentation for xylitol production by Candida tropicalis. J Ind Microbiol Biotechnol 29(1):16–19. https://doi.org/10.1038/sj.jim.7000257

Ko BS, Kim DM, Yoon BH, Bai S, Lee HY, Kim JH, Kim IC (2011) Enhancement of xylitol production by attenuation of intracellular xylitol dehydrogenase activity in Candida tropicalis. Biotech Lett 33(6):1209–1213. https://doi.org/10.1007/s10529-011-0558-z

Krishnan S, Suzana BN, Wahid ZA, Nasrullah M, Munaim MSA, Din MFBM, Li YY (2020) Optimization of operating parameters for xylose reductase separation through ultrafiltration membrane using response surface methodology. Biotechnol Rep 27:e00498

Kourkoutas Y, Bekatorou A, Banat IM, Marchant R, Koutinas AA (2004) Immobilization technologies and support materials suitable in alcohol beverages production: a review. Food Microbiol 21(4):377–397. https://www.sciencedirect.com/science/article/pii/S0740002003001072?casa_token=iCeq_gu_UzsAAAAA:_gPRiJB0tzOpS6aW9fV_FVE9IHuCYw5YKz9xoHR-XgvqFr5gZS7eVCskYjwcsDIy0Q5Ed8cjIP-C

Kumar V, Krishania M, Preet Sandhu P, et al (2018) Efficient detoxification of corn cob hydrolysate with ion-exchange resins for enhanced xylitol production by Candida tropicalis MTCC 6192. Bioresour Technol 251:416–419.https://doi.org/10.1016/j.biortech.2017.11.039

Kwon DH, Kim MD, Lee TH, Oh YJ, Ryu YW, Seo JH (2006) Elevation of glucose 6-phosphate dehydrogenase activity increases xylitol production in recombinant Saccharomyces cerevisiae. J Mol Catal B Enzym 43(1–4):86–89. https://www.sciencedirect.com/science/article/pii/S1381117706002013?casa_token=0_QYD2DMl6EAAAAA:f3r2fhezjWZhIyFL7uGa3fAtkDWqxpU-rsf0z7dpNbpgfC-dsCky_GiRLar1j-oe9l9ZJECBKCod

Liu S, He H, Fu X, Yuan T, Wang Q, Yang G, Zhang H, Ding M, Liao C (2019) Xylitol production from prehydrolysis liquor of Kraft-based dissolving pulp by Candida tropicalis. BioResources 14:21–30

López-Linares JC, Romero I, Cara C et al (2018) Xylitol production by Debaryomyces hansenii and Candida guilliermondii from rapeseed straw hemicellulosic hydrolysate. Bioresour Technol 247:736–743. https://doi:https://doi.org/10.1016/j.biortech.2017.09.139

Lugani Y, Sooch BS (2020) Fermentative production of xylitol from a newly isolated xylose reductase producing Pseudomonas putida BSX-46. LWT 134:109988. https://doi.org/10.1016/J.LWT.2020.109988

Manaf SFA et al (2018) Fractionation of oil palm fronds (OPF) hemicellulose using dilute nitric acid for fermentative production of xylitol. Ind Crop Prod 115:615

Mareczky Z, Fehér A, Fehér C, Barta Z, Réczey K (2016) Effects of pH and aeration conditions on xylitol production by Candida and Hansenula yeasts. Periodica Polytech Chem Eng 60(1):54–59

Martínez EA, Silva SS, e Silva JBA, Solenzal AI, Felipe MG (2003). The influence of pH and dilution rate on continuous production of xylitol from sugarcane bagasse hemicellulosic hydrolysate by C. guilliermondii. Process Biochem 38(12):1677–1683.

Martínez EA, Santos JAF (2012) Influence of the use of rice bran extract as a source of nutrients on xylitol production. Food Sci Technol 32:308–313

Matrawy AA, Khalil AI, Marey HS, Embaby AM (2021) Biovalorization of the raw agro-industrial waste rice husk through directed production of xylanase by Thermomyces lanuginosus strain A3-1 DSM 105773: a statistical sequential model. Biomass Convers Biorefinery 11(5):2177–2189. https://doi.org/10.1016/j.jenvman.2016.08.059

Moraes EDJC, Silva DDV, Dussán KJ, Tesche LZ, de Almeida Silva JB, Rai M, de Almeida Felipe MDG (2020) Xylitol-sweetener production from barley straw: optimization of acid hydrolysis condition with the energy consumption simulation. Waste Biomass Valorization 11(5):1837–1849

Moran LV, Sampath H, Kochunov P, Hong LE (2013) Brain circuits that link schizophrenia to high risk of cigarette smoking. Schizophr Bull 39(6):1373–1381

Mussatto SI (2012) Application of xylitol in food formulations and benefits for health. In: d-Xylitol (pp 309–323). Springer, Berlin, Heidelberg. https://link.springer.com/chapter/10.1007/978-3-642-31887-0_14

Mussatto SI (2014) Brewer's spent grain: a valuable feedstock for industrial applications. J Sci Food Agric 94(7):1264–1275

Mussatto SI, Roberto IC (2002) Xylitol: a sweetner with benefits for human health. Revista Brasileira De Ciências Farmacêuticas 38(4):401–413

Mussatto SI et al (2008) Effect of hemicellulose and lignin on enzymatic hydrolysis of cellulose from brewer's spent grain. Enzyme Microb Technol 43(2):124–129

Narisetty V, Castro E, Durgapal S et al (2021) High level xylitol production by Pichia fermentans using non-detoxified xylose-rich sugarcane bagasse and olive pits hydrolysates. Bioresour Technol 342:126005. https://doi.org/10.1016/j.biortech.2021.126005

Oh EJ, Ha SJ, Kim SR, Lee WH, Galazka JM, Cate JH, Jin YS (2013) Enhanced xylitol production through simultaneous co-utilization of cellobiose and xylose by engineered Saccharomyces cerevisiae. Metab Eng 15:226–234

Pal S, Choudhary V, Kumar A, Biswas D, Mondal AK, Sahoo DK (2013) Studies on xylitol production by metabolic pathway engineered Debaryomyces hansenii. Biores Technol 147:449–455

Pappu SMJ, Gummadi SN (2018) Effect of cosubstrate on xylitol production by Debaryomyces nepalensis NCYC 3413: A cybernetic modelling approach. Process Biochem 69:12–21. https://doi.org/10.1016/j.procbio.2018.03.023

Pereira RS, Mussatto SI, Roberto IC (2011) Inhibitory action of toxic compounds present in lignocellulosic hydrolysates on xylose to xylitol bioconversion by Candida guilliermondii. J Ind Microbiol Biotechnol 38(1):71–78

Pérez-Bibbins B, Salgado JM, Torrado A, Aguilar-Uscanga MG, Domínguez JM (2013) Culture parameters affecting xylitol production by Debaryomyces hansenii immobilized in alginate beads. Process Biochem 48(3):387–397

Pérez-Bibbins B, de Souza Oliveira RP, Torrado A, Aguilar-Uscanga MG, Domínguez JM (2014) Study of the potential of the air lift bioreactor for xylitol production in fed-batch cultures by Debaryomyces hansenii immobilized in alginate beads. Appl Microbiol Biotechnol 98(1):151–161

Pérez-Bibbins B, Torrado-Agrasar A, Salgado JM, Mussatto SI, Domínguez JM (2016) Xylitol production in immobilized cultures: a recent review. Crit Rev Biotechnol 36(4):691–704

Ping Y, Ling HZ, Song G, Ge JP (2013) Xylitol production from non-detoxified corncob hemicellulose acid hydrolysate by Candida tropicalis. Biochem Eng J 75:86–91

Pronk JT, Meesters PJW, Van Dijken JP, Bos P, Kuenen JG (1990) Heterotrophic growth of Thiobacillus acidophilus in batch and chemostat cultures. Arch Microbiol 153(4):392–398

Rafiqul ISM, Sakinah AM (2013) Processes for the production of xylitol—a review. Food Rev Intl 29(2):127–156

Raj K, Krishnan C (2020) Improved co-production of ethanol and xylitol from low-temperature aqueous ammonia pretreated sugarcane bagasse using two-stage high solids enzymatic hydrolysis and Candida tropicalis. Renew Energy 153:392–403. https://doi.org/10.1016/j.renene.2020.02.042

Ramirez OS, Escoto HH (2021). Enhancement of Xylitol production by fed-batch policy designed through stochastic search. Chem Eng Trans 86:1009–1014. https://doi.org/10.3303/CET2186169

Rao LV, Goli JK, Gentela J, Koti S (2016) Bioconversion of lignocellulosic biomass to xylitol: an overview. Biores Technol 213:299–310

Rao VN, Malu TJ, Cheralathan KK, Sakar M, Pitchaimuthu S, Rodríguez-González V, Shankar MV (2021) Light-driven transformation of biomass into chemicals using photocatalysts–Vistas and challenges. J Environ Manage 284:111983. https://doi.org/10.1016/j.biortech.2015.07.106

Rodrigues RCLB, Felipe MGA, Roberto IC, Vitolo M (2003) Batch xylitol production by Candida guilliermondii FTI 20037 from sugarcane bagasse hemicellulosic hydrolyzate at controlled pH values. Bioprocess Biosyst Eng 26(2):103–107. https://link.springer.com/article/10.1007/s00449-003-0332-2

Salgado JM, Rodríguez N, Cortés S, Domínguez JM (2012) Coupling two sizes of CSTR-type bioreactors for sequential lactic acid and xylitol production from hemicellulosic hydrolysates of vineshoot trimmings. New Biotechnol 29(3):421–427. https://www.sciencedirect.com/science/article/abs/pii/S1871678411001609

Santos JC, Carvalho W, Silva SS, Converti A (2003) Xylitol production from sugarcane bagasse hydrolyzate in fluidized bed reactor. Effect of air flow rate. Biotechnol Progress 19(4):1210–1215. https://aiche.onlinelibrary.wiley.com/doi/abs/10.1021/bp034042d?casa_token=zpaExejItK8AAAAA:VwfNTZJqa0de0sBXGCoqclwJQdscUGjtEkYR6BeBVItnM3xpR7iX_nGfckJkxGOg59URGe10rDWz2UkJDw

Santos JC, Pinto ÍR, Carvalho W, Mancilha IM, Felipe MG, Silva SS (2005) Sugarcane bagasse as raw material and immobilization support for xylitol production. App Biochem Biotechnol 122(1):673–683. https://doi.org/10.1385/ABAB:122:1-3:0673

Sarrouh B, Da Silva SS (2013) Repeated batch cell-immobilized system for the biotechnological production of xylitol as a renewable green sweetener. Appl Biochem Biotechnol 169(7):2101–2110

Sarrouh BF, dos Santos DT, da Silva SS (2007) Biotechnological production of xylitol in a three-phase fluidized bed bioreactor with immobilized yeast cells in Ca-alginate beads. Biotechnol J 2(6):759–763. https://doi.org/10.1002/biot.200600207

Silva-Fernandes T, Santos JC, Hasmann F, Rodrigues RCLB, Izario Filho HJ, Felipe MGA (2017) Biodegradable alternative for removing toxic compounds from sugarcane bagasse hemicellulosic hydrolysates for valorization in biorefineries. Biores Technol 243:384–392

Soleimani M, Tabil L (2013) Interaction of medium detoxification/supplementation and cell recycling in fermentative xylitol production. Biocatal Biotransform 31(4):208–216

De Souza Queiroz S, Jofre FM, dos Santos HA, Hernández-Pérez AF, de Almeida Felipe MDG (2021) Xylitol and ethanol co-production from sugarcane bagasse and straw hemicellulosic hydrolysate supplemented with molasses. Biomass Convers Biorefinery 1–10

Su B, Wu M, Lin J, Yang L (2013) Metabolic engineering strategies for improving xylitol production from hemicellulosic sugars. Biotech Lett 35(11):1781–1789

Tamburini E, Costa S, Marchetti MG, Pedrini1 1ì P (2015) High productivity fed-batch fermentation of xylitol from xylose using a hyper-acidophilic Candida tropicalis. Biomolecules 5(3):1979–1989

Tochampa W, Sirisansaneeyakul S, Vanichsriratana W, Srinophakun P, Bakker HH, Chisti Y (2005) A model of xylitol production by the yeast Candida mogii. Bioprocess Biosyst Eng 28(3):175–183

Tochampa W, Sirisansaneeyakul S, Vanichsriratana W, Srinophakun P, Bakker HHC, Wannawilai S, Chisti Y (2015) Optimal control of feeding in fed-batch production of xylitol. Ind Eng Chem Res 54(7):1992–2000. https://doi.org/10.1021/ie5032937

Unrean P, Ketsub N (2018a) Integrated lignocellulosic bioprocess for co-production of ethanol and xylitol from sugarcane bagasse. Ind Crops Prod 123:238–246

Unrean P, Ketsub N (2018b) Integrated lignocellulosic bioprocess for co-production of ethanol and xylitol from sugarcane bagasse. Ind Crop Prod 123:238–246. https://doi.org/10.1016/j.indcrop.2018b.06.071

Ur-Rehman S, Mushtaq Z, Zahoor T, Jamil A, Murtaza MA (2015) Xylitol: a review on bioproduction, application, health benefits, and related safety issues. Crit Rev Food Sci Nutr 55(11):1514–1528

Vallejos ME, Chade M, Mereles EB, Bengoechea DI, Brizuela JG, Felissia FE, Area MC (2016) Strategies of detoxification and fermentation for biotechnological production of xylitol from sugarcane bagasse. Ind Crops Prod 91:161–169
Villarreal MLM, Prata AMR, Felipe MGA, Silva JAE (2006) Detoxification procedures of eucalyptus hemicellulose hydrolysate for xylitol production by Candida guilliermondii. Enzyme Microb Technol 40(1):17–24. https://www.sciencedirect.com/science/article/pii/S0141022906002237?casa_token=KxSxser_zvQAAAAA:0sh3Dcf6HQ40iULIVcTBCshq-qI05v0C3ii20nXAgpHH8RnBuqLDFBby5zetUFT4pCkyaRB9IMNk
Visscher F, Van der Schaaf J, Nijhuis TA, Schouten JC (2013) Rotating reactors–a review. Chem Eng Res Design 91(10):1923–1940. https://www.sciencedirect.com/science/article/pii/S0263876213002980?casa_token=YKv2edqwHl0AAAAA:JGRNhgB4g9HSyO9oJUo2zXxO8LGIALPtAfu1BaV-pBGuB38AXhr0EqPB--mZT4uQZ7Ky9cCfHzu4
Wang M, Chen J, Lin X, Huang L, Li H, Wen C, He Z (2021) High humidity aggravates the severity of arthritis in collagen-induced arthritis mice by upregulating xylitol and L-pyroglutamic acid. Arthritis Res Ther 23(1):1–12
Wannawilai S, Chisti Y, Sirisansaneeyakul S (2017) A model of furfural-inhibited growth and xylitol production by Candida magnoliae TISTR 5663. Food Bioprod Process 105:129–140
Xu Y, Chi P, Bilal M, Cheng H (2019) Biosynthetic strategies to produce xylitol: an economical venture. Appl Microbiol Biotechnol 103(13):5143–5160
Yang BX, Xie CY, Xia ZY, Wu YJ, Gou M, Tang YQ (2020) Improving xylitol yield by deletion of endogenous xylitol-assimilating genes: a study of industrial Saccharomyces cerevisiae in fermentation of glucose and xylose. FEMS Yeast Res 20(8); foaa061
Zahed O, Jouzani GS, Abbasalizadeh S, Khodaiyan F, Tabatabaei M (2016) Continuous co-production of ethanol and xylitol from rice straw hydrolysate in a membrane bioreactor. Folia Microbiol 61(3):179–189
Zhang J, Zhang B, Wang D, Gao X, Hong J (2015) Improving xylitol production at elevated temperature with engineered Kluyveromyces marxianus through over-expressing transporters. Biores Technol 175:642–645
Zhang B et al (2016) Simultaneous fermentation of glucose and xylose at elevated temperatures co-produces ethanol and xylitol through overexpression of a xylose-specific transporter in energineered Kluyveromyces marxianus. Biores Technol 216:227–239
Zhang H, Yun J, Zabed H, Yang M, Zhang G, Qi Y, Qi X (2018a) Production of xylitol by expressing xylitol dehydrogenase and alcohol dehydrogenase from Gluconobacter thailandicus and co-biotransformation of whole cells. Bioresour Technol 257:223–228
Zhang H et al (2018b) Production of xylitol by expressing xylitol dehydrogenase and alcohol dehydrogenase from Gluconobacter thailandicus and co-biotransformation of whole cells. Biores Technol 257(1):223–228
Zheng J, Liu M, Wu F, Zhang L (2021) Enabling easy and efficient hydrogen release below 80 °C from NaBH4 with multi-hydroxyl xylitol. Int J Hydrogen Energy 46(55):28156–28165
Zhuang Y, Dong X, Tao S, Wang Y, Yang W, Zhang L (2021) Origami-based bionic reactor. Ind Eng Chem Res 60(11):4279–4289. https://pubs.acs.org/doi/abs/10.1021/acs.iecr.0c06066?casa_token=leo7zHQ3F-8AAAAA:6eMmHAv6DNM1DSuPQFXUfCavke-a8LEF88v6po1qQJKebYqrO5fg9zwFSnAOf_gqlYiUkmb2pNgV3DYxzQ

Chapter 6
Methods for Xylitol Recovery: Appraisal and Future Perspectives

E. A. Martinez, K. J. Dussán, D. D. V. Silva, R. L. Souza, and A. M. R. Prata

Abstract The recovery of a product obtained by a biotechnological process can represent from 30 to 90% of the total cost of the process, depending, among other factors, on the type of product and the degree of purity necessary to make its use viable. The recovery of xylitol obtained by bioprocess is not yet technologically established, both because of the complexity of the raw materials used in the process and the purity required, as it is a food grade product. There is a range of key operations for the recovery of a bioproduct and, for xylitol, crystallization seems to be of great importance today and, apparently, the most promising. In this chapter, initially an introduction to the recovery and purification of bioproducts in general is presented, aiming to allow a connection with the strategies that will be presented for the case of xylitol. Then, the most relevant works found in the literature so far are discussed, showing the updated panorama. Emphasis was given to crystallization as a

---

E. A. Martinez
Department of Technology, State University of Feira de Santana (UEFS), Feira de Santana, Bahia, Brazil
e-mail: ernesto.amartinez@uefs.br

K. J. Dussán
Department of Engineering, Physics and Mathematics, Institute of Chemistry, São Paulo State University (UNESP), Araraquara, São Paulo, Brazil
e-mail: kelly.medina@unesp.br

D. D. V. Silva
Department of Chemistry and Biological Sciences, Institute of Biosciences, São Paulo State University (UNESP), Botucatu, São Paulo, Brazil
e-mail: debora.virginio@unesp.br

R. L. Souza
Tiradentes University (UNIT), Aracaju, Sergipe, Brazil
e-mail: ranyere_souza@itp.org.br

A. M. R. Prata (✉)
Departament of Biotechnology, Engineering School of Lorena, University of São Paulo (USP), Lorena, São Paulo, Brazil
e-mail: amrprata@usp.br

© The Author(s), under exclusive license to Springer Nature Switzerland AG 2022

M. das G. de Almeida Felipe and A. K. Chandel (eds.), *Current Advances in Biotechnological Production of Xylitol*,
https://doi.org/10.1007/978-3-031-04942-2_6

key operation for obtaining xylitol in commercial form. Finally, some future perspectives regarding the recovery and purification of xylitol obtained by biotechnological process were discussed.

Keywords Xylitol · Downstream · Raw materials · Lignocellulosics · Crystalization · Bioproducts

6.1 Introduction

In recent years, biotechnology is revolutionizing industrial and agricultural practices, improving the quantity and quality of products. In addition, the number of commercial biotechnology products is increasing every year (Ricroch et al. 2022). In the agriculture and agribusiness sector, for example, biotechnology applications play a significant role, from increasing productivity to adding value and diversifying products, while reducing their environmental impact. In the manufacturing sector, biotechnology is used to produce a wide range of bulk and fine chemicals.

Consequently, these recent innovations in the "upstream" were not followed by a similar improvement in the "downstream" process, to the point that the latter currently represents a bottleneck in terms of consumption of time and costs due to low throughput. In some cases, the downstream process represents up to 60% of total manufacturing costs (Straathof 2011). The term "downstream" generally indicates the recovery and purification of a product from a complex mixture (Buyel et al. 2015). Thus, the purification methods chosen must distinguish between molecules that often exhibit only small variations in size, hydrophobicity or charge. Furthermore, the recovery or purification of natural bioproducts, or those originating from a bioreactor, are also highly influenced by the desired standard of specifications required by the market. As the scale increases, the methods for recovery and purification are not only important, but also essential for the integration of competitive processes, therefore, there is a need for the development of efficient and economical processes in the biotechnology industry.

A product recovery and purification methodology generally starts with a clarification step, which corresponds to the separation of cells from liquid phase of fermented broth where, in most cases, lies the product. Xylitol is an example of secreted bioproduct, which means it will be encountered in such a liquid phase. Unit operations involved are mainly centrifugation, filtration and flocculation, according to Kilikian and Pessoa (2021), who present the downstream steps for recovery and purification of biotechnological products. If the target product or biomolecule is not secreted by the cells, supernatant from clarification step can be discarded, and the cell biomass must be disrupted to release the product. A very common operation in this case is high pressure homogenization. After disruption, another clarification step is necessary to eliminate cells fragment and debris.

Purification methods have evolved rapidly in the last decades, with improved techniques aiming at economical, efficient, fast and easy to use methods. In general, at least two different purification steps are generally required to isolate the target product with the required purity. The first part of the purification process is the removal of process-related impurities, that is, species that are not chemically similar to the target molecule (Pfister et al. 2018). They generally include nucleic acids, host cell proteins, lipids, cell culture medium components, salts, etc. that derive from the manufacturing process.

Although a typical bioprocess usually consists of two main parts, it is usual to have 10 to 20 steps for the overall purification process. This fact reflects the complex nature, which in aqueous mixture of cells, intracellular or extracellular products, unprocessed substrate and secondary products of the "upstream" process. From this complex mixture, the target product must be isolated, and both the cost and the losses of the target molecule in the purification process are of fundamental importance for the viability of the process, since the recovery percentage is inversely related to the number of steps and high-resolution operations would represent a higher cost in the process.

Biomolecules that do not require complex purification processes, such as ions, pigments, nucleic acids, polysaccharides, lipids, viruses and even proteins when present in high concentrations, often apply techniques aimed at concentrating or reducing the water content of the solution (Kilikian and Pessoa 2021). Thus, the target molecule precipitation technique is usually followed by re-solubilization in a reduced volume of solvent; filtering the medium through a membrane of porosity such that water molecules and generally inorganic components are separated from the target molecule; lyophilization, which is the removal of water through sublimation, appropriate to the biological nature of the bioproducts; and the crystallization of the biomolecule that promotes, in a single step, the increase in its concentration, purity and stability.

On the other hand, molecules with a high demand for purity, such as proteins, enzymes, antibodies, DNA, etc., require highly complex purification processes. The most efficient methods are based on chromatographic techniques where packed columns or countercurrent systems are used (De Luca et al. 2020; Teal et al. 2000). In this process, a solution containing the molecules to be separated is homogenized in a solvent called eluent or mobile phase, which can be in gaseous or liquid form, and applied to a stationary or fixed phase, immiscible with the mobile phase. Often the stationary phase is accommodated within a column or on a solid surface in the case of paper chromatography (Carta and Jungbauer 2020). The stationary phase can be composed of porous silica, carbohydrate polymers, synthetic organic polymers which are presented in the form of spherical particles with approximately 100 μm in diameter, soaked in solvent, which constitutes most of the stationary phase (~90%) which for this reason is also called a gel.

Chromatography can be divided into two large groups according to the physical state of the mobile phase: liquid and gas. Liquid phase chromatography is of interest in the purification of biomolecules. In liquid chromatography, the biomolecules present in the liquid medium are retained in the stationary phase, which is a bed of porous material, by means of chemical or physical adsorption, partition, or molecular exclusion phenomena. Subsequently, the action of the eluent or liquid mobile phase promotes the gradual removal of previously retained solutes, which will be removed or eluted at different speeds due to the different affinities of the solutes with the stationary phase and the mobile phase. Components that interact more strongly with the stationary phase move more slowly than those that interact weakly and, therefore, are retained for less time in the column. This will result in the differential migration of the sample components, and hence their efficient purification.

If, on the one hand, chromatographic processes promote an increase in the purity of the target molecule, on the other hand, it reduces the yield, since losses are directly proportional to the number of steps in the process and are cumulative. Thus, the sequence of steps in the purification process must be considered, as well as the maximum yield of the target molecule and the level of purity necessary for its application. Ideally, the desired purification should be achieved in a single chromatographic step to avoid product losses (De Luca et al. 2020; Tarafder et al. 2010). The application of more than one chromatographic method, however, may be necessary and efficient, in view of the different fundamentals that govern the separations in each method, which allows the elimination of different impurities in different chromatographic methods.

As mentioned earlier, separation methods are generally performed with numerous steps (eg precipitation, centrifugation, dialysis, ion exchange chromatography, affinity chromatography, etc.) and require complex procedures with high energy and chemical consumption (Camperi et al. 1996; Linke and Berger 2011). It is important to consider that to achieve the purification of products to homogeneity, these steps provide high costs for the final product (Nagarajan 2012; Saxena et al. 2003). Another interesting approach is the isolation of target biomolecules by liquid–liquid extraction. Possible advantages are related to high capacity, greater selectivity and integration between recovery and purification. This technological approach should be considered as an alternative based on a critical view of the previous technology, since currently, in scientific and industrial scope, purification strategies of low cost, fast processing, with high yield and capable of operations in large-scale are essential.

Aqueous two-phase systems (ATPS) have the potential for high-throughput purifications. Although the resolution is not comparable to column chromatography, ATPS can represent a continuous, batch operational purification process, integrated with other techniques (eg extraction process) and highly biocompatible (Albertsson 1986; Cienfuegos et al. 2017; Espitia-Saloma et al. 2014; Santos et al. 2018). ATPS are formed when two hydrophilic, but incompatible components are dissolved in water. Examples of such ATPS are the aqueous solution of two hydrophilic polymers (polyethylene glycol (PEG)/dextran) or a polymer (PEG) and a salt (eg phosphate salt) (Albertsson and Tjerneld 1990). The use of these ATPS has been satisfactorily studied for over 40 years, for the separation and purification of biological molecules, such as DNA, proteins, enzymes, alkaloids, antibiotics, drugs, among others (Asenjo and Andrews 2012; Bora et al. 2005; Kessel 1981; Li et al. 2004; Matos et al. 2014; Mohamed et al. 2014; Rosa et al. 2011), and also compounds of non-organic origin, such as metal ions and antibiotics (Mandal and Mandal 2014; Rogers et al. 1993). The ability to manipulate the properties of the phases is the driving force to achieve a selective separation and is mainly promoted by phase formers, who currently explore different constituents, from the famous polymers and salts, such as the more recently investigated, until ionic liquids, deep eutectic solvents, organic solvents and carbohydrates (Asenjo and Andrews 2012; Freire et al. 2012; Ghazizadeh and Pazuki 2021; Souza et al. 2015).

ATPS are composed of two immiscible aqueous phases that coexist in equilibrium promoted by the addition of water-soluble compounds. Each phase of the system becomes enriched with one of the compounds, originating two aqueous phases of different chemical and physical nature, leading to the migration of biomolecules to one of the phases by affinity. One of the main features of the system is the high water content in the phases, allowing the separation of biomolecules from different sources under non-denaturing conditions (Albertsson 1986; Albertsson and Tjerneld 1990; Johansson 1989).

The development of a biomolecule purification process must therefore be guided by the maximum yield and level of purity necessary for its application. A fundamental tool for the purification process is related to quantification and characterization methods, which will monitor the effective progress in the isolation of the target molecule in each applied step. The characterization of biomolecules is necessary for the selection of appropriate techniques for purification, for the elimination of specific impurities and, at the end of the process, for the determination of their chemical identity, a requirement for the validation of the process. In addition to characterization and quantification, it is necessary to preserve the biological activity of the target biomolecule during the purification process as well as during transport and storage of the final product.

Briefly, some basic rules must be considered for the development of an efficient purification process. Among them, knowledge of the characteristics of the target molecule and its contaminants, to select the most appropriate techniques; the establishment of analytical methodologies for quantification of the target molecule so that each step is monitored and evaluated for the desired level of purification. We also consider the removal of impurities present in a high proportion in the initial stages of the process, based on different fundamentals such as molecular size, hydrophobicity, charge, etc.

6.2 Recovery and Purification of Xylitol

Works dealing with xylitol recovery methods such as chromatographic techniques, membrane separation and crystallization applied either to xylitol solutions obtained by chemical means or to fermented media were already reviewed by Aliakbarian et al. (2012) and an updated literature review of crystallization processes for xylitol recovery from solutions obtained by both ways was provided by Martinez et al. (2015).

After the fermentation process, centrifugation must be carried out to separate the biomass from the medium containing xylitol as a component in higher concentration. To decrease the concentration of substrates such as xylose and arabinose the fermentation time must be increased.

Purification of fermented hydrolysate is more complex than that of synthetic fermentation broth because of the presence of colored compounds, proteins, lignin, cellulose, and hemicellulose derivatives that are not readily fermented by yeasts (Alves et al. 2021a, b). Purification steps are included to lower the concentration of other metabolites such as arabitol, glycerol and ethanol and yeast parts or debris. The purification of fermented media containing xylitol has been studied in the last ten years using methods such as nanofiltration with membranes, adsorption with activated charcoal and on silica gel, ion exchange resins, combined use of ultrafiltration and electrodeionization, liquid–liquid extraction and supercritical technology. Then the fermented liquor must be subjected to concentration to increase the xylitol content to values suitable for crystallization.

Strategies for xylitol purification using activated charcoal, ion-exchange resins, pH adjustment, liquid–liquid extraction, membrane separation, precipitation, supercritical CO_2 extraction and combined methods in the main studies published since 1995 to 2014 were presented and discussed by Martinez et al. (2015). In this chapter, information available in the literature on xylitol purification studies will be updated. In Table 6.1 are summarized the main strategies for xylitol purification published since 2015.

Physical separation technologies using membranes allow the separation of different compounds from a solution by applying a hydrostatic pressure difference between the two sides of a permselective barrier. Thus, the fed solution is divided into a permeate fraction containing all the components that permeated the membrane and a retentate fraction containing all the compounds rejected by the membrane, within part of the solvent. Separation by ultrafiltration (UF) or nanofiltration (NF) is determined by the membrane pore size range with values of 1,000–100,000 Da and 150–1,000 Da, respectively, which is mainly related to its molecular weight cutoff (MWCO), and to a lesser extent in shape molecular, charge and hydrophobicity (Cassano et al. 2018). Membrane techniques offer the advantages of low energy consumption, good separation efficiency, reduced number of steps, and high-quality final products (Desiriani et al. 2017; Alves et al. 2021a, b; Santos et al. 2011).

Affleck (2000) proposed the membrane separation as a method for the recovery of xylitol from fermented broth. Xylitol produced by *Candida tropicalis* from corn fiber hydrolyzate with a xylitol yield factor of 0.6 g g^{-1} of xylose was optimally separated from the impurities with a 10,000 nominal molecular weight cutoff polysulfone membrane, allowing 82.2–90.3% of xylitol contained in the fermentation broth to pass through, while retaining 49.2–53.6% of oligopeptides and peptides. Permeate crystallization led to xylitol crystals with purity up to 90.3%.

Desiriani et al. (2017) evaluated a membrane-based separation process of the synthetic medium containing xylose (2.7 g L^{-1}), xylitol (2.8 g L^{-1}), acetic acid (2.9 g L^{-1}) and cells (4 g L^{-1}) using a combination of a hydrophilic polysulfone UF and polyamide NF to isolate and concentrate xylitol from a fermentation broth of

from 67 to 46%. The purity of xylitol was enhanced from 82 to 93% due to the removal of arabinose from the mixed sugars solution. The selectivity of the membrane towards xylitol was enhanced by the presence of ethanol in the synthetic fermentation solution. Long term filtration (6 h or more) showed the antifouling properties of the blended membranes.

According to Kresnowati et al. (2019), the electrodeionization (EDI) membrane is a combination of electrodialysis and ion exchange system that allows continuous ion transport under electrical current and self-regenerating ion-exchange resins. In their work, ultrafiltration (UF) and electrodeionization (EDI) membranes were used for xylitol purification from either model solution or oil palm empty fruit bunch (OPEFB) hydrolysate fermented by *Debaromyces hansenii.* The validation experiment using the OPEFB hydrolysate fermentation broth showed that UF-EDI membrane was able to remove 99% of microorganism and biomass, 99% pigment, >46% of xylose, and >99% ionic impurities, including >90% acetic acid, with a loss of xylitol about 30–50%.

According to Alves et al. (2021a, b) microfiltration and ultrafiltration did not result in high purification and separation factors but were able to eliminate agglomerates and large (high molar mass) compounds. Thus, a model solution containing the major components of fermentation broths (30 g L^{-1} xylitol, 6 g L^{-1} arabinose, 5 g L^{-1} ethanol, 3 g L^{-1} glycerol, 2 g L^{-1} xylose, 1 g L^{-1} acetic acid, and 0.2 g L^{-1} glucose) was subjected to NF step. Results showed a xylitol purification factor of 3.3, protein separation factor of 8.4, and color removal of 96.0% using a polypepirazine amide (NF) membrane.

Activated carbon has thousands of pores in its structure with great physical or chemical worship capacity through van der Waals or interactions through mainly covalent bonds, respectively. Thus, this material is used to remove color and other impurities from solutions. The majority of attempts at purifying biotechnological xylitol with activated carbon using batch processing are shown in Table 6.1.

Gurgel et al. (1995) used activated charcoal and ion-exchange resins (Amberlite 200C, strong cation-exchange resin and Amberlite 94S weak anion-exchange resin) to clarify xylitol-containing sugarcane bagasse hydrolyzate fermented by *Candida guilliermondii.* Optimal clarification with xylitol loss (19.1%) was obtained using 250 g l^{-1} activated carbon at 80 °C and pH 6.0, under agitation at 100 rpm for 1 h.

The treatment of the corn fiber hydrolyzate fermented by *C. tropicalis* carried out using activated carbon for 1 h (2.5 g per 100 g of solution, pH 6.0, 80 °C, and 100 rpm) was able to more effectively remove colored contaminants (up to 79.5%) and proteins (69.0%), but adsorbed about 25–50% of xylitol (Affleck, 2000). Studies on xylitol (5 to 200 g L^{-1}) adsorption by charcoal from synthetic broths fermented by *D. hansenii*, at 25 and 50 °C, pH 6.0, under magnetic stirring for 1 h were reported by Sampaio et al. (2006). According to the authors, the best clarification results were obtained using 20 g L^{-1} activated charcoal at 25 °C, removing nearly 79% of colored contaminants, 94% of amino acids, and 69% of total proteins, and recovering xylitol almost completely.

In recent studies, the fermentation broth almost became translucent, residual xylose were removed and xylitol recovery were 76.2 and 77.1% for crude glycerol/xylose and pure glycerol/xylose, respectively, after the charcoal treatment (5%) according to Prabhu et al. (2020a, b).

The capacity of xylitol purification in fermented media by activated carbon in fixed bed columns was verified by Cardoso and Forte (2021). The broth composition was total solids 98.1 g L^{-1}; protein 19.2 g L^{-1}; xylitol 15.02 g L^{-1}; xylose 1.44 g L^{-1}; glucose 0.00 g L^{-1}; arabinose 3.38 g L^{-1}; ethanol 5.93 g L^{-1} and glycerol 1.79 g L^{-1}. The process was an alternative for biotechnological xylitol purification presenting values of purification factors for proteins (18.0), colored compounds (reducing more than 99% of the absorbance at 420 nm and 560 nm) and ethanol (20.4) as well as a volumetric productivity of 36.8 g L^{-1} h^{-1} and a retention coefficient of 31.1% of xylitol. Besides, a good degree of separation of total solids in relation to xylitol when the feeding was done by pulse input was observed. However, it did not present separation when using continuous feeding. According to the authors, recovery of up to 83% of the injected xylitol, separation of proteins (93–100% purity), colored compounds (~100% purity) and ethanol (~95% purity) were obtained.

Liquid–liquid extraction (LLE) or solvent extraction consists of the transfer of certain components from one phase to another when immiscible or partly soluble liquid phases are brought into contact. The solutes are separated based on their different solubility in two different liquids. The extraction of xylitol produced by *Candida tropicalis* from hemicellulose hydrolyzate of Meranti wood sawdust, hydrolyzed by xylose reductase (XR), was reported by Mun et al. (2016). Ethyl acetate was employed as solvent in the extraction process. The yield of 78.14% (w/w) in the xylitol extraction was obtained using a reaction time of 60 min, sample to solvent ratio of 1: 4.5 v/v and 5 stages of extraction.

Supercritical fluid extraction (SFE) is the process of separating a component from a solid or liquid matrix using supercritical fluids as the extraction solvent. Due to its easy-to-reach supercritical conditions, carbon dioxide (CO_2) is the most commonly used supercritical fluid, sometimes modified by co-solvents such as ethanol or methanol. Furthermore, CO_2 has low toxicity, is non-flammable and is cheap compared to other fluids. The SFE is simple to operate, has a low extraction temperature and is a clean and ecologically correct technology, in addition to providing clean extracts with high purity (Silva et al. 2020a, b).

Xylitol production from sugarcane biomass (sugarcane straw and bagasse) hemicellulosic hydrolysate using *Scheffersomyces amazonensis* UFMG-HMD-26.3 and the xylitol recovering by supercritical CO_2 extraction was reported by Silva et al. (2020a, b). The concentrated broth containing concentrations of 62.63 g L^{-1} total sugar and 171.55 g L^{-1} xylitol was submitted to supercritical fluid extraction using pressurized CO_2, ethanol as co-solvent, 60 °C and 300 bar. The maximum extraction (40.5%) and xylitol purity (99.6%) were reached using the lowest extraction time (20 min) and highest ratio of sample/solvent (11/9).

Xylitol crystallization

Crystallization is one of the oldest purification techniques known to human beings, used when it is desired to obtain a solid product from a solution. To induce crystallization from a solution, a drive force is needed. The selection of the crystallization method using methods such as solvent evaporation, cooling, salting out or precipitation is dependent on the solid solubility in the solution and saturation slopes with the temperature.

A supersaturated solution is one that contains a solute content above equilibrium, under the same temperature and concentration conditions as the other components. This state of the solution allows the occurrence of crystallization and the properties have marked effects on the process and parameters of the product obtained (Nývlt et al. 1985). The solubility of pure xylitol depends significantly on the temperature and composition of the water–ethanol solvent mixture. Xylitol solubility in water (850 g L^{-1} at 60 °C) is higher than in ethanol 95.3% (200 g L^{-1} at 60 °C). Thus, the use of water–ethanol mixtures (50–50% w/w) contributes to facilitate the xylitol crystallization process from fermentation solutions that contain impurities in undetectable concentrations. The xylitol crystallization in solutions containing ethanol, arabitol and adonitol was studied by Fernandez et al. (1999). The presence of two xylitol isomers, and possible by-products of xylitol secretion by microorganisms, prevent the xylitol crystallization especially at high concentrations of the mixtures.

There are few works in the literature on the xylitol crystallization process (Martinez et al. 2015). The effectiveness of this process depends on the purity of the product obtained by fermentation, which requires several previous purification steps, as can be seen in Fig. 6.1.

The complexity of the crystallization phenomenon consists in the appearance and growth of solid particles in the medium, caused by instability in the solution. This instability can be caused by changes in the physical properties of the solution such as concentration and temperature to form a solid product. A decrease in temperature and an increase in concentration both favor the crystallization.

In the phase equilibrium of a solid–liquid system, two solution states are generally considered, the unsaturated and the saturated. There is, however, another state of solution, supersaturation, which allows the occurrence of crystallization, and whose properties have marked effects on the process and parameters of the product obtained (Nývlt et al. 1985). A supersaturated solution is one that contains a solute content above equilibrium, under the same temperature and concentration conditions as the other components and can be created by cooling, evaporation, addition of anti-solvent or by precipitation (van Rosmalen et al. 2003).

Crystallization occurs through nucleation processes, or the generation of new crystals (nuclei) followed by crystal growth. The nucleation mechanisms can be classified as primary nucleation (homogeneous and heterogeneous) and secondary nucleation originated by crystals, by an intermediate layer or by contact (van Rosmalen et al. 2003). Heterogeneous nucleation occurs in the presence of solid substances foreign to the medium, such as dust, colloids and crystallizer walls, and if the nucleation takes place in a crystalline suspension, then it is called secondary nucleation (Nývlt et al.

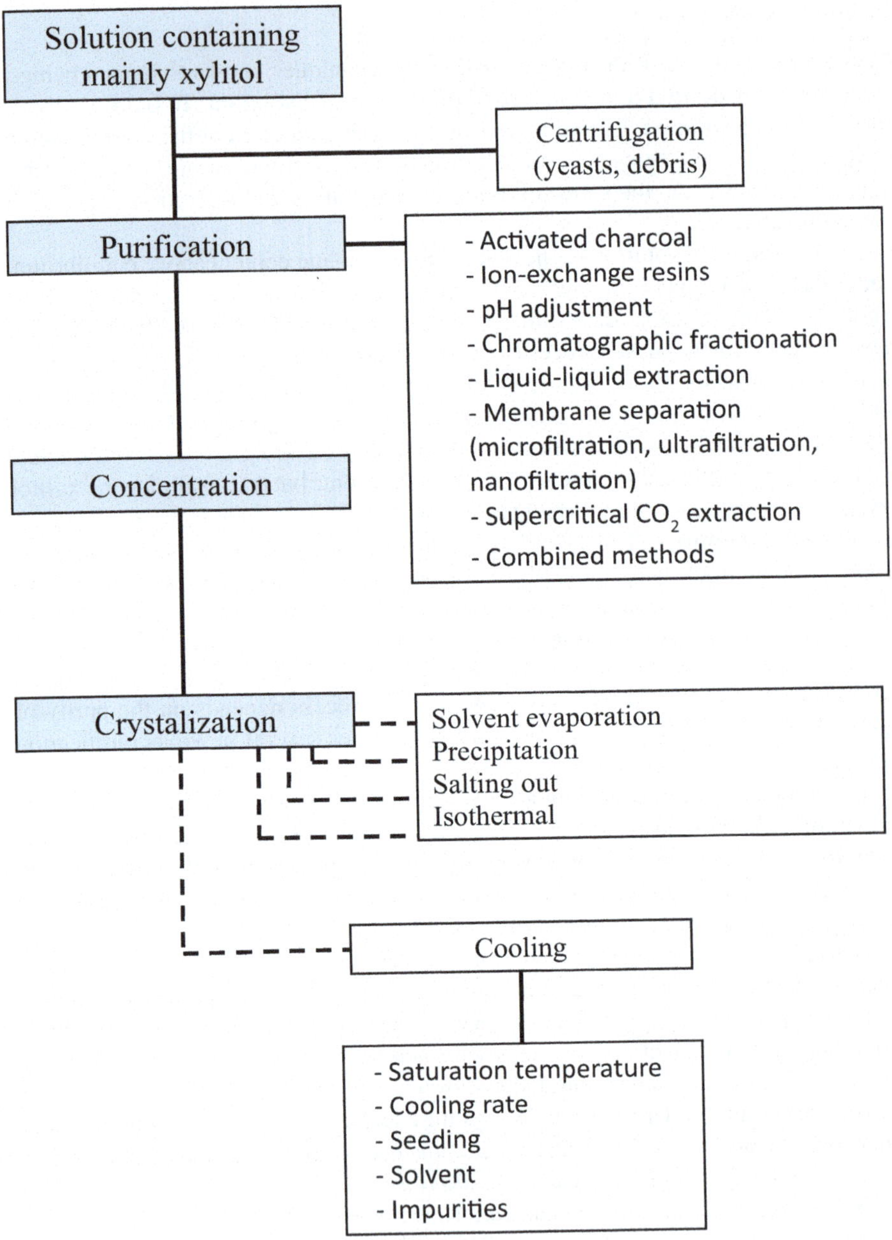

Fig. 6.1 Flow chart of downstream process for xylitol crystals production

2001; van Rosmalen et al. 2003). The apparent secondary nucleation mechanism, where nuclei are introduced into the solution, occurs when a dry crystal is submerged in the supersaturated solution (Mullin 2001; Nývlt et al. 2001; van Rosmalen et al. 2003). On the surface of the crystal there are microcrystals, adhered by electrostatic or capillary forces, generated by friction of the crystals, or by the drying of the mother liquor. After the crystal is submerged in the solution, they begin to detach from the adhered surface, serving as growth nuclei.

The kinetics of primary nucleation, crystal growth and secondary nucleation can be studied using model experiments, on a laboratory scale or in pilot plant crystallizers, operating in continuous or batch mode. As crystallization occurs in them under conditions similar to those of industrial crystallizers, kinetic data are more reliably obtained for its use in chemical engineering, reducing project risk (Nývlt et al. 2001).

The solvent choice in a crystallization process can also affect the most suitable type of process: evaporation, when the solubility practically does not vary with temperature, or it can be by cooling, when the variation is significant. The dependence of the xylitol solubility with temperature is greater than for sucrose (Ullmanns 1998). Xylitol solubility in water and 50% (v/v) isopropanol-water is higher than for glucose and xylose, between 5 and 70 °C; and in mixtures such as 50% (v/v) 2-hydroxyethanolamine acetate-water and 50% (v/v) water–ethanol at temperatures higher than 30 °C and 50 °C, respectively (Junior and Rocha 2021).

Xylitol crystallization studies could be based in part on general methods applied to different sugars, due to the structural and physicochemical similarity between these compounds. However, specific system characteristics, such as high viscosity, complex rheology, instability towards high temperatures, physicochemical similarities between products and contaminants, such as small density differences, make sugar purification one of the most difficult steps in production of pure substances (Weatherley 1994).

The xylitol crystallization from synthetic media fermented or not and fermented hemicellulosic hydrolysates studied since 1990 using solutions with high concentrations of xylitol adding different solvents, in cooling processes and with the use of seeds are presented in Table 6.2.

Heikkilä et al. (1990, 1992, 1997) patented a simple method to produce xylitol from hemicellulosic hydrolyzates and from xylan-containing material by *Debaryomyces hansenii* and *Candida tropicalis*. The xylitol-rich solution (86.5 to 92.2 g/100 g solution) was obtained after xylitol recovered by chromatographic separation from the fermented broth and a concentration step. The solutions were seeded with xylitol crystals (0.04 mm size), and the crystallization solution was cooled using different cooling rates. After centrifugation the recovered crystals had a mean size of 0.37 mm and purity values between 56.5 and 99.4%.

Nurmi et al. (1999) reported that the use of effective stirring enhances nucleation, enabling it to occur spontaneously, preventing solidification of a crystallization mass and allowing nuclei to develop crystal shapes and to grow. These authors carried the experiments using solutions containing sugars and sugar alcohols (93.8–94.4 g

Table 6.2 Xylitol crystallization studies from materials using different solvents, seeds and cooling rate

Starting material	xyl in solution	Seed	CR (°C/min) and T (°C)	Results	References
Xylan-containing material by *Debaromyces hanseni*	92.0 g/100 g water 86.5 g/100 g water	0.03% Seed suspended in isopropanol	0.006 and 45 °C 0.027	L = 0.37 mm and 99.4% pure crystals 99% pure crystals	Heikkilä et al. (1990)
Biomass hemicellulosic hydrolyzates by *Candida tropicalis*	92.0 g/100 g water	0.03%	0.006 and 45	L = 0.37 mm and 99.4% pure crystals	Heikkilä et al. (1992)
Xylan-containing material	91.4 g/100 g water	0.06 g	0.010 and 30	77.0% xyl purity, 30.72% xyl yield and 81.2% crystal purity	Heikkiläet al. (1997)
	91.3 g/100 g water	0.05 g	0.012 and 30	64.3% xyl purity, 31.74% xyl yield and 93.3% crystal purity	
	92.2 g/100 g water	0.05 g	0.007 and 60	56.5% xyl (purity), 29.73% xyl and 68.0% crystal purity	
Solution containing sugars and sugar alcohols	93.8 g/100 g water	10 g	0.042 and 50	67.0% xyl yield	Nurmi et al. (1999)
	94.1 g/100 g water	10 g	0.030 and 50	57.0% xyl yield	
	94.4 g/100 g water	10 g	0.019 and 56	68.0% xyl yield	
Synthetic solution	582 and 730 g/l water	1 g/l	−10, −5 and 0 °C	56% crystallization yield or almost 100% purity at 730 g/l and −5 °C	De Faveri et al. (2002)

(continued)

Table 6.2 (continued)

Starting material	xyl in solution	Seed	CR (°C/min) and T (°C)	Results	References
Sugarcane bagasse hydrolyzate by *C. guilliermondii*	470 g/l water	1 g/l	−10 °C	29% crystallization yield, 92% purity	
Synthetic solution	582, 656 and 730 g/l water	1 g/l	−10, −5.2 and 0	54% crystallization yield and 97% purity at 728 g/l and −6 °C 25% crystallization yield and 100% purity at 583 g/l and −2.4 °C	De Faveri et al. (2004)
Corncob hydrolyzates fermented by *D. hansenii*	0.400 kg/kg 40, 50 and 60% ethanol	1 g/l	−10 or −5 °C during 72 h	98.9 wt % of xyl; 0.4710 g/g yield	Rivas et al. (2006)
Sugarcane hemicellulosic hydrolyzate by *C. guilliermondii*	First crystallization: 935.4 g/l Second crystallization 42.63 g/100 g 52.98 g/100 g 63.59 g/100 g water – ethanol 50–50%	0.1% 0.1% 0.1%	and 50 0.10 and 30 0.25 and 40 0.50 and 50	85% purity and 3% moisture 91.20–94.85% purity and 1.55 – 2.77% moisture	Martinez et al. (2007)
Synthetic solution Wheat straw hemicellulosic hydrolyzate by *C. guilliermondii*	720 g/l 726.5 g/l water – ethanol 50–50%	200 mg	0.2 and 50	51.4% yield and 99.9% purity Using 30% concentrated hydrolysate + 70% xylitol solution: 43.5% yield and 95.9% purity	Canilha et al. (2008)
Hydro alcoholic solution	42.63, 52.98 and 63.59 g/100 g	4 seeds	0.10, 0.25, 0.50 and 40, 50 and 60	$g = 2.44$; $n = 2.44$; $m = 3.44$; $B_N = 4.76 \times 10^{-18}$ kg $kg^{-1}{}_{solv}$	Martinez et al. (2008)

(continued)

Table 6.2 (continued)

Starting material	xyl in solution	Seed	CR (°C/min) and T (°C)	Results	References
Semi synthetic medium by *C. guilliermondii*	First crystallization: 745.3 g xyl/l Second crystallization: 42.63 g/100 g 52.98 g/100 g 63.59 g/100 g water – ethanol 50–50%	0.1% 0.1% 0.1%	0.50 and 50 0.25 and 30 0.25 and 40 0.25 and 50	95% purity and 3.08% moisture 98.5–99.2% purity L = 0.3293 mm; CM = 1800 kg/m^3_{solv}; G: 2.76 × 10^{-8} m/s; dN/dt = 4.58 × 10^7#/ m^3_{solv}s L = 0.3975 mm; CM = 1359.4 kg/m^3_{solv}; G: 5,15 × 10^{-8} m/s; dN/dt = 3.04 × 10^7#/ m^3_{solv}s L = 0.3263 mm; CM = 718.54 kg/m^3_{solv}; G: 3.87 × 10^{-8} m/s; dN/dt = 2.66 × 10^7#/ m^3_{solv}s	Martinez et al. (2009)
Synthetic solution	52.98 g/100 g water – ethanol 50–50%	4 seeds	1–4 and 40	Using hydrophobic silica (0,1%): L = 0.1912 mm; CM = 723.97 kg/m^3_{solv}; G: 1077.4 × 10^{-8} m/s; dN/dt = 44.9 × 10^7#/ m^3_{solv}s Using hydrophilic silica (0,1%): L = 0.2889 mm; CM = 764.96 kg/m^3_{solv}; G: 801.19 × 10^{-8} m/s; dN/dt = 13.3 × 10^7#/ m^3_{solv}s	Martinez et al. (2012)

(continued)

Table 6.2 (continued)

Starting material	xyl in solution	Seed	CR (°C/min) and T (°C)	Results	References
Cashew apple bagasse hemicellulosic hydrolysate by *Kluyveromyces marxianus*	200 g/l ethanol, isopropanol and the ionic liquid 2-hydroxyl-ethylammonium acetate	3 seeds	0.25–0.50	69.7% yield and 84.8% purity	Junior and Rocha (2021)

* CR = cooling rate; xyl = xylitol; CM = crystallization mass; g = order of crystal growth kinetics; n = order of nucleation kinetics; m = apparent order of nucleation kinetics; B_N = system kinetic constant; L = crystal size; G = crystal growth rate; dN/dt = nucleation rate; T = crystallization temperature

of xylitol in 100 g) that were seeded with 10 g ground xylitol and cooled from 50–56 °C to 25–20.5 °C using different rates (0.042–0.019 °C.min^{-1}), thus obtaining crystallization masses with 3.0–3.6 supersaturation and xylitol yields in the range of 57–68%.

De Faveri et al. (2002, 2004) studied xylitol recovery from synthetic solutions and hardwood hemicellulose hydrolyzate fermented by *D. hansenii* containing xylitol and xylose (75 g. l^{-1} xylitol and 24 g. l^{-1} xylose). These broths were evaporated until supersaturation was achieved. After addition of 1 g.l^{-1} xylitol to favor nucleation, crystallization was carried out by cooling at −10, −5, or 0 °C, centrifuging and filtering the solutions. Promising results both in terms of crystallization yield (56%) and purity degree (close to 100%) were obtained from very concentrated synthetic xylitol solutions (730 g l^{-1}) at relatively high temperature (−5 °C), whereas the fermented hydrolyzate gave poorer results (29% crystallization yield and 92% purity) even at lower concentration (470 g.l^{-1}) and temperature (−10 °C), because of very quick impurity precipitation (De Faveri et al. 2002). An initial xylitol supersaturation value of 728 g.l^{-1} in synthetic solutions and cooling temperature of −6.0 °C were the operating conditions under which xylitol purity (97%) and crystallization yield (54%) were simultaneously optimized. However, purity close to 100%, which is needed to meet industrial standards, was predicted for conditions (583 g.l^{-1} and −2.4 °C) leading to unsatisfactory 25% crystallization yield (De Faveri et al. 2004).

Rivas et al. (2006) studied xylitol crystallization from concentrated corncob hydrolyzates previously detoxified with charcoal, fermented by *D. hansenii*, and then subjected to sequential stages of adsorption, concentration, precipitation with hydroalcoholic solutions, and further concentration. At −5 °C an increase in ethanol percentage from 40 to 60% (v/v) resulted in a progressive rise in the crystallization yield from 15.6 to 47.1%, while at −10 °C xylitol content of mother liquors was almost the same, but the crystallization yield was remarkably improved in media treated with 40 and 50% ethanol, because of both a reduction of viscosity and an acceleration of crystallization. Under optimal conditions, crystallization led to regularly shaped, well-formed, and homogeneous crystals, containing 98.9% (w/w) xylitol.

Martinez et al. (2007) studied the downstream process for xylitol produced from sugarcane bagasse hemicellulosic hydrolyzate by *Candida guiliermondii*. Syrup containing 935.4 g. l^{-1} xylitol and 13.1 g.l^{-1} arabinose was first submitted to a preliminary crystallization step using 0.5 °C.min^{-1} cooling rate, 50 °C saturation temperature and four xylitol seeds. Thus, crystals with properties such as 85% purity, −147.81 J.g^{-1} melting heat, 84.67 °C melting point, and 3% moisture content were obtained. These crystals were used in a second crystallization step in 50–50% (w/w) water/ethanol solutions using the solubility values (42.63, 52.98 and 63.59 g.100 mL^{-1}) corresponding to the saturation temperatures (30, 40, and 50 °C) and applying different cooling rates (0.1, 0.25, and 0.50 °C.min^{-1}). Crystals with higher purity (91.20–94.85%), melting point (91.8–94.1 °C) and melting heat (−174.18/−214.62 J.g^{-1}, as an absolute value), and reduced moisture content (1.55 to 2.77%) were obtained. Thus, xylitol crystals with properties close to those of commercial were obtained (purity of 99.8%, heat of fusion of −259.63 J.g $^{-1}$, melting point of 93.61 °C and moisture content of 0.05%).

Canilha et al. (2008) carried out the xylitol crystallization of xylitol purified from wheat straw hemicellulosic hydrolyzate fermented by *C. guilliermondii*. Initially, a solution with commercial xylitol (720 g/l) and a concentrated broth [xylitol (726.5), xylose (4.3), arabinose (3.2), glycerol (16.5), acetic acid (1.0) and lignin derivatives (12.2)] were submitted at 50 °C saturation temperature and 0.2–0.4 °C.min^{-1} cooling rate, and 200 mg of xylitol crystals were added to the system to favor nucleation. In the concentrated broth, the mixing of the medium (0, 30, and 50%) with a commercial xylitol solution (100, 70, and 50%) was carried out. A higher yield (43.5%), close to the value reached with pure xylitol solution (51.4%), was obtained by performing crystallization in a mixture composed of 70% pure xylitol solution and 30% concentrated medium (nucleation temperature of 12 °C).

The determination of the kinetic parameters of xylitol crystallization can be performed by applying the method of Ný vlt et al. (1985). Each data set of the crystal size distribution of the crystalline product [$M(L)$] are converted into the dimensionless z according to the gamma function that describes the dependence of the oversize fraction on the crystal size:

$$M(L) = 100\left(1 + z + \frac{z^2}{2} + \frac{z^3}{6}\right)\exp(-z) \tag{6.1}$$

$$\text{with } z = L_m/Gt_c \tag{6.2}$$

where z is a dimensionless crystal size; G is the crystal growth rate; t_c is the crystallization time and L_m is the crystal mean size.

In this distribution the dependence of $M(L)$ on the dimensionless crystal size (z) is marked by an inflection point corresponding to the maximum value on the differential size distribution where $z = 3$. The substitution of this value into Eq. (6.1) yields the oversize fraction $M(L)$ of 64.7%. The values of mean size (L_m) are then determined from the linear relation of z and L for $z = 3$ corresponding to the maximum value of $M(L)$ for each experiment (Martinez et al. 2008).

Thus, the average crystal growth rate (G) can be then determined by dividing the mean size by three times the crystallization time (Eq. 6.2).

A simple relation for the nucleation rate (dN/dt, in #/m$^3{}_{solv}$ s) can be used:

$$\frac{dN}{dt} = \frac{27GM_c}{2\alpha\rho_c L_m^4} \tag{6.3}$$

where M_c is the xylitol mass/solvent volume (kg/m$^3{}_{solv}$); α is the crystal volume shape factor and ρ_c is the crystal density (kg/m^3).

From the logarithmic form of Eq. (6.3) the c, g/n and B_N values can be estimated by the multilinear correlation function:

$$\log\left(\frac{dN}{dt}\right) = \log(k_N/k_G^{n/g}) + c\log(M_c) + \left(\frac{n}{g}\right)\log(G) \tag{6.4}$$

where k_G and k_N are growth and nucleation rate constants, respectively, n is the order of nucleation kinetics and g the order of the overall crystal growth kinetics.

According to Nýchvlt et al. (1985), the crystal mean size in the model experiments using different saturation temperatures and cooling rates can be determined by:

$$L_m^{(1+3\frac{g}{n})} = 3B_N M_c^{(1-c)(\frac{g}{n})}(tc/3)^{(1-\frac{g}{n})} \tag{6.5}$$

And the system kinetic constant (B_N) can be described by:

$$B_N = \frac{4.5^{\frac{g}{n}} k_G}{(\alpha\rho_c k_N)^{g/n}} = \left(\frac{4.5k_G}{\alpha\rho_c k_N}\right)^{g/n} \tag{6.6}$$

The combined effects of saturation temperature and cooling rate on crystal size and kinetics parameters of the xylitol crystallization process in hydro-alcoholic solution using models-experiments was used by Martinez et al. (2008, 2009, 2012).

Martinez et al. (2008) studied the combined effects of saturation temperature (40, 50 and 60 °C) and cooling rate (0.10, 0.25, and 0.50 °C.min^{-1}) on kinetics of commercial xylitol crystallization in an ethanol/water (50:50% w/w) solution according to model experiments. The best correlation between nucleation and growth rates (in log terms) was obtained when the value of the mass concentration exponent (c) in secondary nucleation kinetics of the solids concentration was 0 (zero). The nucleation model explained 73% of the total variation in the response as a function of the apparent growth rate, and the regression constant or intercept was 14.88. Estimated kinetic parameters such as orders of growth (g = 2.44) and nucleation kinetics (n = 2.44), apparent order of nucleation kinetics (m = 3.44), system kinetic constant ($B_N = 4.76 \times 10^{-18}$ kg.kg$_{solv}^{-1}$) and the mean crystal size calculated were satisfactorily compared with the experimental one (11.3% mean deviation). The kinetic parameters of xylitol crystallization were used to represent the medium size of the crystals obtained after the second crystallization step of the xylitol produced by biotechnological way using *C. guilliermondii* (Martinez et al. 2009). A good fit was verified in the correlation between calculated and experimental crystals size with errors between 2 and 22% in the obtained xylitol crystals. Thus, the impurity concentration of the syrup did not have a significant influence on kinetic parameters.

Martinez et al. (2009) carried out a two-step protocol for xylitol crystallization from a semi-synthetic medium fermented by *C. guilliermondii* centrifuged and concentrated to produce a syrup containing 745.3 g. l^{-1} xylitol. Crystal with 95% purity, 3.08% humidity, −206.81 J.g^{-1} and 91.65 °C melting point were obtained after a first crystallization step, carried out at 1.17 supersaturation, saturation temperature of 50 °C, and a 0.5 °C. min^{-1} cooling rate. To improve the xylitol properties, a second step was carried out using the crystals obtained, which were dissolved in ethanol/water solution (50:50% w/w) at the concentration corresponding to the saturation temperature of 30, 40 and 50 °C. It was used 0.25 °C.min^{-1} cooling rate and seeding of 0.1% xylitol. The highest nucleation rate (4.58 × 10^{7}.m^{-3} s^{-1}) and crystallization time (3,982.8 s) were obtained with the lowest saturation temperature (30 °C), while the highest crystal growth rate (5.15 × 10^{-8} m.s^{-1}) and the largest average size (93.975 × 10^{-4} m) were obtained at 40 °C. The xylitol crystals had 98.5–99.2% purity, 0.38–0 0.63% moisture, −236.8/−244.95 J.g^{-1} heat of melting and 93.9–94.1 °C melting point.

The use of pyrogenic nanometric silica (hydrophobic and hydrophilic) with crystals from 5 to 7 nm in the formation of crystals with heterogeneous nuclei, aiming at more efficient crystallization of xylitol in saturated solution was patented by Martinez et al. (2012). Essays were carried out in 100 g of the hydro-ethanolic solutions (92 ml) of xylitol 99.7%, containing 52.98 g of xylitol in 100 g of water–ethanol solution 50–50% (w/w) at the saturation temperature of 40 °C using a cooling speed of 0.5 °C/min. Four xylitol crystals and 0.10% m/m of pyrogenic (hydrophobic or hydrophilic) nanometric silica were added in the tests without and with the addition of silica, respectively. The use of hydrophobic silica produced a decrease in the average size of the xylitol crystals (L = 0.1912 mm), the xylitol crystallized mass (Mc = 723.97 kg_{xyl}/m^3_{solv}) and the crystals growth rate (G = 1077.41*10^{-8} m/s). However, an increase in the nucleation rate was observed (dN/dt = 44.9 × 10^{7}/$m^3_{solv}s$). With the use of hydrophilic silica (0.0487 g) crystals were produced with an average size (L = 0.2889 mm) and crystal growth rate (G = 764.96 × 10^{-8} m/s) lower than that in the cooling crystallization of xylitol in water–ethanol mixture (50–50% w/w) with xylitol crystals as seeds (L = 0.3310 mm). On the other hand, there was an increase in the mass of xylitol crystals produced (Mc = 801.19 kg/m^3_{solv}) and in the nucleation rate (dN/dt = 13.3 × 10^{7}/$m^3_{solv}s$).

After nine years, xylitol was biotechnologically produced by *Kluyveromyces marxianus* ATCC36907 using the cashew apple bagasse hemicellulosic hydrolysate (CABHH) by Júnior and Rocha (2021). After concentration, the fermented hydrolyzate contained 200 g. L^{-1} of xylitol, 9.4 g. L^{-1} of xylose and 26.4 g. L^{-1} of glucose. Xylitol recovery and purification was performed using ethanol, isopropanol, 2-hydroxyl-ethylammonium acetate (2-HEAA)], cooling rates of 0.25 and 0.50 °C.min^{-1} and three xylitol crystals as seeds. Higher yield (69.7%) and purity (84.8%) values were obtained using 50% (v/v) of isopropanol as anti-solvent and higher cooling rate (0.5 °C. min^{-1}).

6.3 Conclusion and Future Perspectives

For some years now, xylitol has been considered an important input in the food, dental-pharmaceutical industries, and more recently, its use in the cosmetic industry stands out (Ahuja et al. 2020). As a result, xylitol has been placed in the select group of 12 top sugar-derived building blocks that can be produced from biomass (Werpy and Petersen 2004; Prabhu et al. 2020a, b), and from it can be produced many chemicals such as glycerol, glycol, polyester resins, heat transfer agent, hydraulic fluids (GVR 2021).

More recently, its use has been expanded for new areas as the production of xylitol-based plasticizer (Hou et al. 2021) and a natural deep eutectic solvent as the absorption medium of NH_3, which is a safe and environmentally friendly material used for pollution control (Liu and Jie 2021).

With this increased interest in using xylitol for different applications, mainly as a low-cost substitute for petrochemicals, xylitol is fast gaining ground over, and with the higher xylitol demand, the higher need to improve recuperation and purification processes increases. Therefore, apart from the search for new xylitol producing yeasts (wild-type or genetically modified), adjustments in the fermentation process (increasing yield and productivity), coupling to existing bio-production processes implementing new biorefineries, now the grand challenge is to overcome the barriers imposed by the difficulty of xylitol recovery from fermentation broth.

It is well known that the downstream process for bio-products is the most expensive stage compared to upstream. Since the last century, analytical separation techniques have been carried out by classical methods such as precipitation, distillation and extraction. However, the complexity of the media resulting from fermentation processes and the reduction in separation and purification costs have forced scientists to search for new, cheaper and more economically viable separation techniques. To choose a separation technique, besides considering economic and accessibility criteria, the physical and structural properties of the molecules to be separated or the characteristics of the matrix in which they are found are important factors. Typically, the xylitol production process by fermentation generates a complex stream from the point of view of recovery and purification of the product of interest, requiring several steps, which would increase the final cost of the product (Júnior and Rocha 2021). It is important to highlight that it is mandatory a high degree of purification, especially when xylitol is used in food or pharmaceutical industries.

In this way, unconventional separation techniques appear as an option for separation and purification stages. In the hybrid processes, the reaction or fermentation is combined with separation techniques, in the same equipment, such as distillation, crystallization, extraction, adsorption, pervaporation, and membrane separation, to achieve certain advantages, such as reducing the cost of capital or overcoming some limitations that are imposed by the equilibrium, which cannot be obtained when using conventional processes (Antony et al. 2021; Jatoi et al. 2021; Li et al. 2021; Anshul et al. 2022).

Mardawati et al. (2020) and Jain and Sanjoy (2021) reported that xylitol could be recovered from fermented broth and be purified applying different methods or combinations of methods, which cite chromatography, ion exchange resins, membrane filtration, nanofiltration and crystallization.

Several studies have reported that crystallization is still the most used process, with many variations and combinations (Delgado et al. 2021; Swart et al. 2021; Zhou et al. 2021; Prabhu et al. 2020a, b; Mardawati et al. 2020). However, other methods have been evaluated as an alternative for recovery and purification to obtain xylitol-rich crystals and higher yield. For example, the combination of ultrafiltration and nanofiltration (Kresnowati et al. 2017; Alves et al. (2021a, b); a combination of carbonation, activated charcoal and ion-exchange (Kumar et al. 2019) or xylitol recuperation using supercritical CO_2, integrating the CO_2 produced during fermentation to the xylitol purification step in a biorefinery (Silva et al. 2020a, b).

It is really important to consider xylitol production combined with other bio-products or biofuel production in biorefineries when thinking about the future. For this reason, it is indispensable deal with the recovery process in parallel for both products. Recent works brought this kind of concerns as Galan et al. (2021), who had evaluated integrated production of xylitol and sorbitol from different lignocellulosic biomass, and Swart et al. (2021), evaluating xylitol and xylooligosaccharides co-production from brewers spent grains, both trying to develop platforms with positive cost-effective into a biorefinery concept.

For the biorefineries to become increasingly viable and contribute to the development of a sustainable production chain (Volmer et al. 2020), it must incorporate product recovery processes that consider the analysis of Life Cycle Assessment, carbon, water, and land-use footprint, strengthening interaction with approaches of circular economy and 4.0 industry.

References

Affleck RP (2000) Recovery of xylitol from fermentation of model hydrolysate using membrane technology. Master of science thesis, State University of Virginia. 104p

Ahuja V, Macho M, Ewe D, Singh M, Saha S, Saurav K (2020) Biological and pharmacological potential of xylitol: a molecular insight of unique metabolism. Foods 9(11):1592. https://doi.org/10.3390/foods9111592

Albertsson PAJG, Tjerneld F (1990) Aqueous two-phase separations. In: Asenjo JA (ed) Separation processes in biotechnology. Marcell Dekker, New York, pp 287–327

Albertsson PA (1986) Partitioning of cell particles and macromolecules. New York: John Wiley & Sons.

Aliakbarian B, De Faveri D, Perego P, Converti A (2012) An assessment on xylitol recovery methods. In: Silva SS, Chandel AK (eds) D-Xylitol. Springer, Berlin, pp. 229–244.

Alves YPC, Antunes FAF, Silva SS, Forte MBS (2021a) From by-to bioproducts: selection of a nanofiltration membrane for biotechnological xylitol purification and process optimization. Food Bioprod Process 125:79–90. https://doi.org/10.1016/j.fbp.2020.10.005

Alves YPC, Antunes FAF, da Silva SS, Forte MBS (2021b) From by- to bioproducts: selection of a nanofiltration membrane for biotechnological xylitol purification and process optimization. Food Bioprod Process 125:79–90. https://doi.org/10.1016/j.fbp.2020.10.005

Anshul Y, Pawan KL, Vinod KS (2022) Membrane distillation crystallization technology for zero liquid discharge and resource recovery: Opportunities, challenges and futuristic perspectives. Sci Total Environ 806(2):150692. https://doi.org/10.1016/j.scitotenv.2021.150692

Antony FM, Pal D, Wasewar K (2021) Separation of bio-products by liquid–liquid extraction. Phys Sci Rev 6(4):20180065. https://doi.org/10.1515/psr-2018-0065

Asenjo JA, Andrews BA (2012) Aqueous two-phase systems for protein separation: phase separation and applications. J Chromatogr A 1238:1–10. https://doi.org/10.1016/j.chroma.2012.03.049

Bora MM, Borthakur S, Rao PC, Dutta NN (2005) Aqueous two-phase partitioning of cephalosporin antibiotics: effect of solute chemical nature. Sep Purif Technol 45:153–156. https://doi.org/10.1016/j.seppur.2004.08.006

Buyel JF, Twyman RM, Fischer R (2015) Extraction and downstream processing of plant-derived recombinant proteins. Biotech. Advances 33:902–913. https://doi.org/10.1016/j.biotechadv.2015.04.010
Camperi SA, Auday RM, delCanizo AN, Cascone O (1996) Study of variables involved in fungal pectic enzyme fractionation by immobilized metal ion affinity chromatography. Process Biochem 31:81–87. https://doi.org/10.1016/0032-9592(95)00020-8
Canilha L, Carvalho W, Giulietti M, Felipe MGA, Almeida e Silva JB (2008) Clarification of a wheat straw-derived medium with ion-exchange resins for xylitol crystallization. J Chem Technol Biotechnol 83:715-751. https://doi.org/10.1002/jctb.1861
Cardoso BS, Forte MBS (2021) Purification of biotechnological xylitol from *Candida tropicalis* fermentation using activated carbon in fixed-bed adsorption columns with continuous feed. Food Bioprod Process 126:86–90. https://doi.org/10.1016/j.fbp.2020.12.013
Carta G, Jungbauer A (2020) Protein chromatography: process development and scale-up. New York: John Wiley & Sons.
Cassano A, Comidi C, Ruby-Figueroa R, Castro-Muñoz R (2018) Nanofiltration and tight ultrafiltration membranes for the recovery of polyphenols from agro-food by-products. Int J Mol Sci 19:351. https://doi.org/10.3390/ijms19020351
Cienfuegos NEC, Santos PL, García AR, Soares CMF, Lima AS, Souza RL (2017) Integrated process for purification of capsaicin using aqueous two-phase systems based on ethanol. Food Bioprod Process 106:1–10. https://doi.org/10.1016/j.fbp.2017.08.005
De Faveri D, Perego P, Converti A, Del Borghi M (2002) Xylitol recovery by crystallization from synthetic solutions and fermented hemi-cellulosic hydrolysates. Chem Eng J 90:291–298. https://doi.org/10.1016/S1385-8947(02)00134-1
De Faveri D, Torre P, Perego P, Converti A (2004) Optimization of xylitol recovery by crystallization from synthetic solutions using response surface methodology. J Food Eng 61:407–412. https://doi.org/10.1016/S0260-8774(03)00148-1
De Luca C, Felletti S, Lievore G, Chenet T, Morbidelli M, Sponchioni M, Cavazzini A, Catani M (2020) Modern trends in downstream processing of biotherapeutics through continuous chromatography: the potential of Multicolumn Countercurrent Solvent Gradient Purification. TrAC, Trends Anal Chem 132:116051. https://doi.org/10.1016/j.trac.2020.116051
Delgado M, Navarro M, Lázaro A, Boyer SAE, Peuvrel-Disdier E (2021) Triggering and acceleration of xylitol crystallization by seeding and shearing: Rheo-optical and rheological investigation. Sol Energy Mater Sol Cells 220:110840. https://doi.org/10.1016/j.solmat.2020.110840
Desiriani R, Kresnowati MTAP, Wenten IG (2017) membrane-based downstream processing of microbial xylitol production. Int J Technol 8:1393–1401. https://doi.org/10.14716/ijtech.v8i8.726
Espitia-Saloma E, Vázquez-Villegas P, Aguilar O, Rito-Palomares M (2014) Continuous aqueous two-phase systems devices for the recovery of biological products. Food Bioprod Process 92:101–112. https://doi.org/10.1016/j.fbp.2013.05.006
Faneer KA, Rohani R, Mohammad AW, Ba-Abbad MM (2017) Evaluation of the operating parameters for the separation of xylitol from a mixed sugar solution by using a polyethersulfone nanofiltration membrane. Korean J Chem Eng 34(11):2944–2957. https://doi.org/10.1007/s11814-017-0186-y
Faneer KA, Rohani R, Mohammad AW (2018) Influence of pluronic addition on polyethersulfone membrane for xylitol recovery from biomass fermentation solution. J Clean Prod 171(995):e1005. https://doi.org/10.1016/j.jclepro.2017.10.075
Fernandez C, Avelino A, Farelo F (1999) Crystallization of xylitol from hydro-alcoholic solutions containing arabitol and adonitol. In: 14th international symposium on industrial crystallization, 12–16 September, Cambridge University
Freire MG, Claudio AFM, Araujo JMM, Coutinho JAP, Marrucho IM, Canongia Lopes JN, Rebelo LPN (2012) Aqueous biphasic systems: a boost brought about by using ionic liquids. Chem Soc Rev 41:4966–4995. https://doi.org/10.1039/c2cs35151j

Galán G, Mariano M, Ignacio EG (2021) Integrated renewable production of sorbitol and xylitol from switchgrass. Ind Eng Chem Res 60(15):5558–5573. https://doi.org/10.1021/acs.iecr.1c00397

Ghazizadeh AF, Pazuki G (2021) Effect of carbohydrates, choline chloride based deep eutectic solvents and salts on the phase behavior of PEG-PPG copolymer ATPSs and partitioning of penicillin G. J Mol Liq 339:117152. https://doi.org/10.1016/j.molliq.2021.117152

Gurgel PV, Mancilha IM, Peçanha RP, Siqueira JFM (1995) Xylitol recovery from fermented sugarcane bagasse hydrolysate. Bioresour Technol 52:219–223. https://doi.org/10.1016/0960-8524(95)00025-A

GVR – Grand View Research (2021) Xylitol building block market size, share & trends analysis report by application, regional outlook, competitive strategies, and segment forecasts, 2019 To 2025. Market Analysis Report. Report ID: GVR1281. https://www.grandviewresearch.com/industry-analysis/xylitol-building-block-market. Accessed 06 Oct 2021

Heikkilä H, Nurmi J, Rahkila L, Töyrylä M (1990) Method for the production of xylitol from mixtures containing xylose. Patent WO 90/08193, January 15

Heikkilä H, Nurmi J, Rahkila L, Töyrylä M (1992) Method for the production of xylitol. U.S. Patent 5,081, 026, January 14

Heikkilä H, Puuppo O, Tylli M, Nikander H, Nygrèn J, Lindroos M, Eroma OP (1997) Method for producing xylitol. Patent WO97/49659, June 23

Hou BY, Ren L, Fu DM, Jiang YY, Zhang, |M. Y., Zhang, H. X. (2021) Novel environmentally sustainable xylitol-based plasticizer: synthesis and application. J Polym Res 28(9):1–12. https://doi.org/10.1007/s10965-021-02694-9

Jain V, Sanjoy G (2021) Biotransformation of lignocellulosic biomass to xylitol: an overview. Biomass Convers Biorefinery 1-19. https://doi.org/10.1007/s13399-021-01904-0

Jatoi AS, Baloch HA, Mazari SA, Mubarak MN, Sabzoi N, Aziz S, Soomro SA, Abro R, Shah SF (2021) A review on extractive fermentation via ion exchange adsorption resins opportunities, challenges, and future prospects. Biomass Conv Bioref https://doi.org/10.1007/s13399-021-01417-w

Johansson G (1989) Affinity partitioning of proteins using aqueous two-phase systems. In: Janson JGR (ed) Protein purification, principles high resolution methods and applications. VCH Publishers, pp 330–345

Júnior JEM, Rocha MVP (2021) Development of a purification process via crystallization of xylitol produced for bioprocess using a hemicellulosic hydrolysate from the cashew apple bagasse as feedstock. Bioprocess Biosyst Eng 44(4):713–725. https://doi.org/10.1007/s00449-020-02480-9

Kessel D (1981) Some determinants of partitioning behavior of lymphoblasts in aqueous biphasic systems. Biochim Biophys Acta (BBA)-Gen Subj 678:245–249. https://doi.org/10.1016/0304-4165(81)90213-0.

Kilikian BV, Pessoa Jr A (2021) Recuperação de produtos obtidos em bioprocessos (downstream processing), in: Schimidell, W. (Organisador), Coleção Biotecnologia Industrial, vol 2, Engenharia Bioquímica. 2.ed. São Paulo: Edgard Blücher. pp 395–454

Kresnowati MTAP, Regina D, Bella C, Wardani AK (2019) Combined ultrafiltration and electrodeionization techniques for microbial xylitol purification. Food Bioprod Process 114:245–252. https://doi.org/10.1016/j.fbp.2019.01.005

Kresnowati MTAP, Desiriani R, Wenten IG (2017) Ultrafiltration of hemicellulose hydrolysate fermentation broth. In: AIP conference proceedings, vol 1818. No. 1. AIP Publishing LLC. https://doi.org/10.1063/1.4976888

Kumar V, Sandhu PP, Ahluwalia V, Mishra BB, Yadav SK (2019) Improved upstream processing for detoxification and recovery of xylitol produced from corncob. Biores Technol 291:121931. https://doi.org/10.1016/j.biortech.2019.121931

Li L, He CY, Li SH, Liu F, Su S, Kong XX, Li N, Li KA (2004) Study on PEG-$(NH_4)_2SO_4$ aqueous two-phase system and distribution behavior of drugs. Chinese J Chem 22:1313–1318. https://doi.org/10.1002/cjoc.20040221120

Li X, Tan S, Luo J, Pinelo M (2021) Nanofiltration for separation and purification of saccharides from biomass. Front Chem Sci Eng 15:837–853. https://doi.org/10.1007/s11705-020-2020-z

Linke D, Berger RG (2011) Foaming of proteins: new prospects for enzyme purification processes. J Biotechnol 152:125–131. https://doi.org/10.1016/j.jbiotec.2010.07.022

Liu B, Jie T (2021) Investigation of glycolic acid natural deep eutectic solvents with strong proton donors for ammonia capture and separation. Ind Eng Chem Res 60(30):11600–11610. https://doi.org/10.1021/acs.iecr.1c01456

Mandal S, Mandal A (2014) Separation of no-carrier-added (TcO4-)-Tc-99m from Mo-99-Tc-99m equilibrium mixture by PEG based aqueous biphasic separation technique using sodium/potassium salts of citric and tartaric acid. J Radioanal Nucl Ch 299:1225–1230. https://doi.org/10.1007/s10967-014-2967-7

Mardawati E, Annazhifah N, Sukri N, Suryadi E, Harahap BM (2020) Physicochemical properties of xylitol crystals from oil palm empty fruit bunches hydrolysate. Int J Adv Sci Eng Inf Technol 10:1646–1653. https://doi.org/10.18517/ijaseit.10.4.11596

Marques Junior JE, Rocha MVP (2021) Development of a purification process via crystallization of xylitol produced for bioprocess using a hemicellulosic hydrolysate from the cashew apple bagasse as feedstock. Bioprocess Biosyst Eng 44:713–725. https://doi.org/10.1007/s00449-020-02480-9

Martinez EA, Almeida e Silva JB, Giulietti M, Derenzo S (2008) Kinetics of the xylitol crystallization in hydro-alcoholic solution. Chem Eng Process 47:2157–2162. https://doi.org/10.1016/j.cep.2007.11.004

Martinez EA, Almeida e Silva JB, Giulietti M, Solenzal AIN (2007) Downstream process for xylitol produced from fermented hydroylysate. Enzym Microb Technol 40:1185–1189. https://doi.org/10.1016/j.enzmictec.2006.09.003

Martinez EA, Almeida e Silva JB, Poço JGR, Giulietti M (2012) Cristalização de xilitol em água-etanol utilizando xilitol e sílica gel (nanoparticulada) pirogênica como semente. Patente BR 10 2012 013327–0 A2.01/06/2012

Martinez EA, Giulietti M, Almeida e Silva JB, Derenzo S, Felipe GA (2009) Batch cooling crystallization of xylitol produced by biotechnological route. J Chem Technol Biotechnol 84:376–381. https://doi.org/10.1002/jctb.2050

Martinez EA, Canettieri EV, Bispo JAC, Giulietti M, de Almeida e Silva JB, Converti A (2015) Strategies for xylitol purification and crystallization: a review, separation science and technology 50(14):2087–2098. https://doi.org/10.1080/01496395.2015.1009115

Martinez EA, Santos JAF (2012) Influence of the use of rice bran extract as a source of nutrients on xylitol production. Food Sci Technol 32(2):308–313. https://doi.org/10.1590/S0101-20612012005000036

Matos T, Johansson HO, Queiroz JA, Bulow L (2014) Isolation of PCR DNA fragments using aqueous two-phase systems. Sep Purif Technol 122:144–148. https://doi.org/10.1016/j.seppur.2013.11.014

Mohamed AS, Ling TC, Muniandy S, Tan YS, Raman J, Sabaratnam V (2014) Recovery and partial purification of fibrinolytic enzymes of Auricularia polytricha (Mont.) Sacc by an aqueous two-phase system. Sep Purif Technol 122:359–366. https://doi.org/10.1016/j.seppur.2013.11.016.

Mullin JW (2001) Crystallization, 4th edn. Oxford: Butterworth-Heinemann.

Mun LW, Rafiqul ISM, Sakimah AMM, Zularisam AW (2016) Purification of bioxylitol by liquid–liquid extraction from enzymatic reaction mixture. Sep Sci Technol 51(14):2369–2377. https://doi.org/10.1080/01496395.2016.1203335

Mussatto SI, Santos JC, Filho WCR, Silva SS (2006) A study on the recovery of xylitol by batch adsorption and crystallization from fermented sugarcane bagasse hydrolysate. J Chem Technol Biotechnol 81:1840–1845. https://doi.org/10.1002/jctb.1614

Nagarajan S (2012) New tools for exploring "old friends—microbial lipases." Appl Biochem Biotechnol 168:1163–1196. https://doi.org/10.1007/s12010-012-9849-7

Nurmi J, Eroma OP, Eriksson K (1999) Method for recovering an organic compound from solutions. U.S. Patent 5980640, November 9

Nývlt J, Sönhel O, Matuchová M, Broul M (1985) The kinetics of industrial crystallization. Prague: Academia.

Nývlt J, Hostomský J, Giulietti M (2001) Cristalização, São Carlos: Ed. da UFSCar.

Pfister D, Nicoud L, Morbidelli M (2018) Continuous biopharmaceutical processes: chromatography, bioconjugation, and protein stability. Cambridge University Press.

Prabhu AA, Thomas DJ, Ledesma-Amaro R, Leeke GA, Medina A, Verheecke-Vaessen C, Coulon F, Agrawal D, Kumar V (2020a) Biovalorisation of crude glycerol and xylose into xylitol by oleaginous yeast *Yarrowia lipolytica*. Microb Cell Fact 19:1–18. https://doi.org/10.1186/s12934-020-01378-1

Prabhu AA, Thomas DJ, Ledesma-Amaro R, Leeke GA, Medina A, Verheecke-Vaessen C, Colon F, Agrawal D, Kumar V (2020b) Biovalorization of crude glycerol and xylose into xylitol by oleaginous yeast *Yarrowia lipolytica*. Microb Cell Fact 19:1–18. https://doi.org/10.1186/s12934-020-01378-1

Ricroch AE, Martin-Laffon J, Rault B, Pallares VC, Kuntz M (2022) Next biotechnological plants for addressing global challenges: the contribution of transgenesis and new breeding techniques. New Biotechnol 66:25–35. https://doi.org/10.1016/j.nbt.2021.09.001

Rivas B, Torre P, Dominguez JM, Converti A, Parajó JC (2006) Purification of xylitol obtained by fermentation of corncob hydrolysates. J Agric Food Chem 54(12):4430–4435. https://doi.org/10.1021/jf053156x

Rogers RD, Bond AH, Bauer CB (1993) Metal Ion separations in polyethylene glycol-based aqueous biphasic systems. Sep Sci Technol 28:1091–1126. https://doi.org/10.1080/01496399308018023

Rosa PAJ, Azevedo AM, Sommerfeld S, Backer W, Aires-Barros MR (2011) Aqueous two-phase extraction as a platform in the biomanufacturing industry: economical and environmental sustainability. Biotechnol Adv 29:559–567. https://doi.org/10.1016/j.biotechadv.2011.03.006

van Rosmalen S, Bruinsma D, Kramer H, Bermingham S, Witkamp G, Derenzo S, Giulietti M, Secker MM, Re MI, Cekinski E (2003) Industrial crystallization and precipitation, IPT-TU DELFT

Sampaio FC, Passos FML, Passos FJV, De Faveri D, Perego P, Converti A (2006) Xylitol crystallization from culture media fermented by yeasts. Chem Eng Process 45:1041–1046. https://doi.org/10.1016/j.cep.2006.03.012

Santos RP, Souza LM, Balieiro AL, Soares CMF, Lima ÁS, Souza RL (2018) Integrated process of extraction and purification of betanin from Opuntia ficus-indica using aqueous two-phase systems based on THF and sodium salts. Sep Sci Technol 53:734–744. https://doi.org/10.1080/01496395.2017.1397022

Santos DT; Sotelo AF, Sarrouh BF, Salaue LM, Palácios NR, Silva SS (2011) Supercritical technology as an alternative for biotechnological xylitol purification. Semina: Ciências Agrárias Londrina 32(2):621–632. https://doi.org/10.5433/1679-0359.2011v32n2p621

Saxena RK, Sheoran A, Giri B, Davidson WS (2003) Purification strategies for microbial lipases. J Microbiol Methods 52:1–18. https://doi.org/10.1016/S0167-7012(02)00161-6

Silva DDV, Dussán KJ, Idarraga A, Grangeiro R, Silva SS, Cardona CA, Quintero J, Felipe MGA (2020a) Production and purification of xylitol by Scheffersomyces amazonenses via sugarcane hemicellulosic hydrolysate. Biofuels Bioprod Biorefin 14(2):344–356. https://doi.org/10.1002/bbb.2085

Silva DDV, Dussán KJ, Idarraga A, Grangeiro L, Silva SS, Cardona CA, Quintero J, Felipe MGA (2020b) Production and purification of xylitol by *Scheffersomyces amazonenses* via sugarcane hemicellulosic hydrolysate. BiofuelsBioproducts Biorefining 14:344–356. https://doi.org/10.1002/bbb.2085

Souza RL, Lima RA, Coutinho JAP, Soares CMF, Lima ÁS (2015) Novel aqueous two-phase systems based on tetrahydrofuran and potassium phosphate buffer for purification of lipase. Process Biochem 50:1459–1467. https://doi.org/10.1016/j.procbio.2015.05.015

Straathof AJJ (2011) 2.57 - the proportion of downstream costs in fermentative production processes. In: Moo-Young M (ed) Comprehensive biotechnology, 2nd edn. Academic Press, Burlington, pp 811–814

Swart LJ, Petersen AM, Bedzo OKK, Görgens JF (2021) Techno-economic analysis of the valorization of brewers spent grains: production of xylitol and xylo-oligosaccharides. J Chem Technol Biotechnol 96(6):1632–1644. https://doi.org/10.1002/jctb.6683

Tarafder A, Aumann L, Morbidelli MJJOCA (2010) The role of ion-pairing in peak deformations in overloaded reversed-phase chromatography of peptides. J Chromatography A., 1217:7065–7073. https://doi.org/10.1016/j.chroma.2010.09.007

Teal HE, Hu Z, Root DD (2000) Native Purification of Biomolecules with Temperature-Mediated Hydrophobic Modulation Liquid Chromatography. Anal Biochem 283:159–165. https://doi.org/10.1006/abio.2000.4640

Ullmanns (1998) Encyclopedia of industrial chemistry, 6th edn. Sugar alcohols-xylitol. Electronic Release

Vollmer NI, Gernaey KV, Mussato SI, Sin G (2020) Surrogate modelling based uncertainty and sensitivity analysis for the downstream process design of a xylitol biorefinery. Computer aided chemical engineering, vol 48. Elsevier, pp. 1663–1668. https://doi.org/10.1016/B978-0-12-823377-1.50278-0

Weatherley LR (1994) Engineering process for bioseperations. Oxford: Butterworth-Heineman.

Werpy T, Petersen G (2004) *Top value added chemicals from biomass: volume I--results of screening for potential candidates from sugars and synthesis gas.* No. DOE/GO-102004–1992. National Renewable Energy Lab., Golden, CO (US). https://doi.org/10.2172/15008859

Zhou H, Lv L, Zhang Y, Ji M, Cen K (2021) Preparation and characterization of a shape-stable xylitol/expanded graphite composite phase change material for thermal energy storage. Sol Energy Mater Sol Cells 230:111244. https://doi.org/10.1016/j.solmat.2021.111244

Chapter 7
Critical Analysis on Xylitol Production Employing Integrated Approaches in Sugarcane and Corn Processing Mills

Daniel D. Durán-Aranguren, María F. Cajiao-Pedraza, Juan A. Ospina-Paz, Juan D. Vásquez-Muñoz, Andrés F. Hernández-Perez, and Rocío Sierra

Abstract Xylitol bioproduction has been studied worldwide, at least during the last three decades, as a potential replacement for the chemical process that has significant drawbacks regarding efficiency and sustainability. Important advances in the efficiency and productivity of this bioprocess have been achieved through developments in bioprocessing and genetic engineering technologies; however, critical bottlenecks still remaining have avoided scale up to commercial scales. Nevertheless, large-scale global production is still mostly based on the catalytic route, while there are only a few initiatives for biotechnological production of this high-added-value sweetener and platform chemical. In this regard, xylitol bioprocess integration into a biorefinery may support its feasibility and sustainability, based on the maximum valorization of the raw material, optimization of energy usage and material fluxes, and diversification of the product portfolio. Xylitol bioprocess can be coupled to existing processes in industrial plants or can be considered a backbone for the design of biorefineries, an approach that will be considered in the present chapter. Herein, xylitol bioproduction technology is reviewed in order to identify gaps and improvement opportunities for its integration into biorefineries. Moreover, alternative processes to obtain other valuable compounds from streams derived from xylitol bioprocess are presented.

Keywords Xylitol · Biorefinery · Process integration · Lignocellulosic biomass · Biomass valorization

D. D. Durán-Aranguren · M. F. Cajiao-Pedraza · J. A. Ospina-Paz · J. D. Vásquez-Muñoz · R. Sierra
Product and Process Design Group, Department of Chemical and Food Engineering, Universidad de Los Andes, Carrera 1 No. 18A-10, 111711 Bogotá, Colombia

A. F. Hernández-Perez (✉)
Department of Biotechnology, Engineering School of Lorena (EEL), University of São Paulo (USP), Estrada Municipal Do Campinho s/n, Lorena, SP 12602-810, Brazil
e-mail: afhernandezmbio@gmail.com

© The Author(s), under exclusive license to Springer Nature Switzerland AG 2022

M. das G. de Almeida Felipe and A. K. Chandel (eds.), *Current Advances in Biotechnological Production of Xylitol*,
https://doi.org/10.1007/978-3-031-04942-2_7

7.1 Introduction

Bioeconomy and circular economy are currently a matter of discussion by national and regional regulatory entities and embraced by corporations and startups as pillars in their strategic plans for the future and as opportunities to update the business models for a changing world. Clear evidence of this revolution is the relevance that ESG (Environmental, Social, and Governance) is reaching for fund investments in several companies and industrial sectors (Ellen MacArthur Foundation 2020).

Biotechnology leads the development of technologies to replace processes based on non-renewable or non-sustainable raw materials and expands the products and services offered to society (Straathof et al. 2019). A recent report from McKinsey defined the current context for biotechnology as the "Bio Revolution", which represents the coupling of bio-based technologies with AI and data-based technologies to potentialize the possibilities and impacts of biotechnology in this new economy (Chui et al. 2020). According to the McKinsey report, up to 60% of the physical inputs currently used in production systems can be produced biologically or replaced using biology (Chui et al. 2020). Moreover, bio innovations may improve the performance and sustainability of already existing productive chains and create new ones. Recent advances in genetic engineering, synthetic biology, and bioprocessing have accelerated knowledge development and the emergence of cell factories for manufacturing a high variety of biofuels, biochemicals, food, and pharmaceutical ingredients (Straathof et al. 2019).

Several examples of startups being created based on the possibility of generating added value from biological products and/or processes can be observed worldwide, as well as medium and large companies incorporating biotechnology within their Research & Development areas, besides those already established and successful bio corporations in several segments, ranging from agriculture, health, pharmaceutics, energy to food and chemicals (Chui et al. 2020). One particular example worth mentioning is the Amyris case, which has been continuously developing high-added-value products from lignocellulosic biomass (mostly sugarcane bagasse), some of which are already manufactured in a scaled-up fashion, such as farnesene, squalene, bisabolol, among others, and by using bioprocessing, synthetic biology, data science, and automation technologies, allowing it to reach high recognition and market value (Hill et al. 2020). Another remarkable example is a recently developed technology to produce bioplastic from CO_2, consisting of three integrated processes: firstly, CO_2 fermentation to ethanol using a technology developed by Lanzatech; followed by dehydration and polymerization to obtain "green" polyethylene, a process performed by Total Energies; and finally, the use of polyethylene to produce cosmetic packaging by L'Óreal (L'Óreal 2020).

Both examples represent a growing trend for a high value and sustainable, productive system based on the integration of various processes, chemical and biological, for the complete use of sustainable raw materials to obtain several products, which can be applied in more than one market segment or niche, allowing companies to diversify both their feedstock and products portfolio. This concept has been denominated Biorefinery, as a "green" parallel to fossil fuels refineries, and it has been particularly used for production facilities that exploit lignocellulosic biomasses.

Several bioproducts have been considered feasible candidates to be jointly produced into biorefineries to attend high volume-low value or low volume-high value markets (IEA Bioenergy 2020). It is within this scenario that bioproducts that can be produced from the hemicellulosic fraction stand out, as it is the case of xylitol, which has been considered since 2004 and recently confirmed as one of the top 12 bio-based chemicals suitable to be part of biorefineries (IEA Bioenergy 2020). Xylitol is a high-value polyol, with a growing market demand based on its important applications in food and pharmaceutical industries, and also it is recognized as a platform chemical since it can be used as a substrate to obtain other valuable compounds, such as glycols, xylaric acid and polymers (Hernández-Pérez et al. 2019; Delgado Arcaño et al. 2020).

Xylitol properties and applications are deeply discussed in other chapters of the present book; however, an interesting analysis of the increasing industrial interest in xylitol can be illustrated through the observation of the patent publication profile. As evidenced in Fig. 7.1a, continuous growth in patents published related to xylitol was observed at least until 2015, when the number of patents published per year remains relatively stable, in values around 1800. To gain insight into the industrial fields interested in xylitol, Fig. 7.1b summarizes the ten most-cited subclasses of the International Patent Classification (IPC), which account for approximately 70% of the total IPC subclasses. Among them, applications related to medical, dental, and

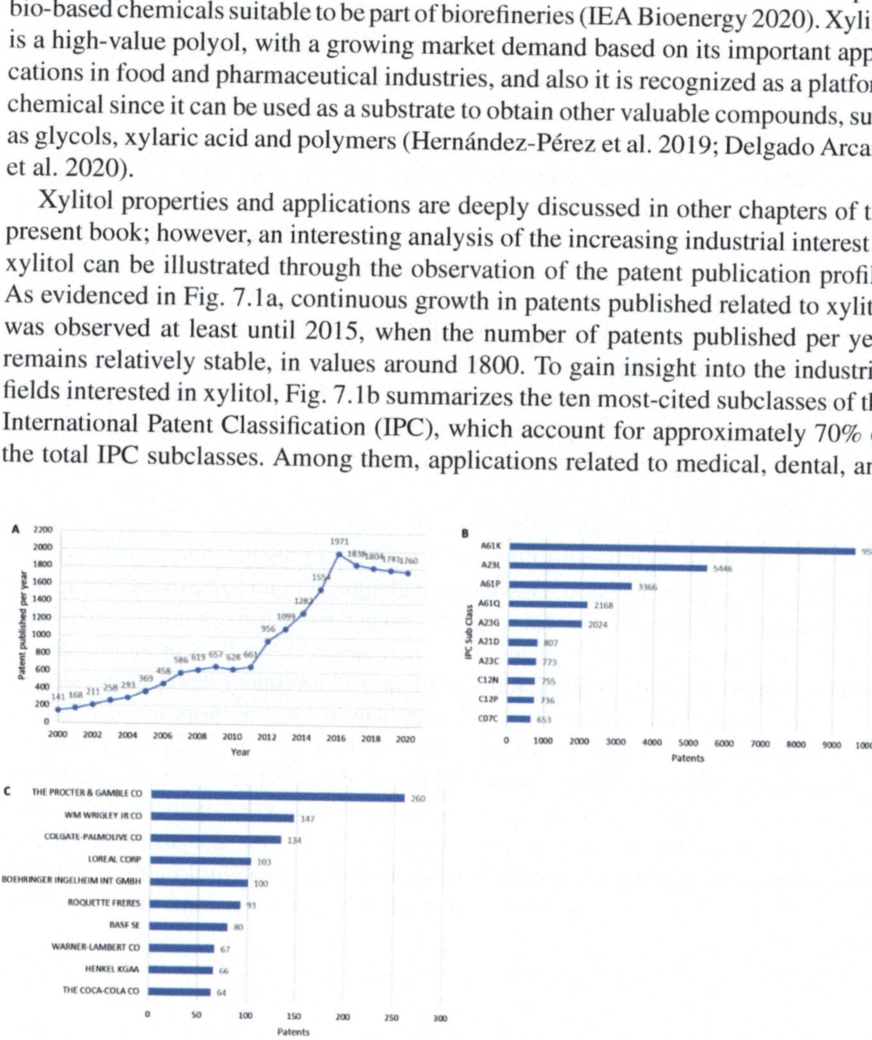

Fig. 7.1 Profile of xylitol-related patent publication. **a** Patent publication in the period 2000–2020; **b** top 10 most cited International Patent Classification (IPC); **c** top 10 standardized assignees. The patent search was performed in Patsnap Analytics software using the query TAC: (xylitol) on October 9th, 2021

therapeutically purposes represent the most common IPCs (A61K, A61P), which is coherent with the fact that pharmaceutical and personal care companies were among the main assignees, such as Boehringer Ingelheim, Warner-Lambert, Procter and Gamble and Colgate Palmolive (Fig. 7.1c). IPCs related to food and beverages formulations (A23L, A23G, A21D, A23P) can also be highlighted (Fig. 7.1b) and, consequently, assignees in these market segments as WM Wrigley Jr and The Coca-Cola Company (Fig. 7.1c). Xylitol application as an ingredient in personal care formulations is also observed with IPC Sub Class A61Q (Fig. 7.1b) and companies like L'Oreal Fig. 7.1c. It is important to notice that Fig. 7.1b shows only a trend on the most frequently cited IPC and their relationship with industrial segments interested in xylitol since it is important to consider that one patent can belong to various IPC subclasses.

Currently, xylitol production at commercial scale is mostly based on the catalytic hydrogenation of xylose, purified from the hemicellulosic fraction of lignocellulosic biomasses, which involves high energy requirements and has important efficiency and environmental drawbacks (Rueda et al. 2015; Hernández-Pérez et al. 2019; Delgado Arcaño et al. 2020; Ahuja et al. 2020). Biotechnological production of xylitol from lignocellulosic biomass has been studied worldwide over the last three decades since it represents the possibility of replacing the chemical process with a potentially more efficient and sustainable process. Xylitol bioproduction is based on the natural capacity of xylose-fermenting microorganisms to reduce this pentose into xylitol, which avoids the need for complex xylose purification and costly catalysts, and high temperature and pressure conditions for xylose conversion (Hernández-Pérez et al. 2019).

Developments in bioprospection, bioprocessing, and genetic engineering technologies have led to significant improvements in the efficiency and productivity of xylose-to-xylitol bioconversion. However, challenges remain to be overcome to take xylitol bioproduction into large scale, not only related to the fermentation step but also to upstream and downstream operations, which can be low efficient and/or energy and water-intensive. In this regard, integration into a biorefinery has been considered a strategy to improve the feasibility and sustainability of the bioprocess, based on the integral use of all fractions of lignocellulosic biomass to obtain other valuable products and to optimize energy usage and material fluxes between unit operations. This perspective allows for utilizing all the components in the lignocellulosic feedstock to obtain added-value products or services. Within this scenario, this chapter aims to critically review and discuss existing literature on xylitol bioproduction to identify possible gaps, opportunities, and challenges for integrating this bioprocess into biorefineries.

7.2 Context of Global Xylitol Production

Xylitol chemical production from lignocellulosic biomass has been occurring since the 1970s, mostly concentrated in Asia and Europe (Delgado Arcaño et al. 2020). According to the IEA Bioenergy report (IEA Bioenergy 2020), xylitol production at a commercial scale has a global capacity of approximately 190.000 tons per year and potential for continuous growth. Among the most important players, companies like Futaste Pharmaceutical, Dupont Danisco, Roquette Frères, and Cargill can be highlighted.

Particularly, Dupont/Danisco (which recently assigned a partnership with IFF for the ingredients segment) is one of the main xylitol producers outside China and holds the brand Xivia (IFF Nutrition and Biosciences 2021). This company has been producing xylitol from wood biomass since 1975 in Finland, and in 2012 it presented the Dupont Wood-based Integration concept as a circular economy process with lower environmental impacts (until 90% lower) and lower energy requirements (85% lower) than conventional xylitol producing-processes based on corncob (DuPont Nutrition & health 2012). The process is integrated with pulp and paper plants, using waste side streams as feedstocks, such as black liquor, and energy production.

Another initiative of xylitol chemical production integrated into other processes that is worth mentioning is the Fortress Global case. This Canadian company, a producer of dissolved pulp, purchased the S2G Biochemicals company in 2018 and announced the intention to build a 2000 ton per year demonstration plant for xylitol production by catalytic hydrogenation of C5 sugars, using the hemicellulose residual stream from the cellulose mill (S2G Biochem 2016). The proprietary technology was developed by S2G Biochemicals and Mondelez International and intended to produce not only xylitol but also glycols (ethylene glycol, propylene glycol, and others) that can be drop-in replacements for petrochemicals (S2G Biochem 2016).

Regarding xylitol production by biotechnological routes, biotech companies from China have also been reported as xylitol producers, such as Hangzhou Shouxing Biotech Co. Ltd., Thomson Biotech (Xiamin) Pte. Ltd., and Yucheng Luijan Biological Ltd. (Ravella et al. 2012; Ahuja et al. 2020). However, it was not possible to obtain information about the processes performed by these companies. On the other hand, the American company ZuChem announced in 2020 the first commercial sale of xylitol obtained by fermentation (ZuChem 2020). According to a press release, the production plant is a joint venture between ZuChem and a China's company called Harbin Yimei Bioengineering Technology Co., which resulted in the brand Sweet Appeal Natural Products, and it is based on the biotechnological route developed by ZuChem scientists in cooperation with the United States Department of Agriculture (USDA) and the University of Illinois. The bioprocess is intended to work with a wide range of non-GMO feedstocks (for example, corn, hardwood, and bagasse), using a recombinant microorganism that is able to tolerate high concentrations of furfural and other inhibitors and to achieve yields near to 100% of theoretical and titers higher than 100 g/L (ZuChem 2020). Further information on the production capacity or current operational state of the ZuChem plant was not found.

Another initiative that is worth mentioning is the case of Virdia and Stora Enso. In June 2014, Stora Enso, a leading global provider of renewable material from wood and biomass, purchased the US-based company Virdia, and in September, announced a demonstration plant for xylose production from sugarcane bagasse (Stora Enso 2015). Virdia had a biorefinery project, according to a 2018 lecture, which involves obtaining various products from the biomass fractions, particularly xylitol from the hemicellulose (Stora Enso 2016). In this regard, Virdia has a US patent granted in 2021, with a priority date of 07 Jan 2015, which describes xylitol production by fermentation of sugars refined from lignocellulosic biomasses, mainly xylose, and using a recombinant strain of *Escherichia coli* (Jansen et al. 2021). However, Stora Enso announced in Jan 2021 that they decided to close the Virdia operations (Stora Enso 2021) permanently. No further statements on the advancement of the Virdia's biorefinery and/or xylitol process were found.

The startup Creatus Biosciences can also be highlighted, which is a spin-off of the University of British Columbia that has been working since 2014 to develop engineered strains to be used in mixed sugar biorefineries (Creatus Biosciences 2021), and it has a patented granted in the US in November 2018 for xylitol producing recombinant strains of *Metschinkowia* genus (Luo et al. 2019). It is worth mentioning also a patent granted in Europe, China, and the United States to Roquette Frères, a known xylitol producer by chemical pathway, in which a recombinant strain for xylitol production using glucose as a substrate is described (Defretin et al. 2018). Another interesting patent was filed in 2018 by Sinopec Shangai Eng, a subsidiary of China's petrochemical company Sinopec, which describes an industrial strain of recombinant *Saccharomyces cerevisiae* for high-yield production of xylitol and ethanol by co-fermenting xylose and glucose (Tang et al. 2018).

As previously presented, initiatives at a large scale for integration of xylitol with other processes have been observed for chemical technologies, such as the cases of Dupont/Danisco and Fortress Global, but not for biotechnology routes. Currently, some research is focused on the technical-economical assessment, looking for the most consolidated and/or scalable bioprocess, and using integration in biorefineries as a pillar for feasibility and sustainability, as well as a way to attract investment to take this bioprocess on a large scale (Mountraki et al. 2017; Hernández-Pérez et al. 2019). For that reason, the following section points out the gaps and challenges of the xylitol bioproduction process for its integration in biorefineries.

7.3 Process Alternatives for the Biotechnological Production of Xylitol and Their Influence on the Selection of Biorefinery Pathways

Xylitol bioproduction from lignocellulosic biomasses has been studied as a potentially feasible and sustainable route to replace the current chemical technologies. However, several drawbacks keep xylitol bioprocess not profitable as a stand-alone plant. Opportunities for improvement can be identified from the existing process alternatives reported in the literature for the xylitol biotechnological production route, which are summarized in Table 7.1.

One of the most fundamental technical aspects of the process that affects the yields of sugar extraction, and consequently of xylitol, is the composition of the raw material. The amount of structural carbohydrates limits how much glucose and xylose can be obtained. Additionally, the lignin content limits the access to cellulose and hemicellulose. Table 7.2 shows the approximate ranges of composition that have been reported in the literature for corn stover, sugarcane bagasse, and sugarcane straw.

It is important to note that the cellulose, hemicellulose, and lignin content is similar, to some extent, among the three biomasses, which are characterized for being rich in these constitutive polymers. Sugarcane mills could benefit from the same bolt-on infrastructure by using mixtures of bagasse and straw. Moreover, if some processing mills share the production of both sugarcane and corn, it would also be possible to use mixtures of all three feedstocks. Nonetheless, the use of feedstock mixtures may imply additional logistics to transport these materials to the biorefinery. From an economic perspective, the biorefinery location affects the selection of the raw material, based on the different agro-industrial biomasses generated in the surroundings. It has been stated that logistic expenses related to obtaining the raw material represent about 34% of the total production costs, followed by the pretreatment cost with 17%, boiler energy, and utilities and storage at 8.5% and 5%, respectively (Chandel et al. 2012). Regardless of the use of biomass mixtures, it would be important to implement cogeneration systems that use a fraction of the raw material to produce energy to fuel the process through incineration or gasification.

Since lignin removal enhances the accessibility to structural carbohydrates during the hydrolysis and, therefore, facilitates the solubilization of hemicellulose and cellulose (Philippini et al. 2019), a biorefinery using mixtures of sugarcane and corn biomasses could control the different proportion of the feedstocks in order to keep lignin content as lowest as possible. Moreover, an alkaline pretreatment step can be implemented for delignification, which produces a solid fraction with high cellulose contents and represents the possibility of further valorizing the solubilized lignin (Raj and Krishnan 2020). The recovery of alkalis would require spending more energy (Venkateswar Rao et al. 2016); however, these black liquors could be recycled into the process to remove more lignin which reduces water consumption in the biorefinery.

Once lignin is removed, the breakdown of structural carbohydrates to obtain fermentable sugars can be performed by different alternatives such as hydrothermal treatments, dilute acid hydrolysis, or enzymes. Dilute acid hydrolysis has been one of the most studied alternatives in bioprocesses aiming at xylitol production since it is considered effective and low cost, although the use of acids as a catalyst is a drawback (Hernández-Pérez et al. 2019). Hydrothermal treatments, such as steam explosion, also result in high yields of xylose (Hernández-Pérez et al. 2019) and have the advantage of avoiding the use of other chemicals and solvents (Aguiar et al. 2021). However, both dilute acid hydrolysis and hydrothermal treatments may also cause the formation or release of potentially toxic compounds to xylitol-producing

Table 7.1 Advantages, disadvantages, and opportunities of different process alternatives reported in the literature for the biotechnological route of xylitol production

	Alternatives	Advantages	Disadvantages	Opportunities for improvement
Biomass composition	Use of a single feedstock	– Well-known yields of sugars	– Fixed amounts of lignin that could make the pretreatment stage mandatory	– Use of mixtures from nearby processing plants, avoiding logistic expenses and their environmental impacts – Implementation of energy-efficient cogeneration systems
	Use of mixtures	– It is possible to use mixtures that decrease the overall lignin content – Same bolt-on infrastructure that could use mixtures of bagasse and straw	– Biomass composition would depend on plant location and season, making it highly variable	
Pretreatment	No pretreatment	– Avoids the use of other chemicals – Less water consumption	– Yields of sugars could be reduced due to lignin interference	– Reduction of water consumption by recycling alkalis – Recuperation of lignin
	Alkaline pretreatment (NaOH, H_2O_2, aqueous ammonia)	– It is possible to reuse alkalis from waste streams and recycle them into the process	– It is necessary to recover alkalis, which requires energy consumption	
Hydrolysis	Hydrothermal treatments	– The use of water at high temperatures allows the solubilization of hemicellulose – Avoids the use of other chemicals and solvents – The material is hydrolyzed in short times – High yields of sugars are obtained with no need for additional chemicals to use and recycle	– A high concentration of furans and other inhibitors can be obtained – Requires excessive amounts of water and energy	– Integration of high–temperature streams into other parts of the process – Recuperation of sugar oxidation byproducts – Water recycling

(continued)

Table 7.1 (continued)

	Alternatives	Advantages	Disadvantages	Opportunities for improvement
	Acid hydrolysis	– High yields of sugars can be obtained – Even higher yields of sugars can be obtained when using biomasses that retain more water (low lignin content) – It is possible to retrieve neutralizing agents that precipitate, which can be recycled into the process or be used in fertilizers and materials	– Inhibitor formation (furans, aldehydes, and phenolic acids) – At high biomass loads, part of the material could remain without reacting, and the retrieval of the hydrolysate could be challenging – It is necessary to neutralize the hydrolysate to avoid sugar degradation – After hydrolysis, toxic compounds from hydrolysis will also be concentrated with evaporation – It is necessary to implement detoxification steps that increase costs and generate waste	– Use of other inorganic and organic acids that have been reported in the literature, such as phosphoric and acetic acid – Use other organic acids (citric, acetic, fumaric, and oxalic) that produce fewer furans, avoid equipment corrosion, and can be recovered from waste streams without consuming high amounts of energy – Integration of high-temperature streams into other parts of the process
	Enzymatic hydrolysis	– Produces fermentable sugars without producing undesired fermentation inhibitors – It operates at low temperatures (~50 °C) in comparison to thermal treatments and acid hydrolysis – Could allow working at high solid loadings (>15%) at special process conditions, which could increase sugar yields considerably – If enzymes can be recovered effectively, the process could be performed several times with the same hydrolyzing agent	– Requires lignin removal prior to the enzymatic treatment to improve digestibility. Without pretreatment, enzymes are not able to access effectively into the lignocellulosic substrate – The use of enzymes requires special process conditions (with customized equipment, adequate feeding and process control, and the incorporation of combined enzymatic cocktails) – It is necessary to separate and recycle enzymes, which is a complex process to carry out due to the presence of solids in the system – Requires less water than acid hydrolysis and thermal treatments	– Implement simultaneous saccharification and fermentation to reduce water and energy consumption – Reduce costs and material usage by recycling enzymes – Study the combination of enzymes from different microorganisms to formulate optimal enzymatic cocktails – Integrate energy from the sugar's concentration into other parts of the process

(continued)

Table 7.1 (continued)

	Alternatives	Advantages	Disadvantages	Opportunities for improvement
Fermentation	Batch	– It can be appropriately used for preliminary studies	– It is not preferred at an industrial scale since it has low productivity – Requires time-outs during cleaning and recharging	– Use of yeasts with high tolerance to inhibitors and ethanol – Study the use of co-cultures either simultaneously or by adding first a yeast that ferments glucose to ethanol and later a yeast that converts xylose into xylitol – Study cell recycling as a method to increase xylitol yields – Recycle excess water after fermentation. This possibility would require wastewater treatment units
	Fed-batch	– Preferred at industrial scale since it increases process productivity – Sugar yields are enhanced when glucose is fermented first and then xylose in a second bioreactor	– Requires time-outs during cleaning and recharging	
	Continuous	– Preferred at industrial scale since it increases process productivity – Increases volumetric productivity due to fewer time-outs during cleaning and recharging – More automation and elimination of the cleaning step	– It is necessary to separate ethanol before the xylitol production reactor, which could interfere with xylose fermentation	
Separation and purification	Crystallization	– Produces uniform crystals that facilitate its handling during the following purification steps	– Requires low temperatures (−15 to 0 °C), which implies high energy consumption – Requires the use of ethanol to produce prismatic crystals	– It is possible to reuse solvents, water streams and integrate residual energy – It is possible to explore the use of spray drying, which has been demonstrated to achieve high purity crystals
	Liquid–liquid extraction	– It is possible to recycle solvents – High purity xylitol can be obtained	– Requires the use of solvents like ethyl acetate in high amounts to perform the extraction – Xylitol losses during clarification and extraction could exist	
	Supercritical CO_2 extraction	– High purity xylitol (95–99%) can be obtained – Uses a green extraction process with CO_2, which can be recycled	– Low extraction efficiency	

Table 7.2 Chemical composition of corn stover, sugarcane bagasse, and sugarcane straw reported in the literature

Residue	Corn stover	Sugarcane bagasse	Sugarcane straw
Cellulose %	37–40	41–46	35–40
Hemicellulose %	24–26	25–31	26–29
Lignin %	7–19	19–23	19–23
Extractives %	3–5	1–9	6–8
Protein %	3–4	~1	~3
Ash %	3–7	1–3	2–8
Reference	Tao et al. (2013), Saini et al. (2015), Glińska et al. (2021), Mensah et al. (2021)	Szczerbowski et al. (2014)	Szczerbowski et al. (2014)

Percentages are shown on a dry weight basis; NR: Not reported

microorganisms, such as furans, organic acids, and phenolic compounds, depending on the process conditions, mainly temperature and time, furans can be formed (Raj and Krishnan 2020; Aguiar et al. 2021). In this alternative, the biorefinery would have the possibility of using the energy from the high-temperature residual streams and recycling water in other parts of the process. An interesting opportunity for research is the use of other inorganic and organic acids that have been reported in the literature, such as phosphoric and acetic acid (Bajpai 2022). The use of organic acids (e.g., citric, acetic, fumaric, oxalic) could be beneficial for a biorefinery since they produce less furans, avoid equipment corrosion, and can be easily recovered and recycled from waste streams without consuming high amounts of energy (Venkateswar Rao et al. 2016).

Furthermore, when using acid hydrolysis, it is necessary to neutralize the hydrolysate to avoid the degradation of fermentable sugars and reach the required pH for fermentation. This requirement makes it necessary to use substances such as calcium oxide, calcium carbonate, sodium hydroxide, and potassium hydroxide (Roberto et al. 1991; Dominguez et al. 1997). If these neutralizing agents can be precipitated, it is possible to recycle these substances into the process (de Beer et al. 2014, 2015), but also, it opens the possibility to elaborate additional products for the biorefinery by producing gypsum-based materials such as binders or plasterboards (Kamarou et al. 2021; Erbs et al. 2021). The target market for the gypsum application must consider plant location to reduce transport costs. Therefore, if there are crops that need fertilizers nearby or if there is an urban core that needs building materials, the market decision might change.

Regarding the use of enzymatic hydrolysis, it must be noted that this alternative allows for a different biorefinery configuration compared to the ones that could use the other hydrolysis methods. The use of enzymes produces sugar-rich hydrolysates without producing undesired fermentation inhibitors (Raj and Krishnan 2020), which means that there will not be valorization of furfural, phenols, and other byproducts.

Besides, the solid fraction from hydrolysis would have a cellulose-rich composition that could be further valorized into the biorefinery (Fig. 7.1) without the interference of other byproducts. This particular method would reduce the overall energy consumption requirements of the biorefinery since it is carried out at relatively low temperatures (~50 °C) in contrast to the other hydrolysis alternatives (Rafiqul and Sakinah 2013). Moreover, the use of enzymes at special process conditions (with customized equipment, adequate feeding and process control, and the incorporation of combined enzymatic cocktails) could allow working at high solid loadings (>15%), which could increase sugar yields considerably (Mussatto et al. 2021). It is important to mention that it is necessary to separate and recycle enzymes in this option, which is a complex process to carry out due to the presence of solids in the system (Rafiqul and Sakinah 2013). The high cost of enzymes makes it necessary to implement their recycling (Liguori and Faraco 2016). However, if enzymes can be recovered effectively, the process could be performed several times with the same hydrolyzing agent.

After hydrolysis, using any of the alternatives presented, an evaporation unit should be included to increase the concentration of sugars for fermentation. During concentration, some toxic compounds (derived from thermal treatments or acid hydrolysis) like furfural will increase their concentration, and others like 5-hydroxymethyl furfural will be selectively removed during vacuum evaporation (Mussatto and Roberto 2004). As a result, the order in which hydrolysates should be neutralized, detoxified, and concentrated is still a point of discussion that should be further studied. Moreover, the high temperature used for concentration and the retrieved water represent also opportunities for energy integration and water recycling.

Regarding the fermentation step, three operation modes can be used: batch, fed-batch, and continuous, as presented in Table 7.1. From these options, fed-batch and continuous configurations have been more widely implemented at the industrial level in other bioprocesses, and they may be more attractive for the biorefinery they allow to obtain high xylitol productivity (Vallejos and Area 2017; Baptista et al. 2021). Yeasts from the genus *Candida*, *Debariomyces*, and *Kluyveromyces* have been widely used for xylitol production since they are able to assimilate pentose sugars (Baptista et al. 2021). Particularly, yeasts from the genus *Kluyveromyces* are attractive for their application in biorefineries since they can produce both xylitol and ethanol by consuming C5 and C6 sugars. The co-production of ethanol and xylitol has been an attractive alternative due to the potential economic benefits that joint production could provide in a biorefinery scenario (Unrean and Ketsub 2018).

As a final step of the process, it is necessary to separate and purify xylitol. Crystallization has been the most common method to purify xylitol since it produces uniform crystals (Albuquerque et al. 2014). Crystallization requires the use of low temperatures (−15 to 0 °C) to cause the nucleation of the crystals with the help of solvents like ethanol to reduce the time of crystallization (Seppälä et al. 2010; Delgado Arcaño et al. 2020). Precipitated crystals are later separated by centrifugation and filtration and dried at 30 °C (De Faveri et al. 2002; Sampaio et al. 2006; Seppälä et al. 2010). Another method to separate xylitol is liquid–liquid extraction with solvents like ethyl acetate (Mussatto et al. 2005). Liquid–liquid extraction has been used at an industrial level, but this separation process could result more expensive than crystallization due to the type and amount of solvents needed for the process (Misra et al. 2011). Furthermore, the use of supercritical fluid extraction with CO_2 has been reported to achieve high purity of xylitol (95–99%), using ethanol as co-solvent (Silva et al. 2020). The use of supercritical CO_2 is a novel approach which should be further explored since uses a green extraction process in which both CO_2 and ethanol can be recycled (Silva et al. 2020). Nonetheless, in the context of a biorefinery, the downstream separation of xylitol gives the opportunity to reuse solvents and integrate residual energy and water streams. An interesting option that should be further explored to purify xylitol is the use of spray drying, which has been reported in the literature and could also result on high quality crystals (Mardawati et al. 2020).

7.4 Insights on Alternative Valorization Pathways That Could Be Integrated with the Xylitol Biotechnological Process Under the Biorefinery Concept

Sugarcane and corn residues result from the production of starch, syrups, and other food and feed derivatives. These feedstocks have been strongly linked to the production of alcoholic beverages and both first- and second-generation ethanol (Cortes-Rodríguez et al. 2018; Dionísio et al. 2021). The literature is abundantly documented on several examples of ethanol-producing biorefineries using yeast's well-known fermentation process (Dias et al. 2009; Li et al. 2020; Aguiar et al. 2021; Ntimbani et al. 2021). In this case, the primary production process (foods and feeds) is frequently complemented by producing ethanol from the sugar present in juices or hydrolysates, followed by the valorization of the remnant solid residues by hydrolyzing remnant cellulose or by producing energy (through incineration, gasification, or anaerobic digestion).

Nonetheless, since the biotechnological xylitol production process has as a central step the production of hemicellulosic hydrolysates from lignocellulosic biomass (Hernández-Pérez et al. 2019), the integration of this process into different configurations of biorefineries requires having the C5 sugar platform at its core. Examples of biorefineries that produce xylitol and other value-added products from sugarcane bagasse are listed in Table 7.3. To the best of our knowledge, examples of biorefineries from corn stover that specifically combine xylitol production and other byproducts have not been reported in the literature yet.

In order to maximize valorization of lignocellulosic biomass, other process alternatives can be implemented besides xylitol production to obtain other valuable products through: (i) extracting other substances from the feedstock before the hydrolysis step without affecting the hemicellulose content, (ii) diversifying the use of the C5 and C6 hydrolysates to ferment other substances apart from xylitol, (iii) using the remnant solid obtained after hydrolysis to obtain other products or energy, and (iv)

Table 7.3 Examples of biorefineries reported in the literature that produce xylitol and other value-added products from sugarcane bagasse and straw

1st step	2nd step	3rd step	4th step	5th step	Reference
Production of fermentable sugars through acid hydrolysis of sugarcane bagasse	Fermentation of sugar-rich liquid hydrolysate into xylitol with *C. tropicalis*	Simultaneous saccharification and fermentation to ethanol of the solid residue from hydrolysis with *S. cerevisiae* and commercial enzymes	–	–	Unrean and Ketsub (2018)
Production of xylose and furfural through acid hydrolysis of sugarcane bagasse	Enzymatic hydrolysis of the solid residue from hydrolysis to produce glucose	Fermentation of the glucose-rich hydrolysate into ethanol with *S. cerevisiae*	–	–	Mesa et al. (2014)
Production of fermentable sugars through acid and steam hydrolysis of sugarcane bagasse	Fermentation of sugar-rich liquid hydrolysate into xylitol with *Kluyveromyces marxianus*	Enzymatic hydrolysis of the solid residue from hydrolysis to produce glucose	Fermentation of the glucose-rich hydrolysate into ethanol by recycling cells of *Kluyveromyces marxianus*, previously used for xylitol production	–	Dasgupta et al. (2017)
Production of fermentable sugars through acid hydrolysis of sugarcane bagasse	Fermentation of sugars into xylitol by *Williopsis Saturnus*	Fermentation of the solid residue from hydrolysis to produce xylanases and fungal biomass with *Aspergillus terreus*	Extraction of cell oil from fungal biomass	Production of biodiesel using microbial lipids	Kamat et al. (2013)
Production of sugar cane juice and fermentation of juice into ethanol with *S. cerevisiae*	Production of fermentable sugars through acid hydrolysis of sugarcane bagasse and straw	Fermentation of C5 sugars into ABE using bacteria from the genus *Clostridium*	Fermentation of C6 sugars into ethanol with *S. cerevisiae*	Cogeneration of steam and electricity using remnant bagasse, straw, and solid residue from hydrolysis	Pereira et al. (2015)

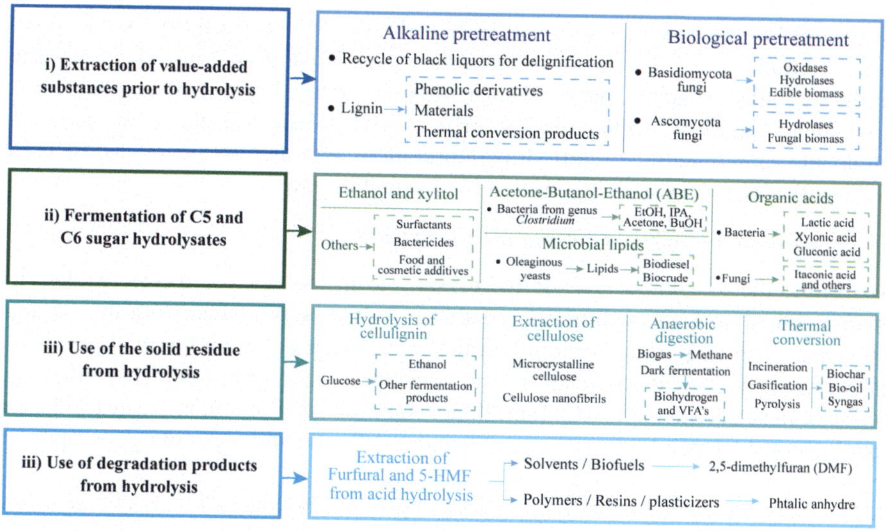

Fig. 7.2 Alternatives for biomass valorization under the biorefinery concept and considering xylitol production as the main process

using of degradation products from hydrolysis. These options are summarized in Fig. 7.2 and discussed next.

(i) Extraction of value-added substances prior to hydrolysis

Considering that the lignin content in sugarcane and corn residues could interfere with hydrolysis, it is possible to implement an alkaline pretreatment process to remove this resistant polymer and use it as another value-added product. The recycle (three to five reuse cycles) of sodium hydroxide and hydrogen peroxide in the process as black liquors to delignify sugarcane bagasse and corn stover has been previously reported as an alternative which also reduces water consumption considerably (Rocha et al. 2014; Alencar et al. 2017). The separated lignin can be used as an end-product itself, or it can be used to obtain phenolic derivatives (e.g., vanillin) or materials (e.g., composites). One example is the synthesis of formaldehyde resins from sugarcane bagasse, as reported by (Raj et al. 2020). These resins can be used as replacements for oil-derived phenolic resins to produce adhesives. Alternatively, it can be thermally converted into energy, syngas, or other chemicals and fuels.

On the other hand, lignin can be coupled in a C5 biorefinery using lignin-degrading fungus. Instead of the aforementioned alkaline pretreatment, a biological pretreatment with white-rot fungi (e.g., fungi from the genus *Ceriporiopsis*, *Pleurotus*, and *Ganoderma*) can be implemented. This route takes advantage of the lignocellulolytic enzymes secreted by these fungi as an added-value product (Cui et al. 2012; Manavalan et al. 2012; Dong et al. 2013). Corn stover and sugarcane bagasse have been used widely as substrates for white-rot fungi, increasing the accessibility to the lignocellulosic matrix. In addition, edible fungal protein-rich biomass can also be obtained in the case of *Pleurotus* (El-Sayed et al. 1994). Similarly, Ascomycota fungi from the genus *Aspergillus*, *Penicillium*, and *Trichoderma* have been used to produce cellulases and xylanases when grown in sugarcane bagasse (Rana et al. 2014; Boonyuen et al. 2014; Valladares-Diestra et al. 2021). However, these fungi are not edible and cannot degrade lignin. It is important to note that fungi can also be used in the solid fraction obtained after hydrolysis (cellulignin).

(ii) Fermentation of C5 and C6 sugar hydrolysates

Apart from the well-known use of hydrolysates from sugarcane and corn residues to produce bioethanol and xylitol, other substances can be produced from the C5 and C6 sugars. It is important to note that the same methodologies for the hydrolysis of the residues, their detoxification, and concentration can be used. The following options can be coupled with a C5 biorefinery since different products can be produced simultaneously depending on market demand and the specific needs of a geographical region.

One option is to use these hydrolysates to produce organic acids. Lactic acid has been obtained from sugarcane bagasse and corn stover hydrolysates with bacteria such as *Lactobacillus casei* and *Bacillus coagulans* (Sakdaronnarong et al. 2014; Ahring et al. 2016). Lactic acid can be further transformed into polylactic acid, a versatile, environmentally friendly polymer that has been incorporated into products as a replacement for non-renewable plastics (Sakdaronnarong et al. 2014). Besides lactic acid, other organic acids such as xylonic acid and gluconic acid have also been explored using microorganisms such as *Gluconobacter oxydans* and *Paraburkholderia sacchari* (Zhang et al. 2016; Bondar et al. 2021). Both substances can be used in the food industry as refreshing flavor additives in wines and juices (Bondar et al. 2021). These organic acids have been used as additives in alkaline conditions to increase cement resistance, can be used as green solvents, and are precursors of other chemical substances like polyamides (Bondar et al. 2021). Similarly, itaconic acid (which can be used in detergent builders, surfactants, and polyester resins) has been reported to be produced from sugarcane bagasse from the fermentation with the Ascomycota fungi *Aspergillus terreus* (Nieder-Heitmann et al. 2018).

Another alternative is the use of hydrolysates to produce Acetone-Butanol-Ethanol (ABE). Bacteria from the genus *Clostridium* such as *Clostridium acetobutylicum* and *Clostridium saccharoperbutylacetonicum* have been used for that purpose due to their ability to convert both C5 and C6 sugars into ABE (da Conceição Gomes et al. 2019). The use of these bacteria is an interesting alternative since they are able to ferment different types of sugars and their mixtures (Zetty-Arenas et al. 2021). Hydrolysates can be obtained using conditions similar to the ones widely studied for xylitol production. These bacteria convert the sugars present in the media into butyric and acetic acid, which are later converted into solvents like ABE and, in some cases, isopropanol (da Conceição Gomes et al. 2019). For example, using sugarcane bagasse and the bacteria *Clostridium acetobutylicum* DSM 6228, a final concentration of 9.1, 4.5, and 0.6 g/L of butanol, acetone, and ethanol were obtained, respectively (da

Conceição Gomes et al. 2019). In another study, where *Clostridium saccharoperbutylacetonicum* was used to ferment sugarcane hydrolysates mixed with sugarcane molasses, the final concentrations of butanol and ABE were 7.8 and 9.8 g/L (Zetty-Arenas et al. 2021). Higher yields (169 L ABE/ton sugarcane straw) and productivities (0.14 g ABE/h*L) were obtained using sugarcane straw hydrolysates during their simultaneous saccharification and fermentation in comparison to separated hydrolysis and fermentation (Pratto et al. 2020). Moreover, corn stover hydrolysates have been reported as substrates for ABE with yields of around 40–65 g ABE/kg of fresh corn stover (Zhang et al. 2019).

Oleaginous yeasts, which can metabolize xylose and other pentoses, can also be used to produce lipids. These microbial lipids are of great importance since they are susceptible to being transformed into biodiesel and other fuels through the esterification of the fatty acids. In sugarcane bagasse, the engineered oleaginous yeast *Rhodotorula mucilaginosa* IIPL32 has been reported to use C5 sugars to produce microbial lipids (Bandhu et al. 2019). Also, the mutant yeast *Trichosporon dermatis* and the fungus *Mucor circinelloides* have been reported to convert both C5 and C6 sugars into lipids (Carvalho et al. 2019; Sun et al. 2021). Lipids produced from *Lipomyces starkeyi* using the C5 fraction from corn stover hydrolysates have been used to produce a hydrocarbon blend (biocrude) by submitting the lipid-rich cell yeasts to hydrothermal liquefaction, which was proposed as an alternative to reduce the cost of extracting lipids (Collett et al. 2019). In another study, hydrolysates from corn stover were converted into lipids (13.8 g/L of cell biomass with 51% of lipids) by *Rhodosporidium toruloides* under a biorefinery concept where the cellulose-rich residue from hydrolysis was used as carbon nanofibers for supercapacitors (Wang et al. 2020). Other yeasts from the genus *Cryptococcus* and *Mortierella* have also been reported to convert hydrolysates from corn stover (Fang et al. 2016; Gong et al. 2016; Xu et al. 2019). Once again, the use of these specialized yeasts in sugarcane and corn stover residues to produce lipids has shown to use similar unit operations in the processes and close enough operating conditions to those required for xylitol production. Specifically, the unit operations of biomass pretreatment, hydrolysis (acid or enzymatic), detoxification, and fermentation (Xu et al. 2019; Wang et al. 2020; Sun et al. 2021) are the same. One main difference for lipids production is the necessity of downstream units for oil recovery from cells, which may include the use of organic solvents.

Other applications that have been reported in the literature to valorize hydrolysates from sugarcane bagasse and straw include the production of surfactants, pigments, food additives, and bactericides. Hydrolysates from sugarcane bagasse and straw have been used to produce rhamnolipids with a high emulsifier index, using various yeast like *Cutaneotrichosporum mucoides* and bacteria such as *Pseudomonas aeruginosa* (Lopes et al. 2017; Marcelino et al. 2019; Jiménez et al. 2020). Moreover, red pigments with application in the food industry have also been obtained from sugarcane hydrolysate using the glucose-consuming filamentous fungi *Monascus ruber* (Terán Hilares et al. 2018). In another study, *Lasidiplodia theobromae* CCT 3966 was grown on sugarcane straw hydrolysate. These fungi produced β-glucan, a fiber that can be used in cosmetics and with several health benefits as a food additive

(Abdeshahian et al. 2020). Finally, pediocin PA-1 (a bacteriocin) was obtained from sugarcane bagasse hydrolysate using the bacteria *Pediococcus pentosaceous,* which can eradicate food pathogens such as *Listeria monocytogenes* (Kuniyoshi et al. 2021). Bactericides obtained this way could be a sustainable alternative to replace oil-derived germicides.

(iii) Use of the solid residue from hydrolysis

The solid fraction that results from hydrolysis is still rich in cellulose and lignin. One option to valorize this stream is to extract the remnant cellulose to produce microcrystalline cellulose (Katakojwala and Mohan 2020) or cellulose nanoscale materials like cellulose nanofibrils (Ehman et al. 2020). On the other hand, this cellulose-rich residue from hydrolysis can be again hydrolyzed to obtain C6 sugars, a substrate for ethanol (and other added-value products) via fermentation. One interesting option is the fermentation of cellulose with the microorganism *Aureobasidium pullulans.* This fungus usually converts starch into the biopolymer pullulan; however, a recent study showed how sugarcane bagasse hydrolysates, rich in cellulose, can be used as a substrate for this microorganism (Terán Hilares et al. 2019).

Another possibility is to obtain energy from the remnant fraction from hydrolysis. Biogas production has been considered a core processing step in sugarcane biorefineries since it has low energy requirements and can produce energy that can be used in several units like hydrolysis, sugars concentration, and ethanol fermentation (Fuess et al. 2021). The mesophilic co-digestion of corn stover with dairy manure resulted in yields of biogas of ~500 mL biogas/g TS (~65% of methane content), which could be attributed to the cellulose content of corn stover, which is rapidly degraded and has been demonstrated to be selective at high retention times (Yue et al. 2013). In another example, the digestion of corn stover using an eluent from a mesophilic anaerobic digester resulted in methane yields around ~190–220 mL CH_4/g VS (Wang et al. 2019). It is important to note that it is possible to use either the hydrolysates or the remnant solid residue from hydrolysis since most anaerobic bacteria can hydrolyze cellulignin by themselves. Also, the digestate from anaerobic processes can be recovered and used as fertilizer.

Besides that, it is possible to obtain other products in anaerobic conditions. Dark fermentation, which occurs in conditions that inhibit methanogenic bacteria, such as solid loadings where volatile fatty acids are produced in a relatively short time (<3 days) (Murali et al. 2017), results in the production of hydrogen. Sugarcane bagasse and corn stover have been used to produce hydrogen (~230 mL H_2/g Carbohydrate) (de Sá et al. 2020; Rodríguez-Valderrama et al. 2020). Several volatile fatty acids, like acetic, propionic, butyric, and valeric acids, have been obtained from the dark fermentation of corn stover and sugarcane bagasse (Murali et al. 2017). These acids are important molecules with a high market value that can be transformed into multiple chemical substances and biofuels. The production of volatile fatty acids can also be done using rumen fluid or manure as a source of inoculum.

It is important to consider that several sugarcane and corn processing mills use part of the bagasse and other residues as sources of energy through their direct combustion or incineration, which can feed heat and power units (Nieder-Heitmann et al.

2018). Moreover, it is possible to enhance the products and the amount of energy obtained from this alternative by using gasification. A gas fraction (59.6 mol %) rich in hydrogen and methane was obtained with H_2/CO ratios of 1.1–1.2 using residues from the hydrolysis of corn stover (Howe et al. 2017). The tar formed (11.8 mol %) contained polyaromatic hydrocarbons, phenol, and benzene, and the remnant fraction was constituted of char (Howe et al. 2017). Similarly, the production of bio-oil and carbon microspheres was reported from sugarcane bagasse through pyrolysis of hydrolysis residues for one hour at 400 and 500 °C in a nitrogen atmosphere (Sakdaronnarong et al. 2018). Higher oil yields were obtained at higher temperatures, and conversely, higher carbon yields were obtained at lower temperatures. Nonetheless, the use of pyrolysis requires the addition of external energy sources since the decomposition reactions at these conditions are endothermic, contrary to gasification.

(iv) Use of degradation products from hydrolysis

Furans, which result from the oxidation of glucose and xylose, are also molecules with added value that can be recovered. Furfural can be used as a solvent, and it is a precursor of sustainable biofuels and chemicals (Teng et al. 2020). In a study, xylose from corn stover hydrolysates was converted into furfural and later into phthalic anhydride used in resins and plasticizers (Giarola et al. 2016).

7.5 Conclusions

Xylitol bioproduction has a great potential to be integrated into biorefineries as a high-added-value alternative for the hemicellulosic fraction of lignocellulosic biomasses rich in C5 carbohydrate, particularly xylose. Integral utilization of the structural fractions of biomasses allows obtaining several products in the same production plant, which represent a possibility to meet different market segments, with diverse value and with the ability to adapt to the changes on demands from the market. At the same time, an integrated biorefinery can take advantage of similar biomass compositions to diversify the raw materials that can be used, which may favor logistic costs of biomass collection and transportation, as well to overcome seasonal production. Successful integration of this bioprocess with other processes depends on utilizing, reusing, and/or recycling streams, such as cold water, heat, or vapor, aiming to reduce the energy and water requirements that increase the bioprocess's operational cost and reduce its sustainability.

Some aspects of the bioprocess affecting economic feasibility and environmental sustainability should be also considered, such as: (a) pretreatment selection and performance, since this step has high energy requirements and environmental challenges; (b) elimination or intensification of detoxification procedures, whose implementation at large scale can be difficult and costly; (c) use of fed-batch or continuous operation for fermentation to increase productivity, to overcome problems related to the presence glucose and potential toxic compounds in the hydrolysate, and to avoid the frequent pauses of batch processes, as well to evaluate strategies of co-fermentation to produce concomitantly ethanol; (d) evaluation of alternative technologies to xylitol purification besides crystallization, such as spray drying and supercritical CO_2, which can bring potential gains in performance and sustainability; (e) strategies to use various byproducts generated during some stages of the bioprocess that can be further processed to become co-products, allowing to introduce more products to the biorefinery which can contribute with the economic feasibility and also to favor the sustainability by reducing residues.

References

Abdeshahian P, Ascencio JJ, Philippini RR et al (2020) Utilization of sugarcane straw for production of β-glucan biopolymer by Lasiodiplodia theobromae CCT 3966 in batch fermentation process. Bioresour Technol 314:123716. https://doi.org/10.1016/j.biortech.2020.123716

Aguiar A, Milessi TS, Mulinari DR et al (2021) Sugarcane straw as a potential second generation feedstock for biorefinery and white biotechnology applications. Biomass Bioenergy 144:105896. https://doi.org/10.1016/j.biombioe.2020.105896

Ahring BK, Traverso JJ, Murali N, Srinivas K (2016) Continuous fermentation of clarified corn stover hydrolysate for the production of lactic acid at high yield and productivity. Biochem Eng J 109:162–169. https://doi.org/10.1016/j.bej.2016.01.012

Ahuja V, Macho M, Ewe D et al (2020) Biological and pharmacological potential of xylitol: a molecular insight of unique metabolism. Foods 9:1592. https://doi.org/10.3390/foods9111592

Albuquerque TL de, da Silva IJ, de Macedo GR, Rocha MVP (2014) Biotechnological production of xylitol from lignocellulosic wastes: a review. Process Biochem 49:1779–1789. https://doi.org/10.1016/j.procbio.2014.07.010

Alencar BRA, Reis ALS, de Souza R de FR et al (2017) Recycling the liquid fraction of alkaline hydrogen peroxide in the pretreatment of corn stover. Bioresour Technol 241:928–935. https://doi.org/10.1016/j.biortech.2017.06.022

Bajpai P (2022) Overview of key pretreatment methods. In: Lignocellulosic biomass in biotechnology. Elsevier, pp 97–130

Bandhu S, Bansal N, Dasgupta D et al (2019) Overproduction of single cell oil from xylose rich sugarcane bagasse hydrolysate by an engineered oleaginous yeast Rhodotorula mucilaginosa IIPL32. Fuel 254:115653. https://doi.org/10.1016/j.fuel.2019.115653

Baptista SL, Romaní A, Domingues L (2021) Biotechnological advancements, innovations and challenges for sustainable xylitol production by yeast. In: Encyclopedia of mycology. Elsevier, pp 420–427

Bondar M, da Fonseca MMR, Cesário MT (2021) Xylonic acid production from xylose by Paraburkholderia sacchari. Biochem Eng J 170:107982. https://doi.org/10.1016/j.bej.2021.107982

Boonyuen N, Manoch L, Luangsa-ard JJ et al (2014) Decomposition of sugarcane bagasse with lignocellulose-derived thermotolerant and thermoresistant Penicillia and Aspergilli. Int Biodeterior Biodegrad 92:86–100. https://doi.org/10.1016/j.ibiod.2014.04.013

Carvalho AKF, Bento HBS, Reis CER, De Castro HF (2019) Sustainable enzymatic approaches in a fungal lipid biorefinery based in sugarcane bagasse hydrolysate as carbon source. Bioresour Technol 276:269–275. https://doi.org/10.1016/j.biortech.2018.12.118

Chandel AK, Antunes FAF, de Arruda PV et al (2012) Dilute acid hydrolysis of agro-residues for the depolymerization of hemicellulose: state-of-the-art. In: da Silva SS, Chandel AK (eds) D-Xylitol. Springer-Verlag, Berlin Heidelberg, pp 39–61

Chui M, Evers M, Manyika J et al (2020) The bio revolution. Innovations transforming economies, societies, and our lives

Collett JR, Billing JM, Meyer PA et al (2019) Renewable diesel via hydrothermal liquefaction of oleaginous yeast and residual lignin from bioconversion of corn stover. Appl Energy 233–234:840–853. https://doi.org/10.1016/j.apenergy.2018.09.115

Cortes-Rodríguez EF, Fukushima NA, Palacios-Bereche R et al (2018) Vinasse concentration and juice evaporation system integrated to the conventional ethanol production process from sugarcane – heat integration and impacts in cogeneration system. Renew Energy 115:474–488. https://doi.org/10.1016/j.renene.2017.08.036

Creatus Biosciences (2021) The right platform for bioprocess development. https://creatusbio.com

Cui Z, Shi J, Wan C, Li Y (2012) Comparison of alkaline- and fungi-assisted wet-storage of corn stover. Bioresour Technol 109:98–104. https://doi.org/10.1016/j.biortech.2012.01.037

da Conceição Gomes A, Rodrigues MI, de França Passos D et al (2019) Acetone–butanol–ethanol fermentation from sugarcane bagasse hydrolysates: utilization of C5 and C6 sugars. Electron J Biotechnol 42:16–22. https://doi.org/10.1016/j.ejbt.2019.10.004

Dasgupta D, Ghosh D, Bandhu S, Adhikari DK (2017) Lignocellulosic sugar management for xylitol and ethanol fermentation with multiple cell recycling by Kluyveromyces marxianus IIPE453. Microbiol Res 200:64–72. https://doi.org/10.1016/j.micres.2017.04.002

de Beer M, Doucet FJ, Maree JP, Liebenberg L (2015) Synthesis of high-purity precipitated calcium carbonate during the process of recovery of elemental sulphur from gypsum waste. Waste Manag 46:619–627. https://doi.org/10.1016/j.wasman.2015.08.023

de Beer M, Maree JP, Liebenberg L, Doucet FJ (2014) Conversion of calcium sulphide to calcium carbonate during the process of recovery of elemental sulphur from gypsum waste. Waste Manag 34:2373–2381. https://doi.org/10.1016/j.wasman.2014.07.010

De Faveri D, Perego P, Converti A, Del Borghi M (2002) Xylitol recovery by crystallization from synthetic solutions and fermented hemicellulose hydrolyzates. Chem Eng J 90:291–298. https://doi.org/10.1016/S1385-8947(02)00134-1

de Sá LRV, Faber M de O, da Silva AS et al (2020) Biohydrogen production using xylose or xylooligosaccharides derived from sugarcane bagasse obtained by hydrothermal and acid pretreatments. Renew Energy 146:2408–2415. https://doi.org/10.1016/j.renene.2019.08.089

Defretin S, Gerard T, Heysen A et al (2018) Production of xylitol from glucose by a recombinant strain

Delgado Arcaño Y, Valmaña García OD, Mandelli D et al (2020) Xylitol: a review on the progress and challenges of its production by chemical route. Catal Today 344:2–14. https://doi.org/10.1016/j.cattod.2018.07.060

Dias MOS, Ensinas AV, Nebra SA et al (2009) Production of bioethanol and other bio-based materials from sugarcane bagasse: integration to conventional bioethanol production process. Chem Eng Res Des 87:1206–1216. https://doi.org/10.1016/j.cherd.2009.06.020

Dionísio SR, Santoro DCJ, Bonan CIDG et al (2021) Second-generation ethanol process for integral use of hemicellulosic and cellulosic hydrolysates from diluted sulfuric acid pretreatment of sugarcane bagasse. Fuel 304:121290. https://doi.org/10.1016/j.fuel.2021.121290

Dominguez JM, Cao N, Gong CS, Tsao GT (1997) Dilute acid hemicellulose hydrolysates from corn cobs for xylitol production by yeast. Bioresour Technol 61:85–90. https://doi.org/10.1016/S0960-8524(97)84703-7

Dong XQ, Yang JS, Zhu N et al (2013) Sugarcane bagasse degradation and characterization of three white-rot fungi. Bioresour Technol 131:443–451. https://doi.org/10.1016/j.biortech.2012.12.182

DuPont Nutrition & health (2012) Xivia xylitol white paper. https://www.dupontnutritionandbiosciences.com/content/dam/dupont/amer/us/en/products/nutrition-health/documents/2e_XIVIA_White_Paper.pdf

Ehman NV, Lourenço AF, McDonagh BH et al (2020) Influence of initial chemical composition and characteristics of pulps on the production and properties of lignocellulosic nanofibers. Int J Biol Macromol 143:453–461. https://doi.org/10.1016/j.ijbiomac.2019.10.165

El-Sayed SA, Zaki MT, Abou El-Khair AW (1994) Bioconversion of sugarcane bagasse into a protein-rich product by white rot fungus. Resour Conserv Recycl 12:195–200. https://doi.org/10.1016/0921-3449(94)90007-8

Ellen MacArthur Foundation (2020) Unlocking the value of the circular economy. https://ellenmacarthurfoundation.org/articles/unlocking-the-value-of-the-circular-economy

Erbs A, Nagalli A, Carvalho KQ de et al (2021) Development of plasterboard sheets exclusively from waste. J Build Eng 44:102524. https://doi.org/10.1016/j.jobe.2021.102524

Fang H, Zhao C, Chen S (2016) Single cell oil production by Mortierella isabellina from steam exploded corn stover degraded by three-stage enzymatic hydrolysis in the context of on-site enzyme production. Bioresour Technol 216:988–995. https://doi.org/10.1016/j.biortech.2016.06.051

Fuess LT, Cruz RBCM, Zaiat M, Nascimento CAO (2021) Diversifying the portfolio of sugarcane biorefineries: anaerobic digestion as the core process for enhanced resource recovery. Renew Sustain Energy Rev 147:111246. https://doi.org/10.1016/j.rser.2021.111246

Giarola S, Romain C, Williams CK et al (2016) Techno-economic assessment of the production of phthalic anhydride from corn stover. Chem Eng Res Des 107:181–194. https://doi.org/10.1016/j.cherd.2015.10.034

Glińska K, Gitalt J, Torrens E et al (2021) Extraction of cellulose from corn stover using designed ionic liquids with improved reusing capabilities. Process Saf Environ Prot 147:181–191. https://doi.org/10.1016/j.psep.2020.09.035

Gong Z, Zhou W, Shen H et al (2016) Co-utilization of corn stover hydrolysates and biodiesel-derived glycerol by Cryptococcus curvatus for lipid production. Bioresour Technol 219:552–558. https://doi.org/10.1016/j.biortech.2016.08.021

Hernández-Pérez AF, de Arruda PV, Sene L et al (2019) Xylitol bioproduction: state-of-the-art, industrial paradigm shift, and opportunities for integrated biorefineries. Crit Rev Biotechnol 39:924–943. https://doi.org/10.1080/07388551.2019.1640658

Hill P, Benjamin K, Bhattacharjee B et al (2020) Clean manufacturing powered by biology: how Amyris has deployed technology and aims to do it better. J Ind Microbiol Biotechnol 47:965–975. https://doi.org/10.1007/s10295-020-02314-3

Howe DT, Taasevigen D, Garcia-Perez M et al (2017) Steam gasification of a thermally pretreated high lignin corn stover simultaneous saccharification and fermentation digester residue. Energy 119:400–407. https://doi.org/10.1016/j.energy.2016.12.094

IEA Bioenergy (2020) Bio-based chemicals: a 2020 update. IEA Bioenergy

IFF Nutrition & Biosciences (2021) XIVIA xylitol. https://www.dupontnutritionandbiosciences.com/product-range/sweeteners/xivia.html

Jansen R, Lawson JA, Carden AT et al (2021) Method for producing xylitol by fermentation

Jiménez IM, Chandel AK, Marcelino PRF et al (2020) Comparative data on effects of alkaline pretreatments and enzymatic hydrolysis on bioemulsifier production from sugarcane straw by Cutaneotrichosporon mucoides. Bioresour Technol 301:122706. https://doi.org/10.1016/j.biortech.2019.122706

Kamarou M, Korob N, Kwapinski W, Romanovski V (2021) High-quality gypsum binders based on synthetic calcium sulfate dihydrate produced from industrial waste. J Ind Eng Chem 100:324–332. https://doi.org/10.1016/j.jiec.2021.05.006

Kamat S, Khot M, Zinjarde S et al (2013) Coupled production of single cell oil as biodiesel feedstock, xylitol and xylanase from sugarcane bagasse in a biorefinery concept using fungi from the tropical mangrove wetlands. Bioresour Technol 135:246–253. https://doi.org/10.1016/j.biortech.2012.11.059

Katakojwala R, Mohan SV (2020) Microcrystalline cellulose production from sugarcane bagasse: sustainable process development and life cycle assessment. J Clean Prod 249:119342. https://doi.org/10.1016/j.jclepro.2019.119342

Kuniyoshi TM, Mendonça CMN, Vieira VB et al (2021) Pediocin PA-1 production by Pediococcus pentosaceus ET34 using non-detoxified hemicellulose hydrolysate obtained from hydrothermal

pretreatment of sugarcane bagasse. Bioresour Technol 338:125565. https://doi.org/10.1016/j.biortech.2021.125565
L'Óreal (2020) Lanzatech, Total and L'Oréal announce the production of the first cosmetic plastic bottle. https://www.loreal.com/en/press-release/sustainable-development/lanzatech-total-and-loral-announce-a-worldwide-premiere-the-production-of-the-first-cosmetic/
Li W-C, Zhang S-J, Xu T et al (2020) Fractionation of corn stover by two-step pretreatment for production of ethanol, furfural, and lignin. Energy 195:117076. https://doi.org/10.1016/j.energy.2020.117076
Liguori R, Faraco V (2016) Biological processes for advancing lignocellulosic waste biorefinery by advocating circular economy. Bioresour Technol 215:13–20. https://doi.org/10.1016/j.biortech.2016.04.054
Lopes V dos S, Fischer J, Pinheiro TMA et al (2017) Biosurfactant and ethanol co-production using Pseudomonas aeruginosa and Saccharomyces cerevisiae co-cultures and exploded sugarcane bagasse. Renew Energy 109:305–310. https://doi.org/10.1016/j.renene.2017.03.047
Luo Z, Jansen van Vuuren HJ, DeBono AG, Ferguson AT (2019) Xylitol producing metschnikowia species
Manavalan T, Manavalan A, Thangavelu KP, Heese K (2012) Secretome analysis of Ganoderma lucidum cultivated in sugarcane bagasse. J Proteomics 77:298–309. https://doi.org/10.1016/j.jprot.2012.09.004
Marcelino PRF, Peres GFD, Terán-Hilares R et al (2019) Biosurfactants production by yeasts using sugarcane bagasse hemicellulosic hydrolysate as new sustainable alternative for lignocellulosic biorefineries. Ind Crops Prod 129:212–223. https://doi.org/10.1016/j.indcrop.2018.12.001
Mardawati E, Rialita T, Suryadi E et al (2020) The evaluation of spray drying process condition on the characteristics of xylitol powder from oil palm empty fruit bunches. Ind J Teknol Manaj Agroind 9:17–24. https://doi.org/10.21776/ub.industria.2020.009.01.3
Mensah MB, Jumpah H, Boadi NO, Awudza JAM (2021) Assessment of quantities and composition of corn stover in Ghana and their conversion into bioethanol. Sci Afr 12:e00731. https://doi.org/10.1016/j.sciaf.2021.e00731
Mesa L, Morales M, González E et al (2014) Restructuring the processes for furfural and xylose production from sugarcane bagasse in a biorefinery concept for ethanol production. Chem Eng Process: Process Intensif 85:196–202. https://doi.org/10.1016/j.cep.2014.07.012
Misra S, Gupta P, Raghuwanshi S et al (2011) Comparative study on different strategies involved for xylitol purification from culture media fermented by Candida tropicalis. Sep Purif Technol 78:266–273. https://doi.org/10.1016/j.seppur.2011.02.018
Mountraki AD, Koutsospyros KR, Mlayah BB, Kokossis AC (2017) Selection of biorefinery routes: the case of xylitol and its integration with an organosolv process. Waste Biomass Valoriz 8:2283–2300. https://doi.org/10.1007/s12649-016-9814-8
Murali N, Fernandez S, Ahring BK (2017) Fermentation of wet-exploded corn stover for the production of volatile fatty acids. Bioresour Technol 227:197–204. https://doi.org/10.1016/j.biortech.2016.12.012
Mussatto S, Roberto I (2004) Kinetic behavior of Candida guilliermondii yeast during xylitol production from highly concentrated hydrolysate. Process Biochem 39:1433–1439. https://doi.org/10.1016/S0032-9592(03)00261-9
Mussatto SI, Santos JC, Filho WCR, Silva SS (2005) Purification of xylitol from fermented hemicellulosic hydrolyzate using liquid-liquid extraction and precipitation techniques. Biotechnol Lett 27:1113–1115. https://doi.org/10.1007/s10529-005-8458-8
Mussatto SI, Yamakawa CK, van der Maas L, Dragone G (2021) New trends in bioprocesses for lignocellulosic biomass and CO2 utilization. Renew Sustain Energy Rev 152:111620. https://doi.org/10.1016/j.rser.2021.111620
Nieder-Heitmann M, Haigh KF, Görgens JF (2018) Process design and economic analysis of a biorefinery co-producing itaconic acid and electricity from sugarcane bagasse and trash lignocelluloses. Bioresour Technol 262:159–168. https://doi.org/10.1016/j.biortech.2018.04.075

Ntimbani RN, Farzad S, Görgens JF (2021) Techno-economic assessment of one-stage furfural and cellulosic ethanol co-production from sugarcane bagasse and harvest residues feedstock mixture. Ind Crops Prod 162:113272. https://doi.org/10.1016/j.indcrop.2021.113272

Pereira LG, Dias MOS, Mariano AP et al (2015) Economic and environmental assessment of n-butanol production in an integrated first and second generation sugarcane biorefinery: fermentative versus catalytic routes. Appl Energy 160:120–131. https://doi.org/10.1016/j.apenergy.2015.09.063

Philippini RR, Martiniano SE, Chandel AK et al (2019) Pretreatment of sugarcane bagasse from cane hybrids: effects on chemical composition and 2G sugars recovery. Waste Biomass Valoriz 10:1561–1570. https://doi.org/10.1007/s12649-017-0162-0

Pratto B, Chandgude V, de Sousa R et al (2020) Biobutanol production from sugarcane straw: defining optimal biomass loading for improved ABE fermentation. Ind Crops Prod 148:112265. https://doi.org/10.1016/j.indcrop.2020.112265

Rafiqul ISM, Sakinah AMM (2013) Processes for the production of xylitol—a review. Food Rev Int 29:127–156. https://doi.org/10.1080/87559129.2012.714434

Raj A, Devendra LP, Sukumaran RK (2020) Comparative evaluation of laccase mediated oxidized and unoxidized lignin of sugarcane bagasse for the synthesis of lignin-based formaldehyde resin. Ind Crops Prod 150:112385. https://doi.org/10.1016/j.indcrop.2020.112385

Raj K, Krishnan C (2020) Improved co-production of ethanol and xylitol from low-temperature aqueous ammonia pretreated sugarcane bagasse using two-stage high solids enzymatic hydrolysis and Candida tropicalis. Renew Energy 153:392–403. https://doi.org/10.1016/j.renene.2020.02.042

Rana V, Eckard AD, Teller P, Ahring BK (2014) On-site enzymes produced from Trichoderma reesei RUT-C30 and Aspergillus saccharolyticus for hydrolysis of wet exploded corn stover and loblolly pine. Bioresour Technol 154:282–289. https://doi.org/10.1016/j.biortech.2013.12.059

Ravella SR, Gallagher J, Fish S, Prakasham RS (2012) Overview on commercial production of xylitol, economic analysis and market trends. D-Xylitol. Springer, Berlin Heidelberg, Berlin, Heidelberg, pp 291–306

Roberto IC, Felipe MGA, Lacis LS et al (1991) Utilization of sugar cane bagasse hemicellulosic hydrolyzate by Candida guilliermondii for xylitol production. Bioresour Technol 36:271–275. https://doi.org/10.1016/0960-8524(91)90234-B

Rocha GJM, Nascimento VM, da Silva VFN et al (2014) Contributing to the environmental sustainability of the second generation ethanol production: delignification of sugarcane bagasse with sodium hydroxide recycling. Ind Crops Prod 59:63–68. https://doi.org/10.1016/j.indcrop.2014.05.002

Rodríguez-Valderrama S, Escamilla-Alvarado C, Magnin J-P et al (2020) Batch biohydrogen production from dilute acid hydrolyzates of fruits-and-vegetables wastes and corn stover as co-substrates. Biomass Bioenergy 140:105666. https://doi.org/10.1016/j.biombioe.2020.105666

Rueda C, Calvo PA, Moncalián G et al (2015) Biorefinery options to valorize the spent liquor from sulfite pulping. J Chem Technol Biotechnol 90:2218–2226. https://doi.org/10.1002/jctb.4536

S2G Biochem (2016) Crossing the chasm to commercialization strategic partnerships. In: 2016 BIO Latin America conference, Sao Paulo, Brazil. https://www.bio.org/sites/default/files/legacy/bio org/docs/27_11h15_CP_S2G.pdf

Saini JK, Saini R, Tewari L (2015) Lignocellulosic agriculture wastes as biomass feedstocks for second-generation bioethanol production: concepts and recent developments. 3 Biotech 5:337–353. https://doi.org/10.1007/s13205-014-0246-5

Sakdaronnarong C, Pipathworapoom W, Vichitsrikamol T et al (2018) Integrative process for a sugarcane bagasse biorefinery to produce glucose, bio-oil and carbon microspheres. Process Saf Environ Prot 116:1–13. https://doi.org/10.1016/j.psep.2018.01.006

Sakdaronnarong C, Srimarut N, Lucknakhul N et al (2014) Two-step acid and alkaline ethanolysis/alkaline peroxide fractionation of sugarcane bagasse and rice straw for production of polylactic acid precursor. Biochem Eng J 85:49–62. https://doi.org/10.1016/j.bej.2014.02.003

Sampaio FC, Passos FML, Passos FJV et al (2006) Xylitol crystallization from culture media fermented by yeasts. Chem Eng Process: Process Intensif 45:1041–1046. https://doi.org/10.1016/j.cep.2006.03.012

Seppälä A, Meriläinen A, Wikström L, Kauranen P (2010) The effect of additives on the speed of the crystallization front of xylitol with various degrees of supercooling. Exp Therm Fluid Sci 34:523–527. https://doi.org/10.1016/j.expthermflusci.2009.11.005

Silva DDV, Dussán KJ, Idarraga A et al (2020) Production and purification of xylitol by Scheffersomyces amazonenses via sugarcane hemicellulosic hydrolysate. Biofuels, Bioprod Biorefin 14:344–356. https://doi.org/10.1002/bbb.2085

Stora Enso (2015) Demonstration plant for bagasse to xylose underway. https://www.storaenso.com/en/newsroom/news/2015/10/demonstration-plant-for-bagasse-to-xylose-underway

Stora Enso (2016) Virdia bio-refinery project. In: Louisiana agricultural technology and management conference. https://www.laca1.org/Presentations/2016/Sugarcane/Virdia Bio-Refinery Project - Bob Jansen.pdf

Stora Enso (2021) Stora Enso will close its Virdia operations in the United States. https://www.storaenso.com/en/newsroom/regulatory-and-investor-releases/2021/1/stora-enso-will-close-its-virdiaoperations-in-the-united-states

Straathof AJJ, Wahl SA, Benjamin KR et al (2019) Grand research challenges for sustainable industrial biotechnology. Trends Biotechnol 37:1042–1050. https://doi.org/10.1016/j.tibtech.2019.04.002

Sun F, Gu Z, Zhou Q et al (2021) Highly efficient microbial lipid synthesis from co-fermentation of enzymatic hydrolysate of sugarcane bagasse by a Trichosporon dermatis mutant. Ind Crops Prod 171:113975. https://doi.org/10.1016/j.indcrop.2021.113975

Szczerbowski D, Pitarelo AP, Zandoná Filho A, Ramos LP (2014) Sugarcane biomass for biorefineries: comparative composition of carbohydrate and non-carbohydrate components of bagasse and straw. Carbohydr Polym 114:95–101. https://doi.org/10.1016/j.carbpol.2014.07.052

Tang Y, Yang B, Chen D et al (2018) Industrial Saccharomyces cerevisiae strain for high-yield production of xylitol and ethanol by co-fermenting xylose and glucose and construction method

Tao L, Templeton DW, Humbird D, Aden A (2013) Effect of corn stover compositional variability on minimum ethanol selling price (MESP). Bioresour Technol 140:426–430. https://doi.org/10.1016/j.biortech.2013.04.083

Teng X, Si Z, Li S et al (2020) Tin-loaded sulfonated rape pollen for efficient catalytic production of furfural from corn stover. Ind Crops Prod 151:112481. https://doi.org/10.1016/j.indcrop.2020.112481

Terán Hilares R, de Souza RA, Marcelino PF et al (2018) Sugarcane bagasse hydrolysate as a potential feedstock for red pigment production by Monascus ruber. Food Chem 245:786–791. https://doi.org/10.1016/j.foodchem.2017.11.111

Terán Hilares R, Resende J, Orsi CA et al (2019) Exopolysaccharide (pullulan) production from sugarcane bagasse hydrolysate aiming to favor the development of biorefineries. Int J Biol Macromol 127:169–177. https://doi.org/10.1016/j.ijbiomac.2019.01.038

Unrean P, Ketsub N (2018) Integrated lignocellulosic bioprocess for co-production of ethanol and xylitol from sugarcane bagasse. Ind Crops Prod 123:238–246. https://doi.org/10.1016/j.indcrop.2018.06.071

Valladares-Diestra KK, Porto de Souza Vandenberghe L, Soccol CR (2021) A biorefinery approach for enzymatic complex production for the synthesis of xylooligosaccharides from sugarcane bagasse. Bioresour Technol 333:125174. https://doi.org/10.1016/j.biortech.2021.125174

Vallejos ME, Area MC (2017) Xylitol as bioproduct from the agro and forest biorefinery. In: Food bioconversion. Elsevier, pp 411–432

Venkateswar Rao L, Goli JK, Gentela J, Koti S (2016) Bioconversion of lignocellulosic biomass to xylitol: an overview. Bioresour Technol 213:299–310. https://doi.org/10.1016/j.biortech.2016.04.092

Wang F, Xu F, Liu Z et al (2019) Effects of outdoor dry bale storage conditions on corn stover and the subsequent biogas production from anaerobic digestion. Renew Energy 134:276–283. https://doi.org/10.1016/j.renene.2018.10.093

Wang X, Xu Q, Cheng J et al (2020) Bio-refining corn stover into microbial lipid and advanced energy material using ionic liquid-based organic electrolyte. Ind Crops Prod 145:112137. https://doi.org/10.1016/j.indcrop.2020.112137

Xu H, Zhao N, Yao H et al (2019) Lipid production from corn stover by a cost-efficient system featuring ammonium carbonate-steam explosion and recirculating enzymatic hydrolysis. Biomass Bioenergy 120:387–395. https://doi.org/10.1016/j.biombioe.2018.11.020

Yue Z, Chen R, Yang F et al (2013) Effects of dairy manure and corn stover co-digestion on anaerobic microbes and corresponding digestion performance. Bioresour Technol 128:65–71. https://doi.org/10.1016/j.biortech.2012.10.115

Zetty-Arenas AM, Tovar LP, Alves RF et al (2021) Co-fermentation of sugarcane bagasse hydrolysate and molasses by Clostridium saccharoperbutylacetonicum: effect on sugar consumption and butanol production. Ind Crops Prod 167:113512. https://doi.org/10.1016/j.indcrop.2021.113512

Zhang C, Chen H, An N et al (2019) Impact of corn stover harvest time and cultivars on acetone-butanol-ethanol production. Ind Crops Prod 139:111500. https://doi.org/10.1016/j.indcrop.2019.111500

Zhang H, Liu G, Zhang J, Bao J (2016) Fermentative production of high titer gluconic and xylonic acids from corn stover feedstock by Gluconobacter oxydans and techno-economic analysis. Bioresour Technol 219:123–131. https://doi.org/10.1016/j.biortech.2016.07.068

ZuChem (2020) Sweet Appeal^{TM} Natural Products announces first commercial sale of xylitol from new-state-of-the-art fermentation facility. https://www.zuchem.com/news/

Chapter 8
Techno-Economic Analysis of Xylitol Production in Stand-Alone and Integrated Biorefineries

Sara Piedrahita-Rodríguez and Carlos Ariel Cardona Alzate

Abstract Xylitol is 5-carbon alcohol, which has multiple applications in industry, mainly pharmaceutical and food. Because of this, the demand for xylitol is currently on the rise. Research on the raw materials that can be processed to obtain this product has been based mainly on the stand-alone process. However, integrating this process into the design of biorefineries could be a sustainable alternative from the techno-economic perspective. This chapter aims to analyze xylitol production from its stand-alone process and compare it in technical and economic terms with integration in biorefineries. For this, initially, three potential raw materials and their viability will be analyzed to obtain the C5 sugar platform so that the best economic alternative is taken to the production of xylitol in both scenarios. The integrated process in biorefineries enables the integral use of the raw material from the other components present in it and reduces costs without sacrificing yields. Therefore, it is concluded that this new processing alternative can technologically complement xylitol, which creates a new world of possibilities that is worth analyzing from the pillars of sustainability.

Keywords Biorefinery · Biotechnological processes · C5 sugar · C6 sugar · Platform · Xylitol

8.1 Introduction

Xylitol is an alcohol obtained from 5-carbon sugars, used mainly in the food and pharmaceutical industry. Due to its properties, the world demand for xylitol remains stable and with a tendency to grow (Rafiqul and Sakinah 2013). The industrial production of xylitol is based mainly on the fermentation route from biomass C5 sugars. However, the catalytic route is also another alternative for xylitol production through the catalytic degradation of xylose.

S. Piedrahita-Rodríguez · C. A. Cardona Alzate (✉)
Departamento de Ingeniería Química, Instituto de Biotecnología y Agroindustria, Universidad Nacional de Colombia, Manizales,Caldas, Colombia
e-mail: ccardonaal@unal.edu.co

© The Author(s), under exclusive license to Springer Nature Switzerland AG 2022

M. das G. de Almeida Felipe and A. K. Chandel (eds.), *Current Advances in Biotechnological Production of Xylitol*,
https://doi.org/10.1007/978-3-031-04942-2_8

Lignocellulosic materials are convenient to obtain the necessary substrate for the production of xylitol. The hemicellulose fraction can be transformed to the C5 platform (xylose), and later converted into xylitol. However, the other fractions of these materials should be used to generate value-added products. Some studies have attempted to propose that the cellulose fraction brought to the C6 platform (glucose) can be used as a substrate with genetically modified microorganisms capable of fermenting both platforms towards the generation of xylitol (Cheng et al. 2014). This phenomenon would be carried out either by diauxic growth or by fermenting the two sugars platforms independently. Based on the ability of microorganism to be adapted or modified to obtain the desired products, biotechnology has been interested in this approach, especially if it can be complemented with other processes or take advantage of the fractions that are not transformed in some way (Baptista et al. 2021).

However, the use of genetically modified microorganisms to obtain xylitol from C5 and C6 fractions, could be expensive, relatively challenging in biotechnology techniques, and difficult to control. For this reason, proposing other schemes for obtaining value-added products from the other fractions of the raw material (mainly cellulose and lignin) becomes a possible alternative. For example, through the concept of biorefinery, a scheme could be proposed that allows the integral use of lignocellulosic raw material, integrating one or more additional processing lines to the xylitol production process. Nevertheless, it is necessary to demonstrate that these schemes could be more sustainable.

The purpose of this chapter is to compare in technical and economic terms the xylitol production process as a stand-alone alternative and integrated into biorefineries, according to the most appropriate design strategies. Therefore, the current known process, routes, kinetics, raw materials, among other aspects are described. Finally, the comparison of different scenarios will allow identifying the benefits of integrating this process within biorefineries and highlighting the challenges that would be immersed in considering this configuration.

8.2 Xylitol Production

Xylitol production can include catalytic as well as biotechnological pathways through fermentation. Industrially, this product is obtained by catalytic reduction of D-xylose. As a raw material, lignocellulosic material is special to generating hemicellulose sugar (xylose), generally through an acid or enzymatic hydrolysis. However, the other fractions of the raw material (especially cellulose and lignin) should be also used. Last years, different attempts are found to generate xylitol from the C6 sugar of cellulose (glucose) (Cheng et al. 2014; Kricka et al. 2015). It will increase the overall yield to obtain xylitol. Other approaches consider using these fractions to obtain different value-added products through the design of integrated processes, such as biorefineries (Dávila et al. 2016; Felipe Hernández-Pérez et al. 2019).

This section will mention the routes for obtaining xylitol, the raw materials most used for this purpose, and some of the technologies currently applied in the industry.

In turn, a brief description of the main applications of xylitol will be given. Finally, at the end of this section, xylitol's potential is discussed, not only as a stand-alone process but also to find a way to propose strategies to integrate this process with others through the biorefinery concept.

8.2.1 Routes and Raw Materials

The catalytic route for obtaining xylitol consists of the hydrogenation of xylose in the presence of a metallic catalyst and hydrogen gas under high pressure and temperature. This process is expensive because the substrate (xylose) must be as pure as possible to achieve high conversions to xylitol (Granström et al. 2007). The catalysts used are usually Nickel, Ruthenium, and Rhodium. The selectivity of the process is sensitive to changes in temperature, so that variable must be selected and controlled in the process (Rosales-Calderon and Arantes 2019). Some studies have shown that at temperatures above 140 °C, by-products can be generated that affect the overall xylitol production performance (Yadav et al. 2012). However, other experiments at moderate conditions (less than 100 °C and 20 bar) achieved a selectivity of more than 97% of xylose towards xylitol (Baudel et al. 2005; Rosales-Calderon and Arantes 2019). The main weakness of applying this process on an industrial and large scale is the high costs of the technology and the reduced overall efficiency. Besides, obtaining xylose from the lignocellulosic material (pretreatment stage), generates additional costs. Also, the limitation generated by the production of by-products is important since it inhibits reaching a high conversion towards xylitol (Rosales-Calderon and Arantes 2019). The xylitol purification process for catalytic hydrogenation is the stage that represents a high level of economic costs. The catalyst used in this case must be removed by filtration (for example, nickel) so that the hydrogenated solution is purified through a process of crystallization and evaporation (Rafiqul and Sakinah 2013). The diagram in Fig. 8.1 shows a simplified scheme of the production of xylitol.

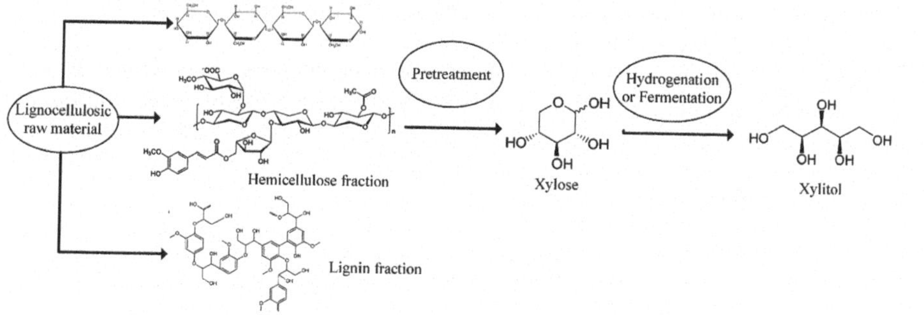

Fig. 8.1 Chemical and biochemical production of xylitol. Simplified scheme

On the other hand, the biotechnological route for obtaining xylitol has gained special importance due to different factors. This route is mainly based on obtaining xylitol through xylose fermentation by the action of several strains. In addition, the generation of xylose is also considered a pretreatment stage of the raw material (either by acid or enzymatic hydrolysis). Nevertheless, it is not necessary to have a highly pure xylose liquor to carry out the process (Ko et al. 2006; Silvério da Silva and Kumar Chandel 2012). This stage consists mainly of providing higher accessibility of the microorganism to the substrate. This alternative for obtaining xylitol could be more sustainable compared to catalytic technology. In economic terms, the pretreatment stage does not require additional purification units of the xylose, which reduces the overall costs of the process (Suzane et al. 2011). Moreover, in environmental terms, it is a safer process (Silvério da Silva and Kumar Chandel 2012). The stage of purification of xylitol obtained by fermentation is the most complex because the product of interest is not highly concentrated, besides to the fact that the fermentation medium has a multicomponent composition. For the separation, crystallization processes are used, which include centrifuges, ethanol precipitation, evaporation, among other unit operations. Also, the design of the purification stage is conditioned to the type of microorganism used, since in many cases, a filtration stage can work to separate the cellular biomass followed by a treatment of the obtained liquor to remove unwanted substances. Nevertheless, for other microorganisms, additional inputs or even different equipment may be required to purify the fermenter outlet streams (Rivas et al. 2006; Sampaio et al. 2006).

As mentioned above, obtaining xylitol by fermentative route depends on the strain to be used. It is possible to obtain this product from microorganisms such as fungi, bacteria, and yeasts. Even genetically modified microorganisms have also proven to be versatile and competitive for the production of xylitol. Table 8.1 shows some of the microorganisms studied by several authors, in which using raw material of lignocellulosic origin or the platform of C5 or C6 directly, xylitol is obtained by fermentation. With this review, it is possible to note that xylitol yields vary between 0.25 and 0.90, with concentrations even higher than 20 g/L. However, due to the control of these strains, their high cost, and the configurations that must be considered in the equipment and processes, many studies have only reached a laboratory scale. In this sense, it is still pending to analyze the processes on a larger scale and comprehensively evaluate their viability.

Xylitol production worldwide is at 200,000 tons per year. The selling price of this product is around $ 4,300 per ton ($ 4.3 per kilogram) (Rosales-Calderon and Arantes 2019). Because this product can be obtained from biomass, the energy and value-added industries consider it one of the main promising chemicals for this type of raw material (Gérardy et al. 2020).

Table 8.1 Microorganisms capable to produce xylitol by fermentative process

Microorganism	Raw material	Xylitol yield (g/g sugar platform) and concentration (g/L)	References
Candida tropicalis	Sugarcane bagasse	32.0 g/L	Vallejos et al. (2016)
	Xylose	22.0 g/L (0.22 g/g)	Martins et al. (2018)
	Rice straw	31.1 g/L (0.71 g/g)	Huang et al. (2011)
Candida guillermondii	Sugarcane straw	16.20 g/L	Hernández-Pérez et al. (2016)
	Rapeseed straw hydrolysate	24.7 g/L (0.65 g/g)	López-Linares et al. (2018)
	Sorghum forage hydrolysate	15.45 g/L (0.35 g/g)	Camargo et al. (2015)
Candida intermedia	Corncob hydrolysate	34.6 g/L (0.40 g/g)	Wu et al. (2018)
Candida mogii	Xylose	0.62 g/g	Sirisansaneeyakul et al.(1995)
Debaromyces hansenii UFV-170	Xylose	37 g/L (0.76 g/g)	Sampaio et al. (2008)
Hansenula polymorpha	Xylose and Glycerol (5%)	58 g/L (0.62 g/L)	Suryadi et al. (2000)
Kluyveromyces marxianus	Cashew apple bagasse	0.49 g/g glucose	Valderez et al. (2014)
	Xylose	24 g/L (0.90 g/g)	Mueller et al. (2011)
Pichia sp.	Xylose	25 g/L (0.58 g/g)	Rao et al. (2007)
Pichia stipitis	Xylose	0.52 g/g	Neeru et al. (2013)
Corynebacterium sp.	Xylose and arabinose	31.0 g/L	Dhar et al. (2016)
Scheffersomyces amazonensis	Xylose	34.2 g/L (0.75 g/g)	Cadete et al. (2016)

8.3 Stand-Alone Process, Using Biomass as Raw Material

As has been seen so far, there are several technologies for obtaining xylitol, in which it can be grouped the catalytic route by hydrogenation and, on the other hand, fermentation (with fungus, yeast, or bacteria). However, many of these configurations vary mainly in the substrate used or the raw material. For the case of the catalytic hydrogenation of xylose, it requires that the liquor of this sugar be as pure as possible. Due to this, the only possible configuration of this process is based on the C5 platform for obtaining xylitol (Yadav et al. 2012). On the other hand, the fermentation route depends on the microorganism used, so the type of substrate is conditioned. It is then possible to find various scenarios for obtaining xylitol by fermentation. Table 8.2

Table 8.2 Characteristics of the processing schemes to obtain xylitol by fermentation pathway

Case	Platform	Characteristics
1	Xylose	It requires a pretreatment stage for xylose production and separation. This process can be carried out by dilute acid hydrolysis or enzymatic hydrolysis. Additionally, non-usable streams (fractions of lignocellulosic raw material and streams not transformed to xylitol) must be managed for their final disposal or use
2	Xylose + Glucose	The raw material pretreatment stage can be simultaneous to obtain the two necessary platform fractions. Higher yields are achieved in the production of xylitol. However, specialized strains capable of transforming both platforms simultaneously (diauxic growth) or individuals are required. It can generate high costs in the acquisition of the strain since it mainly requires genetic modification. Similar to case 1, non-usable streams require a management stage for final disposal or use
3	Xylose + Glucose*	It can be considered the first approach to biorefineries (it is the simplest case since at least two products are obtained from a single raw material). Initially, the pretreatment stage can be simultaneous to obtain the sugar platforms, similar to case 2. However, a separation process is required to obtain both fractions and take them to independent processing lines. Another configuration may consider individual pretreatment for each fraction but may result in further energy and economic costs. Contrary to cases 1 and 2, the unusable streams are considerably reduced. Therefore, the management for final disposal may be different or even require fewer inputs and energy expenditure. By obtaining at least two products, the process can become more profitable than the previous cases (Coral Medina et al. 2018; Unrean and Ketsub 2018)
4	Xylose** + Glucose**	This case can be studied as a variation of case 3. In this sense, xylose and glucose platforms are divided to processing them in two different processing lines. One of them is xylitol production and the others can be other products. The complexity of this configuration will depend on the products to be obtained in addition to xylitol. In this case, the design of biorefineries gains strength since various tools can be applied to optimize the process. Higher profitability would be expected by breaking up the raw material and bringing it to various value-added products. Similar to case 3, waste streams can be reduced and maximized if integrated into the process

* Obtaining other products than xylitol
** Use of only a fraction of the platforms to xylitol and the remainder to a different product(s)

shows each of these cases and their main characteristics, and Fig. 8.2 summarizes each of these processing alternatives.

To consider as a base case the production of xylitol in a stand-alone process, the analysis that will be carried out below and in detail will be for Case 1, in which only the xylose fraction is transformed into xylitol by fermentation. However, it

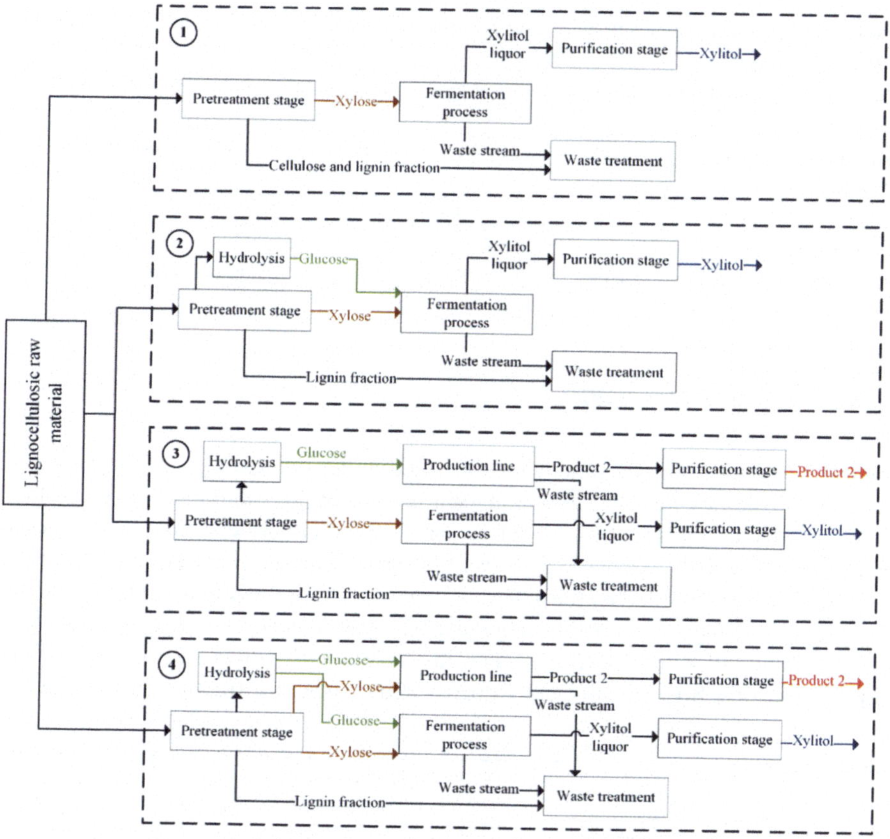

Fig. 8.2 Processing schemes of lignocellulosic raw materials towards obtaining xylitol. (1) Case 1, only use of platform C5 to obtain xylitol. (2) Case 2, use of platform C5 and C6 to obtain only xylitol. (3) Case 3, use of platform C5 towards xylitol and obtaining other value-added products from platform C6. (4) Case 4, partial use of platforms C5 and C6 towards xylitol and the remainder to obtain other value-added products

will start from 3 raw materials of lignocellulosic origin, and its viable conversion to the xylose platform will be evaluated in techno-economic terms and based on the composition of each raw material. This comparative information will identify specific characteristics of the raw material to obtain the C5 platform. First, using the Aspen Plus V.9 tool (Aspen Technologies Inc., USA), a simulation of the process was carried out, and from the mass and energy balances and the configuration of the scheme, the technical and economic analyzes were carried out. Sugarcane bagasse, Pinus patula, and orange peel were considered raw materials for this study, whose compositions can be seen in Table 8.3. Next, the simulated process (similar for each case), the technical and economic analysis methodology, and the results obtained for the xylitol production process based on xylose through the most viable alternative to obtain the xylose platform will be described.

Table 8.3 Chemical compositions of raw materials evaluated in this work for xylose platform production (%w/w)

Raw material	Cellulose	Hemicellulose	Lignin	Extractives	Ash	References
Sugarcane bagasse	46.7	23.62	19.71	8.79	1.13	Aristizábal et al. (2015)
Pinus patula	44.78	23.75	20.22	11.0	0.25	García (2016)
Orange peel	23.88	14.15	5.10	26.56	2.10	Ortiz-Sanchez (2019)

8.3.1 Process Description

Figure 8.3 shows the equipment for the stand-alone processing scheme based on lignocellulosic raw material to obtain xylitol, used in the simulation in Aspen Plus V.9 (Aspen Technologies Inc., USA). In part A, the pretreatment of the raw material is evidenced until it reaches the xylose platform. The dry raw material (less than 10% moisture) is divided by particle size reduction equipment to homogenize it and facilitate its transformation. Subsequently, the raw material is taken to hydrolysis pretreatment with dilute sulfuric acid (1.25% w/v) for 17 min at 120 °C and a solid–liquid ratio of 1:8 (Mussatto and Roberto 2005), where a fraction of cellulose is

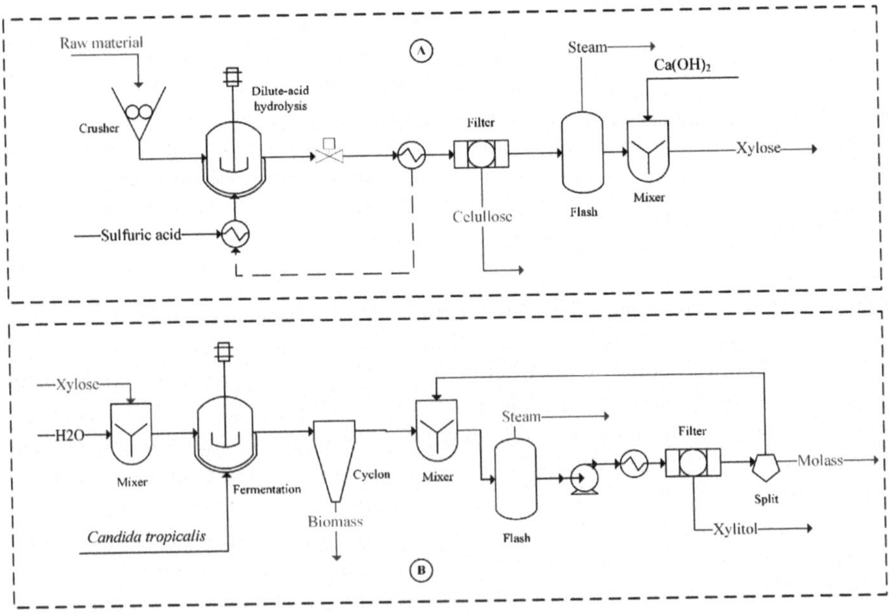

Fig. 8.3 Production of xylitol by fermentation. Stand-alone process. **a** Pretreatment stage (evaluation of three different raw materials), **b** Xylitol production stage

not transformed and a xylose liqueur is obtained. Liquid fraction goes through a purification process, concentrating the xylose to 70 g/L using a flash evaporator at 121 °C and 1 bar of pressure (Dávila 2015), and the concentrated xylose fraction is ready to be transformed into xylitol. In part B, the purify processing of the product is shown. The xylose stream is taken to a fermenter with the necessary nutrients for the metabolism of the *Candida guillermondii* strain. The fermentation conditions are 30 °C and 200 rpm (Dávila 2015). The liquid stream undergoes a filtration process to recover the cellular biomass. The xylitol obtained is taken to a concentration and purification stage.

8.3.2 Technical Analysis (Methodology and Results)

The processes simulations were carried out using Aspen Plus V.9 software (Aspen Technologies Inc., USA). With this software, it is possible to obtain the mass and energy balances of the process. For this case, the Non-Random Two-Liquid (NRTL) thermodynamic method was applied since it allows modeling systems with liquid phases and moderate pressures (<10 bar). For the case of the vapor phase, the Hayden-O'Connell equation of state was used. In order to obtain comparative data with other processes, the technical analysis was carried out by calculating the mass and energy indicators shown in Table 8.4. To homogenize the information and to be able to compare the results, a scale of 1000 kg/day of processing raw materials was selected.

Where, $\dot{m}_{Product,i}$: mass flow of product, (i) [kg/h], $\dot{m}_{Rawmaterial}$: mass flow of raw material, [kg/h], M_p: mass of product (i) obtained in an specific Wp, [kg/h], $M_{Rawmaterial}$: mass of Raw material obtained in an specific Wp, [kg/h], Wp: working period of the process, $\dot{Q}$: Thermal requirements of the process, [MJ/h], $\dot{W}$: Power requirements of the process, [MJ/h], $LHV_{products}$: Lower heating value of products, $LHV_{rawmaterials}$: Lower heating value of raw materials.

Table 8.5 shows the results of the technical indicators calculated for each raw material. The xylose yield is similar for sugarcane bagasse and pinus patula raw materials,

Table 8.4 Mass and energy indicators used for technical analysis of stand-alone process and biorefinery case of study (Alonso-Gómez et al. 2020)

Indicator	Equation
Product yield (Y_P)	$Y_P = \frac{\sum \dot{m}_{Product,i}}{\dot{m}_{Rawmaterial}}$
Process mass intensity index (PMI)	$PMI = \frac{\sum_{i=1}^{N} \dot{m}_i^{in}}{\sum \dot{m}_{Product,i}}$
Mass loss index (MLI)	$MLI = \frac{\sum_{i=1}^{N} \dot{m}_i^{in} - \dot{m}_{Product}}{\sum \dot{m}_{Product,i}}$
Overall energy efficiency (η)	$\eta = \frac{\sum \dot{m}_{Product} * LHV_{products}}{(\dot{m}_{Rawmaterial} * LHV_{Rawmaterial}) + \dot{Q}_{Total} + \dot{W}_{Total}}$
Specific energy Consumption (S_{EC})	$S_{EC} = \frac{\dot{Q}_{Total} + \dot{W}_{Total}}{\dot{m}_{Rawmaterial}}$

Table 8.5 Technical results for each case of xylose platform production

Indicator	Sugarcane bagasse	Pinus patula	Orange peel
Xylose yield (Y_P)	0.53 g/g	0.51 g/g	0.25 g/g
Process mass intensity index (PMI)	0.40 g/g	0.38 g/g	0.15 g/g
Mass loss index (MLI)	39%	40%	45%
Overall energy efficiency (η)	16.4%	17.8%	5.7%
Specific energy Consumption (S_{EC})	125.2 kW/kg	189.1 kW/kg	93.3 kW/kg

due to the high hemicellulose content, instead of orange peel waste. Nevertheless, orange peel waste had a better specific energy consumption but very low energy efficiency in terms of energy indicators. Sugarcane bagasse presented a similar behavior to pinus patula, and it could be a great candidate to obtain xylose products. The xylose production can be compared in terms of mass loss index, showing that the less value is obtained using sugarcane bagasse as raw material. In this sense, the best raw material based on technical aspects to obtain xylose platform is Sugarcane bagasse.

8.3.3 Economic Analysis

The economic analysis was carried out using the Aspen Process Economic Analyzer (APEA) tool, for which the material and energy balances and the equipment used in the simulations served as initial data for the calculation of the production cost. Additionally, the economic values shown in Table 8.6 were taken into account. The economic analysis made it possible to size and calculate the equipment cost to obtain the capital, operation, and production values, following the methodology given by (Peters et al. 2003) (Peters et al. 2003). Besides, the economic indices for the Colombian context were considered, which are tax rate of 25%, annual interest rate of 12.1%, and 0.93 USD/h as operator-supervisor labor cost (Ortiz-Sanchez et al. 2020).

The distribution of the main costs for the three case studies is shown in Fig. 8.4. In all cases, it is evident that for the production of xylose, the investment cost is the one that represents the highest value for the calculation of the profit margin. In fact, for the cases of sugarcane bagasse and patula pine, the profit margin is positive. However, in the case of orange peel, the profit margin is negative, which indicates that this alternative is the least viable to consider in obtaining xylose. Therefore, with these results and those analyzed in terms of mass and energy, the most promising raw material selected as a stand-alone process for xylitol production is sugarcane bagasse. Subsequently, the results of the xylitol production stage are then analyzed in technical–economic terms.

Table 8.6 Economic values for assessing stand-alone and the biorefinery

Item	Value	References
Sugarcane bagasse	0.015 USD/kg	Aristizábal et al. (2015)
Pinus patula	0.012 USD/kg	García (2016)
Orange peel	0.022 USD/kg	Ortiz-Sanchez (2019)
Sulfuric acid	0.35 USD/kg	García-Velásquez and Cardona (2019)
NaOH	0.012 USD/kg	Coral Medina et al. (2018)
Xylose	0.412 USD/kg	Ortiz-Sanchez (2019)
Microorganism	7.0 USD/kg	Coral Medina et al. (2018)
Xylitol	3.0 USD/kg	
Ethanol	0.88 USD/kg	Poveda-Giraldo et al. (2021)
LP stem	7.91 USD/ton	
MP stem	8.07 USD/ton	
Cooling water	0.04 USD/m^3	
Electricity	0.03 USD/MJ	

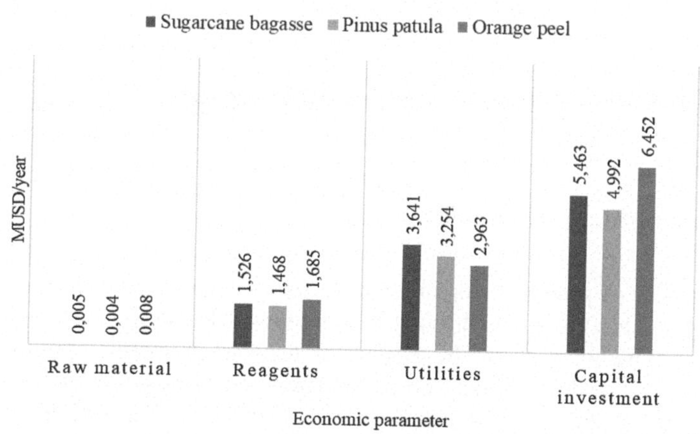

Fig. 8.4 Economic results for xylose production from 3 different raw materials

8.3.4 *Tecno-Economic Results for Best Stand-Alone Alternative, to Produce Xylitol*

The techno-economic results of the production of xylitol from sugarcane bagasse are shown in Table 8.7. These results correspond to the entire process, that is, taking into account both the xylose production stage already evaluated and the xylitol production. Compared with processes reported by other authors, the simulation yielded consistent results. In fact, Vallejos et al. 2016 (Vallejos et al. 2016), reported similar xylitol yields

Table 8.7 Techno-economic results for xylitol stand-alone production from sugarcane bagasse

Technical indicator	Result	Economic parameter	Result
Xylitol yield (Y_P)	0.18 g/g sugarcane bagasse	Revenues (mUSD/year)	26.540
Process mass intensity index (PMI)	0.26 g/g sugarcane bagasse	Raw material (mUSD/year)	0.005
Mass loss index (MLI)	41%	Reagents (mUSD/year)	2.153
Overall energy efficiency (η)	22.4%	Utilities (mUSD/year)	4.251
Specific energy Consumption (S_{EC})	184.1 kW/kg	Capital investment (mUSD/year)	7.891
		Profit margin (%)	12.245

for sugarcane bagasse. However, yields could be improved if xylitol production from the C6 glucose platform is integrated and analyzed. In economic terms, the costs obtained are related to the production scale and the Colombian context indicators that were considered.

8.4 Xylitol Production as Part of an Integrated Biorefineries

8.4.1 Biorefinery Scenarios

The integration of processes for biomass recovery is an alternative for obtaining various value-added products through biorefineries. Thanks to the optimization of the mass and energy streams, these schemes achieve competitive products in the market and achieve the minimum possible generation of waste. In turn, it has been shown that the viability of the processes is considerably improved if the process streams are integrated towards the generation of more products and the main one. In the alcohol world, many biorefinery schemes have been designed. Especially for xylitol, they can be summarized in Table 8.8. Processes have been designed that take advantage of the hydrolysates of raw materials such as corn fiber, spent pulp of Colombian Andes Berry, sugarcane bagasse, among others.

Considering the information shown in Table 8.8, biorefinery 4 was selected as a scheme to be evaluated and compared with the results obtained in obtaining xylitol (stand-alone). Figure 8.5 shows the outline of the process considered for its evaluation. In this case, the sugar streams from the raw material pretreatments (sugarcane bagasse) are the platforms for obtaining xylitol and ethanol. Both processes are carried out through fermentation (xylitol from *Candida tropicalis* and ethanol from

Table 8.8 Biorefineries with xylitol production

Biorefinery	Raw material	Other products	Comments	References
1	Spent pulp of Colombian Andes Berry	Ethanol, phenolic compounds extract, electricity	Xylitol from C5 platform, using *Candida guillermondii*	Dávila et al. (2017)
2	Olive tree pruning	Ethanol, xylitol, antioxidants, electricity	Xylitol from C5 platform, using *Meyerozyma guilliermondii*	Susmozas et al. (2018)
3	Brewer spent grain	Ethanol, PHB	Xylitol from C5 platform, using *Candida guillermondii*	Dávila et al. (2016)
4	Sugarcane bagasse	Ethanol	Xylitol from C5 platform, using *Candida tropicalis*	Unrean and Ketsub (2018)
5	Sugarcane bagasse	Citric acid, glutamic acid, electricity	Xylitol from C5 platform, by catalytic pathway	Özüdoğru et al. (2019)
6	Rye straw	Ethanol	Xylitol from C5 platform, using *Candida* strain	Franceschin et al. (2011)
7	Corn fiber	Ethanol, electricity, biomethane, heat	Xylitol from C5 platform, using *Candida* strain	Fehér et al. (2012)
8	Oil palm empty fruit bunches	Ethanol, Lignin	Xylitol from C5 platform	Coral Medina et al. (2018)

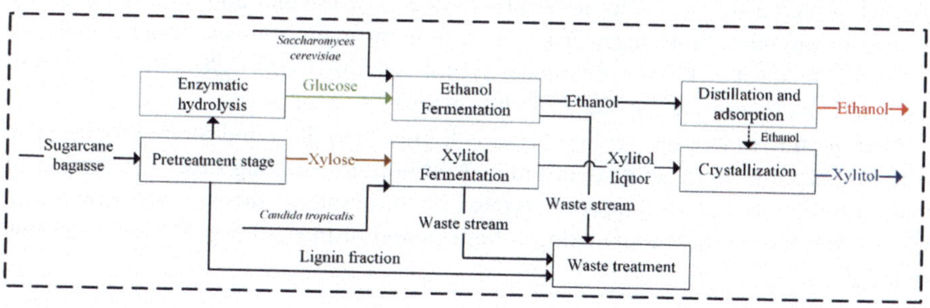

Fig. 8.5 Biorefinery scheme for xylitol and ethanol production from Sugarcane bagasse

Table 8.9 Techno-economic results for biorefinery case from sugarcane bagasse

Technical indicator	Result	Economic parameter	Result
Xylitol yield ($Y_{Xylitol}$)	0.18 g/g sugarcane bagasse	Revenues (mUSD/year)	45.152
Ethanol yield ($Y_{Ethanol}$)	0.35 g/g sugarcane bagasse	Raw material (mUSD/year)	0.005
Process mass intensity index (PMI)	0.26 g/g	Reagents (mUSD/year)	4.588
Mass loss index (MLI)	41%	Utilities (mUSD/year)	5.693
Overall energy efficiency (η)	22.4%	Capital investment (mUSD/year)	9.841
Specific energy Consumption (S_{EC})	184.1 kW/kg	Profit margin (%)	25.025

Saccharomyces Cerevisiae). Then purification stages are considered as distillation, adsorption and crystallization units.

8.4.2 *Techno-Economic Analysis (Methodology and Results)*

Similar to the stand-alone base case of xylitol production, the Aspen Plus V.9 software (Aspen Technologies Inc., USA) was used for the biorefinery. In this way, the material and energy balances were obtained to calculate the mass and energy indicators previously shown in Table 8.4. For this case, same economic values were used to biorefinery assessment.

Table 8.9 shows the results of the biorefinery simulation. Product yields were not affected, comparing the stand-alone processing. The ethanol and xylitol yields are comparable to other biorefinery processes where those processes are considered. For example, Dávila et al. (2016) obtained a xylitol yield of 101.53 kg/ton Brewer's spent grain and an ethanol yield of 32.73 kg/ton Brewer's spent grain.

In economic terms, the profit margin is higher than the stand-alone process, due to the incorporation of another product. Additionally, although the distribution of costs is similar to that for the previous case, the biorefinery allows a more noticeable balance in these values, elucidating a high potential in the integral use of the raw material.

8.5 Conclusion

This analysis reveals that integrating processes through the use of the available platforms of raw material is a techno-economically viable alternative. Lignocellulosic materials have a high potential towards the production of xylitol, so proposing the use of agroindustrial waste could be a solution to environmental problems (related to the final disposal of these wastes), demand for xylitol in the market, and sustainable development of the applied context (region and country). The study of more complex biorefinery schemes would be worth studying to take advantage of all the fractions of the available raw material, which can complement the production of xylitol in several terms.

Acknowledgements The authors express their gratitude for the financial support given by the program: PROGRAMA DE INVESTIGACIÓN RECONSTRUCCIÓN DEL TEJIDO SOCIAL EN ZONAS DE POS-CONFLICTO EN COLOMBIA SIGP Code: 57579 within the research project "COMPETENCIAS EMPRESARIALES Y DE INNOVACIÓN PARA EL DESARROLLO ECONÓMICO Y LA INCLUSIÓN PRODUCTIVA DE LAS REGIONES AFECTADAS POR EL CONFLICTO COLOMBIANO" SIGP Code: 58907. Funded within the framework of Colombia Científica, Contract No FP44842-213-2018. Besides, the authors express their gratitude to Universidad Nacional de Colombia sede Manizales, with the project entitle "ANÁLISIS Y COMPARACIÓN DE RUTAS DE TRANSFORMACIÓN PARA EL APROVECHAMIENTO DE SOSTENIBLE DE MATERIAS PRIMAS LIGNOCELULÓSICAS" Hermes code: 53071.

References

Alonso-Gómez LA et al (2020) 'Performance evaluation and economic analysis of the bioethanol and flour production using rejected unripe plantain fruits (Musa paradisiaca L.) as raw material', food and bioproducts processing. Inst Chem Eng 121:29–42. https://doi.org/10.1016/j.fbp.2020.01.005

Aristizábal et al. (2015) Biorefineries based on coffee cut-stems and sugarcane bagasse: Furan-based compounds and alkanes as interesting products. Biores Technol. Elsevier Ltd, 196: 480–489. https://doi.org/10.1016/j.biortech.2015.07.057

Baptista, S. L. et al. (2021) Metabolic engineering of Saccharomyces cerevisiae for the production of top value chemicals from biorefinery carbohydrates. Biotechnol Adv Elsevier Inc., 47: 107697. https://doi.org/10.1016/j.biotechadv.2021.107697.

Baudel H, de Abreu C, Zaror C (2005) Xylitol production via catalytic hydrogenation of sugarcane bagasse dissolving pulp liquid effluents over Ru/C catalyst. J Chem Technol Biotechnol 80(2):230–233. https://doi.org/10.1002/jctb.1155

Cadete RM et al (2016) The yeast scheffersomyces amazonensis is an efficient xylitol producer. World J Microbiol Biotechnol 32(12):207. https://doi.org/10.1007/s11274-016-2166-5

Camargo D et al (2015) Xylitol bioproduction in hemicellulosic hydrolysate obtained from sorghum forage biomass. Appl Biochem Biotechnol 175(8):3628–3642. https://doi.org/10.1007/s12010-015-1531-4

Cheng H et al (2014) Genetically engineered pichia pastoris yeast for conversion of glucose to xylitol by a single-fermentation process. Appl Microbiol Biotechnol 98(8):3539–3552. https://doi.org/10.1007/s00253-013-5501-x

Coral Medina JD et al (2018) 'Energetic and economic analysis of ethanol, xylitol and lignin production using oil palm empty fruit bunches from a Brazilian factory', J Cleaner Product. Elsevier Ltd 195:44–55. https://doi.org/10.1016/j.jclepro.2018.05.189

Dávila JA (2015) Design of biorefineries for high value added products from fruits - doctoral thesis. Universidad Nacional de Colombia.

Dávila JA, Rosenberg M, Cardona CA (2016) A biorefinery approach for the production of xylitol, ethanol and polyhydroxybutyrate from brewer's spent grain. AIMS Agricul Food 1(1):52–66. https://doi.org/10.3934/agrfood.2016.1.52

Dávila JA, Rosenberg M, Cardona CA (2017) A biorefinery for efficient processing and utilization of spent pulp of Colombian Andes Berry (Rubus glaucus Benth.): experimental, techno-economic and environmental assessment. Biores Technol 223:227–236

Dhar KS, Wendisch VF, Nampoothiri KM (2016) Engineering of corynebacterium glutamicum for xylitol production from lignocellulosic pentose sugars. J Biotechnol Elsevier B.V. 230: 63–71. https://doi.org/10.1016/j.jbiotec.2016.05.011.

Fehér C, Barta Z, Réczey K (2012) Process considerations of a biorefinery producing value-added products from corn fibre. Periodica Polytech, Chem Eng 56(1):9. https://doi.org/10.3311/pp.ch.2012-1.02

Felipe Hernández-Pérez A et al (2019) 'Xylitol bioproduction: state-of-the-art, industrial paradigm shift, and opportunities for integrated biorefineries', CritAl Rev Biotechnol. Taylor & Francis 39(7):924–943. https://doi.org/10.1080/07388551.2019.1640658

Franceschin G et al (2011) Conversion of rye straw into fuel and xylitol: a technical and economical assessment based on experimental data. Chem Eng Res Des 89(6):631–640. https://doi.org/10.1016/j.cherd.2010.11.001

García-Velásquez CA, Cardona CA (2019) Comparison of the biochemical and thermochemical routes for bioenergy production: A techno-economic (TEA), energetic and environmental assessment. Energy 172:232–242. https://doi.org/10.1016/j.energy.2019.01.073

García CA (2016) Hydrogen production through gasification and dark fermentation. - master's thesis, Universidad Nacional de Colombia sede Manizales. http://bdigital.unal.edu.co/54321/1/1053813068.2016.pdf.

Gérardy R et al (2020) continuous flow upgrading of selected C 2 –C 6 platform chemicals derived from biomass. Chem Rev 120(15):7219–7347. https://doi.org/10.1021/acs.chemrev.9b00846

Granström TB, Izumori K, Leisola M (2007) A rare sugar xylitol. Part I: the biochemistry and biosynthesis of xylitol. Appl Microbiol Biotechnol 74(2):277–281. https://doi.org/10.1007/s00253-006-0761-3

Hernández-Pérez AF, de Arruda PV, Felipe M, das G. de A. (2016) Sugarcane straw as a feedstock for xylitol production by Candida guilliermondii FTI 20037. Braz J Microbiol 47(2):489–496. https://doi.org/10.1016/j.bjm.2016.01.019

Huang C-F et al (2011) Development of a yeast strain for xylitol production without hydrolysate detoxification as part of the integration of co-product generation within the lignocellulosic ethanol process. Bioresour Technol. Elsevier Ltd 102(3):3322–3329. https://doi.org/10.1016/j.biortech.2010.10.111

Ko BS, Kim J, Kim JH (2006) Production of xylitol from D-xylose by a xylitol dehydrogenase gene-disrupted mutant of Candida tropicalis. Appl Environ Microbiol 72(6):4207–4213. https://doi.org/10.1128/AEM.02699-05

Kricka W. et al., (2015) Engineering saccharomyces pastorianus for the co-utilisation of xylose and cellulose from biomass. Microbial Cell Factories. ???, 14(1): 61. https://doi.org/10.1186/s12934-015-0242-4.

López-Linares, J. C. et al. (2018) Xylitol production by debaryomyces hansenii and candida guilliermondii from rapeseed straw hemicellulosic hydrolysate. Bioresour Technol Elsevier Ltd, 247: 736–743. https://doi.org/10.1016/j.biortech.2017.09.139

Martins GM et al (2018) 'The isolation of pentose-assimilating yeasts and their xylose fermentation potential', Braz J Microbiol. Sociedade Brasileira De Microbiologia 49(1):162–168. https://doi.org/10.1016/j.bjm.2016.11.014

Mueller M, Wilkins MR, M Banat I (2011) Production of xylitol by the thermotolerant kluyveromyces marxianus IMB strains. J Bioprocess Biotech 01(02). https://doi.org/10.4172/2155-9821.1000102e

Mussatto SI, Roberto IC (2005) Acid hydrolysis and fermentation of brewer's spent grain to produce xylitol. J Sci Food Agric 85(14):2453–2460. https://doi.org/10.1002/jsfa.2276

Neeru C, Chandrajit B, Vidyasagar J (2013) Biological production of xylitol from corn husk and switchgrass by pichia stiptis. Res J Chem Sci 3(3): 58–64. www.isca.in.

Ortiz-Sanchez M, (2019) Evaluation of different biorefinery configurations for orange peel waste processing - master's Thesis. Universidad Nacional de Colombia sede Manizales.
Ortiz-Sanchez M et al (2020) Pre-feasibility analysis of the production of mucic acid from orange peel waste under the biorefinery concept. Biochem Eng J. Elsevier 161(May):107680. https://doi.org/10.1016/j.bej.2020.107680
Özüdoğru HMR et al (2019) Techno-economic analysis of product biorefineries utilizing sugarcane lignocelluloses: Xylitol, citric acid and glutamic acid scenarios annexed to sugar mills with electricity co-production. Ind Crop Prod Elsevier 133(January):259–268. https://doi.org/10.1016/j.indcrop.2019.03.015
Peters M, Timmerhaus K, West R (2003) Plant design and economics for chemical engineers. Fifth ed. New York.
Poveda-Giraldo JA, Piedrahita-Rodríguez S, Cardona-Alzate CA. (2021) Life cycle analysis of biotechnological processes based on the composition of the raw material. Eucalyptus, avocado, and plantain cases in a biorefinery system. Chem Eng Trans 83: 397–402. https://doi.org/10.3303/CET2183067
Rafiqul ISM, Sakinah AMM (2013) Processes for the production of xylitol—a review. Food Rev Intl 29(2):127–156. https://doi.org/10.1080/87559129.2012.714434
Rao RS, Bhadra B, Shivaji S (2007) Isolation and characterization of xylitol-producing yeasts from the gut of colleopteran insects. Curr Microbiol 55(5):441–446. https://doi.org/10.1007/s00284-007-9005-8
Rivas B et al (2006) Purification of xylitol obtained by fermentation of corncob hydrolysates. J Agric Food Chem 54(12):4430–4435. https://doi.org/10.1021/jf053156x
Rosales-Calderon O, Arantes V (2019) A review on commercial-scale high-value products that can be produced alongside cellulosic ethanol. Biotechnol for Biofuels. Biomed Central 12(1):240. https://doi.org/10.1186/s13068-019-1529-1
Sampaio FC et al (2006) Xylitol crystallization from culture media fermented by yeasts. Chem Eng Process 45(12):1041–1046. https://doi.org/10.1016/j.cep.2006.03.012
Sampaio FC et al (2008) Influence of cultivation conditions on xylose-to-xylitol bioconversion by a new isolate of debaryomyces hansenii. Biores Technol 99(3):502–508. https://doi.org/10.1016/j.biortech.2007.01.017
Silvério da Silva S, Kumar Chandel A (2012) D-xylitol. fermentative production, application and commercialization. Springer Verlag Berlin Heidelberg. https://doi.org/10.1007/978-3-642-31887-0
Sirisansaneeyakul S, Staniszewski M, Rizzi M (1995) Screening of yeasts for production of xylitol from d-xylose. J Ferment Bioeng 80(6):565–570. https://doi.org/10.1016/0922-338X(96)87732-4
Suryadi H et al (2000) Polyol production by culture of methanol-utilizing yeast. J Biosci Bioeng 89(3):236–240. https://doi.org/10.1016/S1389-1723(00)88825-8
Susmozas A et al (2018) Designing an olive tree pruning biorefinery for the production of bioethanol, xylitol and antioxidants: a techno-economic assessment. Holzforschung 73(1):15–23. https://doi.org/10.1515/hf-2018-0099
Suzane T, Chandel AK, Branco RDF (2011) Effect of dissolved oxygen and inoculum concentration on xylose reductase production from candida guilliermondii using sugarcane bagasse hemicellulosic hydrolysate. Food Nutr Sci 2:235–240. https://doi.org/10.4236/fns.2011.23033
Unrean P, Ketsub N (2018) Integrated lignocellulosic bioprocess for co-production of ethanol and xylitol from sugarcane bagasse. Ind Crop Prod. Elsevier 123(June):238–246. https://doi.org/10.1016/j.indcrop.2018.06.071
Valderez, M. et al. (2014) Evaluation of dilute acid pretreatment on cashew apple bagasse for ethanol and xylitol production Chem Eng J. Elsevier B.V. 243: 234–243. https://doi.org/10.1016/j.cej.2013.12.099.
Vallejos ME et al (2016) Strategies of detoxification and fermentation for biotechnological production of xylitol from sugarcane bagasse. Ind Crop Prod Elsevier B.V., 91: 161–169. https://doi.org/10.1016/j.indcrop.2016.07.007.

Wu J et al (2018) Single-cell Protein and Xylitol Production by a Novel Yeast Strain Candida intermedia FL023 from Lignocellulosic Hydrolysates and Xylose. Appl Biochem Biotechnol 185(1):163–178. https://doi.org/10.1007/s12010-017-2644-8

Yadav M, Mishra DK, Hwang J (2012) Catalytic hydrogenation of xylose to xylitol using ruthenium catalyst on NiO modified TiO2 support. Appl Catal A Gen Elsevier B.V., 425–426: 110–116. https://doi.org/10.1016/j.apcata.2012.03.007.

Chapter 9
Addressing Key Challenges in Fermentative Production of Xylitol at Commercial Scale: A Closer Perspective

Sreenivas Rao Ravella, David J. Warren-Walker, Joe Gallagher, Ana Winters, and David N. Bryant

Abstract Xylitol has been recognized by the US Department of Energy (DOE) as one of the top 12 value-added chemicals obtained from biomass, with a world market of 200,000 tonnes per year. The global xylitol market is expected to reach a value of US$ 1 Billion by 2026 growing at a compound annual growth rate (CAGR) of 5.8% during 2021–2026. Historically, the commercial xylitol production process has been dependent on the chemical hydrogenation of xylose. Several xylitol production plants, mainly in China that use the chemical process have had to reduce their production capacity to address regulations governing sustainability and environmental standards. In this chapter, key challenges and possible solutions for fermentative xylitol production at commercial scale are discussed in terms of: (1) Feedstock supply for commercial production plants; (2) Industrial biomass pretreatment; and (3) Lessons learned from industrial operations. These are drawn together to identify technology gaps and scaling-up challenges in light of the capital expenditure required to build a state-of-the art xylitol industrial biotechnology (IB) production facility and the potential to reduce climate change impact and contribute towards achieving net-zero targets.

Keywords Xylose · Xylitol commercialisation · Scale-up · Biorefining · Steam explosion · Fermentation

9.1 Introduction

More than 130 counties aim to be climate neutral by 2050, with 14 enshrining this commitment in law, 30 in policy documents, 15 in pledges with the remainder undergoing further discussions (https://www.eciu.net/netzerotracker). In order to facilitate

---

S. R. Ravella (✉) · D. J. Warren-Walker · J. Gallagher · A. Winters · D. N. Bryant
Institute of Biological Environmental and Rural Sciences (IBERS), Aberystwyth University, Aberystwyth SY23 3EE, UK
e-mail: rsr@aber.ac.uk

© The Author(s) 2022
M. das G. de Almeida Felipe and A. K. Chandel (eds.), *Current Advances in Biotechnological Production of Xylitol*,
https://doi.org/10.1007/978-3-031-04942-2_9

the global transition toward climate neutrality, ambitious action plans and objectives related to bioenergy, the circular economy and biorefineries in particular are required. The European Commission emphasised the latter point out in a recently published report on the opportunities afforded from biorefining processes in the EU (EU Biorefinery—outlook to 2030). This EU outlook presented scenarios on how demand and supply for bio-based chemicals and materials could grow to 2030, and actions required to increase the deployment of biorefineries in the EU.

The International Energy Agency (IEA) in, IEA Bioenergy Task 42 "Biorefining in a Circular Economy" identified 4 key areas to classify and describe biorefinery systems: (1) Platforms (e.g. core intermediates such as C5/C6 carbohydrates, syngas, lignin, pyrolytic liquid) (2) Products (e.g. energy carriers, chemicals and material products) (3) Feedstock (i.e. biomass, from dedicated production or residues from forestry, agriculture, aquaculture and other industry and domestic sources and; (4) Processes (e.g. thermochemical, chemical, biochemical and mechanical processes) e.g. (Table 9.1).

Currently, there are more than 224 commercial biorefineries operating across Europe, of which 181 use 'first generation' (1G) feedstock, i.e. sugars, starch or oils. However, comparably fewer, approximately 40, refine lignocellulosic (woody biomass) or "second generation" (2G) feedstock (Hassan et al. 2019). Table 9.2 list examples of commercial 2G biorefineries, their products, capacity and operational status.

Commercially, xylitol, a tooth and diabetic friendly sweetener, is produced by chemical hydrogenation of xylose, a C5 sugar refined from biomass (i.e. birchwood and/or corn cobs etc.) (Makinen 2000; Werpy and Petersen 2004; Ravella et al. 2012). While the scientific literature is awash with fermentative xylitol processes, >900 articles in the past 5 years, using native or engineered yeast (*Candida, Pichia,*

Table 9.1 List of lignocellulosic feedstock stream exploded and processed to release xylose and arabinose in hemicellulosic hydrolysates at Aberystwyth University's BEACON Biorefining Centre, UK (http://www.beaconwales.org)

Feedstock	Xylose: arabinose ratio	Xylose content in biomass (%)
Corn Stover	4.8	25
Willow	6.5	16
Wheat Straw	4.2	30
Sugarcane bagasse	6.5	24
Miscanthus	6.2	25
Brewers spent grain	1.8	20
Oat Hulls	13	24
Birch	11	27
Silver Birch	11	27
Corn cobs	9.0	28
Rye Grass Fibre	3.3	15

Table 9.2 Examples of commercial and demonstration 2nd Generation (2G) lignocellulosic biorefineries

Main investor	Processing technologies	Feedstock	Product	Commercial operations	Location
Beta renewables	Versalis' PROESA® steam explosion pretreatment	Wheat straw and hardwood	2G ethanol 59,000 MT/yr	2013 Deactivated 2021 new venture with Saipem launched	Crescentino (Italy)
Bharat petroleum corporation limited (BPCL)	Tata Projects	Rice and corn straw 488 MT/day	2G ethanol 80 MT/day	Active since/initiated?2022	Bargarh, Odisha, India
Croatian oil and gas company INA	Axens Futurol™ technology http://www.axens.net		2G ethanol 55,000 MT/yr	2020 planning stage	Sisak, Croatia
Assam bio refinery private limited (ABRPL)	Formicobio™ technology—3G technology http://www.chempolis.com	Bamboo 300,000 MT/year	2G ethanol 51,000 MT/yr, furfural 2100 MT/Yr, acetic acid 9000 MT/yr, Biocoal 156,000 MT/yr	2021 planning stage	India
Clariant	Commercial scale Sunliquid® technology, Valmets Biotrac SE process	Wheat/barley Straw/corn Stover 750–800 MT/day	2G ethanol 50,000 MT/year	2022 planning stage	Podari, Romania
Clariant	Demonstration-scale Sunliquid® technology, Valmets Biotrac SE process	Local Agricultural Residues	2G ethanol 150 MT/day	2022 planning stage	Slovakia
Clariant	Pilot-scale Sunliquid® technology, Valmets Biotrac SE process	Wheat/barley/rice straw, corn stover, miscanthus, SCB 12–16 MT/day	2G ethanol pilot plant <3 MT/day	2012 operational	Germany

(continued)

Table 9.2 (continued)

Main investor	Processing technologies	Feedstock	Product	Commercial operations	Location
CTC Sugarcane Technology Centre-Piracicaba, Brazil	Andritz steam explosion pretreatment technology http://www.andritz.com	SCB	Pilot plant for C5 sugar extraction	Deactivated technology for sale	Brazil
LIBERTY Project POET-DSM http://www.poetdsm.com/liberty	Andritz pretreatment technology http://www.andritz.com	285,000 dry tonnes of corn stover	25 million gallons of 2G ethanol	Stopped in 2019, now focusing on R&D to improve process	USA
Bioflex Agroindustrial part of GranBio www.granbio.com	SE pretreatment	Sugarcane straw	2G ethanol 47,000 MT/year	2014 (20 months to construct $265 million plant?)	Brazil
Raízen		SCB	2G ethanol > 31,000 MT/year	2014 technology for sale	Brazil
Fortress Global Enterprises Inc	S2G BioChem and Mondelēz International 2G sugar technology,	Maple, aspen, poplar, birch	Xylitol demonstration 2,000 MT/year commercial > 20,0000 MT/yr	2018 Demonstration plant initiated, commercial plant pending	Sarnia (Ontario, Canada)
DuPont	DWB concept	Wood	Xylose and xylitol	2012 operational	Austria
Sappi	Valmets 'Xylex' technology	Wood	Xylose and Xylitol	2018 operational	South Africa
Oy Karl Fazer Ab		Oat hulls 20,000 MT/yr	Xylitol 4000 MT/yr	2019 construction started	Lahti, Finland
Borregaard Biorefinery	Advanced biorefinery (20 integrated plants)	Wood	2G ethanol, Cellulose fibre and vanillin	1950 operational	Sarpsborg, Norway

(continued)

Table 9.2 (continued)

Main investor	Processing technologies	Feedstock	Product	Commercial operations	Location
SP Energy Technology Center	Bio4Energy	Lignocellulose	Demonstration plant	Operational	Sweden
CLaMber Biorefinery	SE pretreatment and fermentation	Agri waste	Demonstration plant 2G sugars	2015 operational	Spain
ZeaChem Demonstration Biorefinery	Thermal Hydrolysis and fermentation Zea2™	Agri waste and wood 10 tonnes/day	2G ethanol and acetic acid	2014 operational	USA
Stora Enso http://www.storaens.com	B2X technology	SCB	Demonstration plant SCB to xylose	Closed 2021	USA

MT = metric tonnes; SCB = sugarcane bagasse; SE = steam explosion; DWB = Dupont's wood based

Debaromyces etc.) or bacteria, broad commercialisation of an industrial biotechnology (IB) process remains tantalisingly elusive. In the UK and Mexico, the new start-ups ARCITEK Bio Ltd (https://www.arcitekbio.co.uk/) and XiliNat (https://www.xilinat.com/) are on the road to commercialising fermentative processes for xylitol production, but are still at a nascent stage. Recently, however, Sweet Appeal Natural Products LLC announced their first commercial sale of xylitol in China that was produced through a fermentation process using corncobs as a feedstock (https://www.sweet-appeal.com). The focus of this chapter is to explore the current commercial status for IB production of xylitol and highlight the challenges and barriers to market.

In order to supply industrial xylitol manufacture at around 10 k tonne per annum, the amount of biomass required is in the order of >100 k tonnes per annum. Future estimates indicate that up to 40 new second generation biorefineries may be constructed and come on-line in Europe by 2030, thereby significantly increasing the market demand for biomass feedstock (EU Biorefinery—outlook to 2030). While the majority of these biorefineries will focus on products other than xylitol, there may be those where xylose is a side stream that will no doubt be suitable for production of xylitol. Integrating production technologies to manufacture multiple products may very well lead to economic and environmental sustainability gains afforded through the energy recovery, economies of scale and minimisation of waste. Furthermore, a key area that can improve the sustainability of xylitol production and reduce climate change impact will be advances in biomass pretreatment technologies that lower energy demands and maximise 2G sugar release (Silva and Chandel 2012).

As with chemical catalysis, there are several challenges to achieving profitable fermentative xylitol production at commercial scale. These include ensuring a regular supply of biomass and developing an economically viable process including pretreatment, bioconversion of xylose to xylitol and downstream processing (DSP) to obtain a pure product, as well as valorisation of all side streams. Pertinent to fermentative production of xylitol is the potential to use crude xylose hydrolysate streams containing contaminants and/or fermentation inhibitors (e.g. phenolics, organic acids, furfurals). These are formed during pretreatment to extract xylose and purification is a costly processing step essential for chemical catalysis. However, the ability to utilise crude hydrolysates is dependent on the microbial capacity to tolerate or detoxify these as they can significantly impair product titres, rates, and yield. Again, this area has been extensively investigated with tolerance being improved using adaptive evolution or synthetic biology approaches but little headway has been made in the commercialisation of these approaches.

Each of the above steps, or unit operations, require optimisation and integration at pilot and demonstration-scale in order to avoid logistical issues that could result in down-time or failure of commercial manufacturing campaigns. The financial cost associated with integrating unit operations at pre-commercial levels can impede growth of small to medium sized enterprises (SME's) and extend the time taken for the innovative technologies/products to reach market. Across Europe however, several regional or national facilities are available to SME's to help develop laboratory scale fermentation process(es) up to pilot and demonstration-scale. Coupled with financial assistance afforded by schemes such as innovation vouchers, the financial barrier is lower leading towards gaining future investment and validation of processes, such as IB production of xylitol, at a commercially relevant level. Some key challenges and possible solutions in fermentative production of xylitol are discussed in the sections below.

9.2 Feedstock Supply

Availability of lignocellulosic biomass at national and international level needs to be assessed when planning construction of a biorefinery (Martínez-Pérez et al. 2007; Akgul et al. 2012; USDA 2011; Alexander et al. 2015; Hodgson et al. 2016; S2Biom 2016; Dahmen et al. 2019; Schröder et al. 2019; Lüders et al. 2020). The EU funded S2Biom project (https://www.s2biom.wenr.wur.nl/) predicted 476 million tonnes of lignocellulosic biomass will be required to fulfil the needs of all biobased industries in Europe by 2030 (S2Biom 2016). Furthermore, the project predicted at least 1 billion tonnes of lignocellulosic biomass will be produced in Europe on an annual basis by 2050. Zaimes et al (2015) discussed numerous supply chain issues, mostly associated with Life Cycle Analysis (LCA) and outcomes of feedstock choices.

A sustainable biorefining industry is dependent on both sustainable feedstock and a sustainable value chain to ensure that continuity of supply meets demand (Dale 2017). As with a xylitol production facility using chemical conversion, an IB xylitol plant based on fermentation process requires at least 90,000–270,000 dry tonnes of feedstock per annum to produce approximately 10,000–30,000 tonnes of xylitol. Furthermore, losses of up to 50% xylitol can be incurred during downstream processing, purification and crystallization, resulting in more biomass being required than may be projected initially. Table 9.1 gives examples of feedstock that can be utilised to provide a hemicellulosic stream for xylitol production. An important consideration in feedstock choice is not only xylose content and yield per hectare but also the xylose to arabinose ratio (X:A) of the hemicellulose as both the chemical and IB xylitol processes can produce arabitol from arabinose. As a food additive xylitol must meet purity criteria set out in Regulation (EU) 231/2012 amended by Regulation (EU) 724/2013 that state the final product must contain <1% of other polyols. As arabitol is an epimer of xylitol it's separation can confound DSP leading to significant product losses. Microbial biocatlaysts have been engineered to reduce arabitol production during xylitol fermentation offering process improvement benefits, however consumer acceptance of this technology remains to be seen (Yoon et al. 2011).

The biomass supply chain needs to be assessed, on a case-by-case basis. Some refineries can be self-sufficient utilising feedstock produced "in house" for xylitol production. For example, the Finnish corporation, Fazer, is building a xylitol production plant in Lahti that will utilise oat hulls derived from processing of oats in its neighbouring oat milling plant. During the past few years, Fazer has invested approximately 40 million euros in constructing a xylitol manufacturing facility that will utilise a chemical hydrogenation process (https://www.fazer.com/about-us/fazerxylitol/). Sugar manufacturing companies can also provide sugarcane bagasse (SCB), an abundant side-stream, as a feedstock for xylitol production that has a high xylan content and can be readily hydrolysed to xylose. The integration and co-production of microcrystalline cellulose alongside an IB xylitol process using SCB is currently being investigated in the BBSRC Newton-Bhabha Innovate UK project, BIOREVIEW (http://www.bioreviewproject.org/). Another example of an industry that produces a waste stream that can supply xylose for xylitol production is the paper and pulp manufacturing sector. Moreover, recent advances in paper and pulp production, allows recovery of a hemicellulosic C5 stream that can be used for large-scale xylitol production. For example, Danisco® Xylitol branded as XIVIA™ is produced using the DuPont Wood Based integration concept (DWB) (https://www.dupontnutritionandbiosciences.com/).

9.3 Steam Explosion Pre-Treatment

2G biomass is inherently recalcitrant and generally requires application of an appropriate pretreatment for the release and hydrolysis of structural carbohydrates from feedstock (William et al. 2017). Various physico-chemical methods are used to deconstruct the complex lignocellulosic cell-wall matrix of plants (i.e. the crosslinked composite of cellulose, hemicellulose and lignin) into multiple components. These treatments release carbohydrates as monomers or oligomers and increase the surface area of polymers for subsequent hydrolysis by cellulolytic enzyme cocktails (Chandel et al. 2020). Steam explosion (SE) is a widely used, scalable and effective pretreatment adopted in lignocellulosic biorefineries (https://www.valmet.com/). During this pre-treatment, the biomass is subjected to high-pressure steam at a temperature of 160–260 °C enabling water molecules to penetrate the biomass. Following rapid decompression, xylose is liberated from the biomass forming a soluble xylose rich hydrolysate (Walker et al. 2018). SE has a long commercial history, is a scalable technology and in the past few decades has been applied by some commercial lignocellulosic plants (Table 9.2).

Over the past decade several companies have built pilot, demonstration, and commercial SE rigs for lignocellulosic biorefineries (Table 9.2). Valmet (https://www.valmet.com/) have supplied a range of scaled BioTrac pretreatment systems to locations globally; a demonstration-scale system in Straubing, Germany and, a commercial scale system in Romania for 2G bioethanol production, both to Clariant. Valmet's BioTrac system, can process more than 250,000 tons of lignocellulosic feedstock (wheat and barley straw) annually and their SE pilot-scale pretreatment systems have been supplied to India and Sweden. In 2021 Valmet was appointed to rebuild the pretreatment system for RE Energy's biorefinery in Kalundborg, Denmark that will process straw to produce 2G bioethanol and lignin. Furthermore, Valmet supplied a demonstration-scale plant for second generation sugar extraction at the Sappi Ngodwana Mill in South Africa, specifically for the extraction of hemicellulosic sugars and lignin from its dissolving wood pulp (DWP) process (http://www.sappi.com). The C5 sugar stream was used to produce xylitol and furfural production using Plaxica's proprietary "Xylex" technology.

In the UK, Nova Pangea Technologies (http://www.novapangaea.com) employ pilot and demonstration-scale SE pretreatment technology that produces C5 sugar streams as part of their REFNOVA process that also produces both lignin char and glucose from the thermolysis of the remaining pulp. Interestingly, despite considerable commercial effort for fuel production there are no 2G ethanol plants that we are aware of separating C5 streams for the co-production of xylitol, despite the positive technoeconomic effect of the higher xylitol selling price lowering the Payback Selling Price of ethanol from €1.62/kg to €0.79/kg (Bioenergy, IEA—2020; ISBN 978-1-910,154-69-4; De Bari et al. 2017).

9.4 Lessons Learned from Industrial Operations

Information relating to the hurdles encountered during the development and deployment of modern integrated biorefineries can be hard to come by and are often cloaked in industrial secrecy, "know-how", with certain processing requirements being both product and market sector specific. However, there are unit operation and process design considerations that require common solutions for all. For example, Slupska and Bushong (2019) highlight 4 key areas of learning that arose from the commercialisation of cellulosic ethanol at POET/DSM's Project LIBERTY plant, with a starting cost of USD $227 million, which would be equally applicable to a modern-day xylitol production facility. These are related to: Biomass collection; interdependence of unit operations; new operation areas; and saccharification and fermentation. The Borregaard Biorefinery in Norway, is arguably the oldest, commercial integrated biorefinery where 20 manufacturing plants are integrated within one production site and controlled with advanced monitoring systems to produce several by-products (http://www.etipbioenergy.eu/images/Factsheet_Borregaard_final.pdf). In this regard, the challenges associated interdependence of unit operations has been successfully addressed at this facility (Rodsrud et al. 2012).

Some additional lessons learned that have been reported by biorefining operations include:

- Continuous feeding of biomass into pretreatment reactors was challenging using agricultural feedstock as they cannot be handled as easily or in the same manner as wood chips Slupska and Bushong (2019).
- Steam explosion technology is a scalable option for 2G biorefining, however several plants have struggled to operate continuously and have temporarily halted or completely terminated operations and are now looking for new investors (i.e. Beta Renewables, CTC-Piracicaba, Brazil (Table 9.2).

- Commercially available, "off the shelf" equipment may not necessarily integrate easily into new processes. For example, SE systems installed at commercial scale in 2G ethanol plants have required integrating with feedstock collection systems to avoid unnecessary downtime to clear blockages associated with removal of molten net wrapping, used to bale the feedstock, from the reactor (Slupska and Bushong 2019).
- Demonstration plants were useful for Integrated process testing—this is of great value for the development of new technologies, to address integration of processes such as hemicellulose hydrolysate separation or extraction of C5 sugars from 2G lignocellulosic ethanol plants.
- Process changes to current 2G lignocellulosic technology to separate C5 streams depends on further process development, integration of C5 hydrolysate stream separation and concentration technologies for high xylose titres in the hydrolysate.

9.5 CASE STUDY: Challenges of Biomass Pretreatment by Steam Explosion

The BEACON Biorefining Centre (http://www.beaconwales.org) at Aberystwyth University (AU) (Fig. 9.1), have operated batch wise, pilot-scale SE since 2013, investigating multiple biomass feedstock for xylitol production and other industrial

Fig. 9.1 BEACON Pilot plant with pilot-scale steam explosion rig and pilot-scale membrane purification systems (http://www.beaconwales.org)

research applications (Table 9.1). They have applied this pretreatment technology to extract hemicellulosic sugars from multiple biomass feedstocks and demonstrated that a combination of SE and acid treatment is effective for extraction from corn cob powder, SCB and grass fibres.

Through process optimisation studies, xylose recovery yields of up to 90% were achieved while minimising co-production of fermentation inhibitors, such as furan-2-carboxaldehyde (furfural), from hydrothermal dehydration of C5 sugars (Walker et al. 2018). The latter outcome is equally important with respect to maximising C5 yield, as the presence of inhibitors in the hydrolysate represents a xenobiotic challenge for microbial bioconversion of xylose to xylitol during the fermentation process (Ravella et al. 2012; Hernández-Pérez et al. 2019). Depending on microbial tolerance these may need to be eliminated or minimized through additional unit operations, such as over liming to remove furans, or more advanced processes such cross-flow ultrafiltration to separate and isolate xylose from the complex hydrolysate matrix. Each of these solutions bear an economic burden, with lime-based detoxification incurring the additional cost of waste disposal and considerable sugar loss. Indeed, the challenge of pretreatment and integration of multiple unit operations alongside the associated logistics were among the key lessons learned by POET and DSM in the commercialisation of cellulosic ethanol at Project LIBERTY (Slupska and Bushong 2019).

The production of inhibitors formed during the SE pretreatment process can also be feedstock dependent and negatively affect fermentation. Exemplifying this point the *Candida tropicalis* strain isolated at IBERS, AU demonstrated good xylitol productivity using xylose in crude Miscanthus and wheat straw hydrolysates, but not from corn stover where furfural levels where up to 100% greater at equivalent pretreatment severities (Somani et al. 2018; Walker et al. 2018). Taking this into consideration, an industrial biorefinery operating an IB fermentative xylitol production process needs to address feedstock choice, continuity of supply, pretreatment and the challenge of process integration.

During process development at AU, the ability to produce sufficient hydrolysate for larger scale xylitol fermentation, >50 L, using a static, batch fed pilot-scale SE rig, presented a challenge to researchers. The Cambi hydrothermal pretreatment system (http://www.cambi.com) (Fig. 9.1) has a reactor capacity of 30 L, where only 15–16 kg feedstock (<1.5 kg per run) could be processed per day to produce 40–50 L of dilute hydrolysate. As a result, it took several months of work to produce sufficient hydrolysate volume for larger scale fermentation trials (up to 160 L) to provide data relating to the scalability of the process.

In collaboration with Bangor University (BU), UK, (BBSRC Newton-Bhabha Innovate UK project BIOREVIEW), BU were able to increase hydrolysate production using a modified continuous steam refining rig for processing wood chips into fibre. Using this system, it was possible to process 100–200 kg of Miscanthus and SCB per day thereby producing the requisite hydrolysate volume for larger scale fermentations. However, optimising the process required addressing several problems. In agreement with Slupska and Bushong (2019) handling, imbibing and feeding biomass into a continuous dynamic pretreatment process differed between wood chips and agricultural feedstock. The latter was more difficult to process and took substantial development to achieve consistent results. For instance, optimised process parameters for batch SE, that released the majority of xylose from SCB, resulted in near complete destruction of the feedstock in the continuous system. Similarly, due to the larger volumes of material and the physical disruption from milling, recovering liberated xylose from the fibre in larger quantities proved both logistically and technically challenging due to the extremely fine particle size.

Process development for the IB production of xylitol has demonstrated that scaling up from a lab to pilot-scale process involves several unit operations, over and above continuous pretreatment for C5 hydrolysate production, namely; counter current washing of pretreated material to maximise xylose recovery; effective solid and liquid separation; and sugar concentration (Fig. 9.2). Although these processes were performed batch-wise, industrially each operation would need to be integrated as a continuous process (Fig. 9.3) and therefore be subject to the interdependence of unit operation complexities highlighted by Slupska and Bushong (2019). Following, optimisation at pilot-scale, thorough validation at demonstration-scale is required prior to commercial production.

Pretreatment operations in a commercial facility would, by necessity, be performed on a continuous basis to accommodate xylose release from several tonnes

Fig. 9.2 Pilot-scale processing of pretreated biomass at Aberystwyth University biorefining centre. **a** Liquid/solid separation by screw press; **b** counter current washing solids to maximise C5 recovery; **c** recovered hydrolysate; **d** C5 sugar concentration by membrane filtration

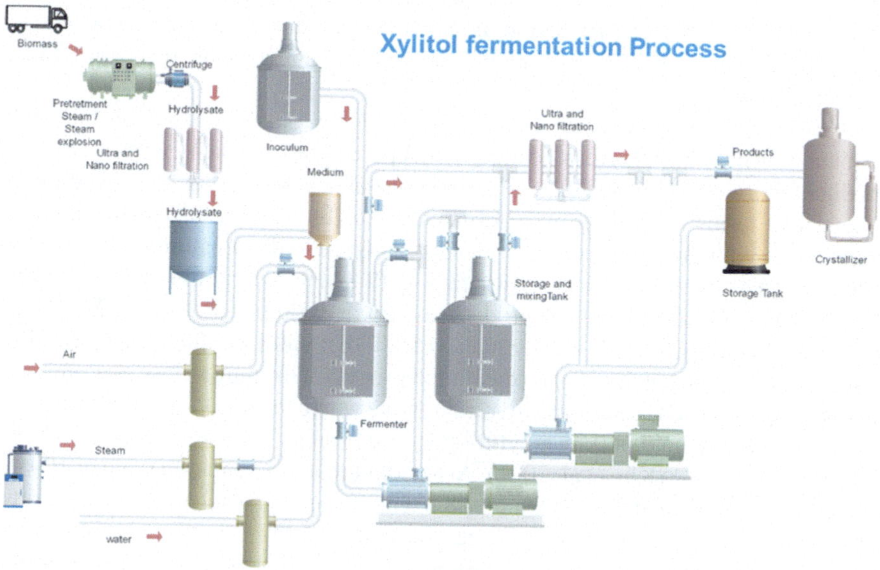

Fig. 9.3 Generalised end-to-end industrial biotechnology xylitol production model

of feedstock per hour. It is important to recognise however, that SE pretreatment systems are big ticket items in terms of CAPEX. Moreover, the transition to a larger-scale, dynamic process represents a new operation area in comparison to static batch-wise processes, thereby presenting financial and logistical scale-up integration challenges.

9.6 Identifying Technology Gaps, Process Changes and Challenges in Scaling-Up

Commercial production of xylitol through a fermentation process involves pretreatment of lignocellulosic biomass, extraction and concentration of a C5 sugar stream (xylose), fermentation of xylose to xylitol using microbes, purification of xylitol from fermentation broth, and finally crystallisation (Fig. 9.3).

Developing a technology process change is a continuous learning experience and these process changes, such as transition from batch-wise, to fed-batch or continuous fermentation, must be evaluated at pilot-scale with further evaluation at demonstration- and commercial- scale. While small to medium sized enterprises (SME's) can access pilot and demonstration-scale biorefining facilities available across Europe and the UK to scale-up sustainable technologies, in global terms, access to such biorefining facilities is limited. The installation, operation and maintenance of pilot/demonstration-scale plants is expensive and requires national and regional investment to assist financing SME access and reduce the fiscal barrier facing fermentative xylitol and other biobased technologies on the road toward commercialization.

Over the past two decades most of the published studies based on lab scale fermentation have been performed using either pure xylose that achieve modest titres, productivity rates and/or yield. Again, one must consider whether these studies are practical in terms of scaling up a fermentative xylitol technology that would be competitive with the incumbent chemically catalysed hydrogenation used commercially. As an example, the Fazer xylitol facility (Table 9.2) currently under construction at Lahti, Finland intends to produce xylitol at 4000 metric tonnes (MT)/year, which if operating for 300 days equals 0.56 T of xylitol product per hour. Now consider a lab-scale batch or fed fermentation process using pure xylose as a substrate that is being proposed as a competitive alternative where the resulting xylitol titre was 187 g L^{-1}, with a yield of 0.75 g xylitol g $xylose^{-1}$ and a volumetric productivity of 3.9 g xylitol L^{-1} h^{-1} (Kim et al. 2002). At a commercial fermentation scale of 100 m^3 production would equal 0.39 MT/h for 48 h with associated discharge, cleaning time and downstream processing costs. Note also that 0.56 MT/h is the amount of final product and conservatively we can assume a figure of 30% product losses occurred during downstream processing. The competitive volumetric productivity target then becomes 0.8 MT/h, approximately double the volumetric productivity developed at lab scale. Obviously, it's possible to increase the amount produced per hour by increasing production capacity, however this would result in increased CAPEX and OPEX with the effect of reducing profitability and potential investor interest.

A fermentative process can use crude hydrolysate and not require xylose purification prior to bioconversion to xylitol, which can offer a technoeconomic advantage over chemical processes. As discussed previously, biomass pretreatment to liberate xylose results in the production of fermentation inhibitors, such as furan-2-carboxaldehyde, and requires the generation and use of inhibitor tolerant yeast and/or bacteria. This area has been widely explored and many reviews are available and in-depth discussion is beyond the scope of this chapter. The key point here is that a costly substrate purification process can be avoided, however improved xylitol productivities are also required to achieve industrial scale–up. In a recent review on *Candida spp.* yeast for xylitol production from agricultural residues and grasses, only one out of 25 studies reported a titre greater than 100 g L^{-1}, productivity of 2.8 g L^{-1} h^{-1} and yield of 0.86 g/g of xylose in 39 h from sugarcane bagasse hydrolysate (West 2021). As discussed in the Fazer competitive scenario, these data suggest that fermentative processes for xylitol production need to be significantly intensified.

Using a combination of commercially realistic production goals, process intensification and the development of high yielding microbial strains, industrially competitive fermentative xylitol production processes could be achieved. For instance, yield has been increased to 1 g xylitol/ g xylose by generating yeast strains deficient in xylitol dehydrogenase that converts xylitol into xylulose for subsequent metabolism in the pentose phosphate pathway (Ko et al. 2006). In an integrated process this would also result in reducing the amount of biomass needing pretreatment in the range of 14–25% with associated OPEX cost savings. However, for process intensification the order of importance to improve production performance are: (1) volumetric productivity of 8–10 g L^{-1} h^{-1}; (2) titre of >10%; and (3) yield of 1:1. Future studies focussing on achieving, or ideally exceeding, these specified targets using crude hydrolysate will help develop a fermentative technology with the commercial potential to bear the costly scale-up development process.

9.7 Capital Investment for Lignocellulosic Biorefineries

Recently, techno-economic analysis (TEA) and LCA for hemicellulosic sugar production from residual 2G feedstock in an integrated small-scale biorefinery was performed (Lopes et al. 2022) (Fig. 9.4). The analysis estimated that production of 2000 tons of xylitol from 30,000 tonnes/year of corn stover, a capital expenditure (CAPEX) of 88.12 million USD was required with an operational cost of 4.66 million USD per year. These financial estimates are in broad agreement with the commercial experience of Oy Karl Fazer Ab (https://www.fazergroup.com/) to finance the reverse integration of a xylitol production plant using oat hulls from their oat milling operations. In 2019, the European Investment Bank (EIB) lent Fazer EUR 40 million in finance towards covering research and development (R&D) and

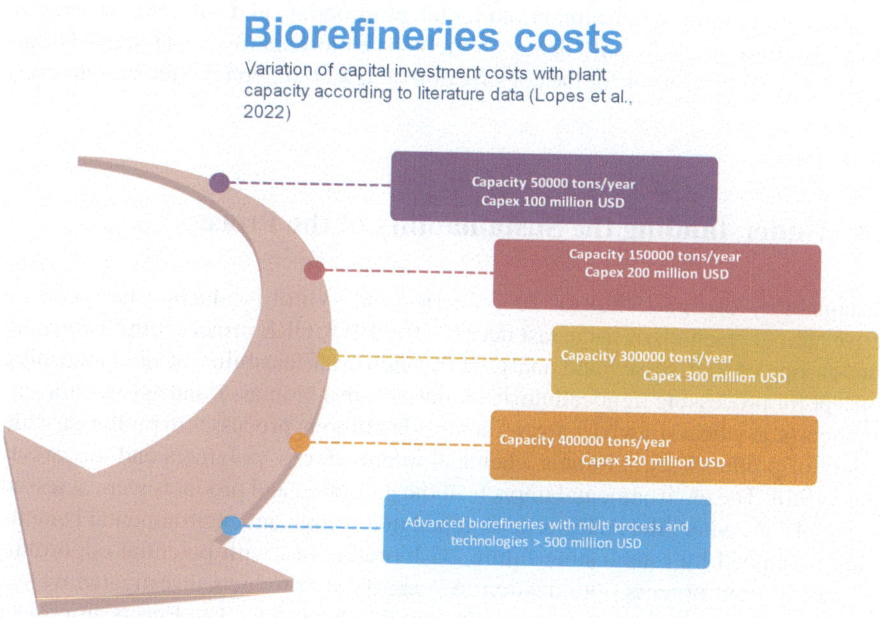

Fig. 9.4 Biorefineries CAPEX cost per tonnes of Lignocellulose biomass processed modified from Lopes et al. 2022

capital expenditure costs, with the total costs approximating to EUR 87 million for a facility to produce 4000 tonnes xylitol per year. In comparison, from 2020 Fortress Global were in the process of operating a 2000 tonnes/year xylitol demonstration unit at its Fortress Specialty Cellulose Mill (FSCM-https://www.fortressge.com/products/) in collaboration with Mondelez International, Inc. (https://www.mondelezinternational.com/). Following successful commissioning and validation of the xylitol demonstration plant, the FSCM has the capacity to supply feedstock for over 20,0000 tonnes per year. Fortress intend to capitalise on this with a further USD $150 million investment for a full-scale plant expected to yield up to $40 million in earnings before interest, taxes, depreciation, and amortization (EBITDA) per year. Within their production process, contaminating alditols, such as arabitol, removed during downstream processing will be valorised by hydrogenation to glycols (IEA Bioenergy: Task 42: Bio-Based Chemicals A 2020 Update).

Another TEA study by Longwen Ou et al. (2020) evaluated the economics of a biorefinery to process 450,000 dry MT of Miscanthus per year to produce sugars and xylitol from the hemicellulose fraction, this refinery cost estimate was with a CAPEX of 379 million USD to 423 million USD based on xylitol from C5 sugars and polyol production from lignin. Franceschin et al., (2011) estimated that a demonstration plant based on 5 MT/h biomass feed rate and 0.374 MT/h xylitol production would cost in the order of 30 million Euros (Fig. 9.5). Whereas the predicted cost of a commercial plant (Fig. 9.6) producing 1.5 MT/h was >100 million Euros.

The recently published study *EU Biorefinery Outlook to 2030* predicts that lignocellulosic biorefining will be important for the bioeconomy and will cost between 81 and 325 million Euros to build demonstration-scale biorefineries, and require investment of 3–13 billion euros for the construction of new commercial scale biorefineries by 2030.

9.8 Understanding the Sustainability of the Process

Sustainability of lignocellulosic biorefineries and xylitol production has been an active area of research over the last decade. The BIOCORE project (http://www.biocore-europe.org/) assessed and analysed the industrial feasibility of the biorefining concept for processing lignocellulosic biomass (forest biomass), and agricultural co-products (e.g. wheat straw, rice straw) alongside different processes to produce a wide variety of products, e.g. biofuels, chemical intermediates, polymers and the sweetener xylitol. The environmental impacts of the processes and products were assessed through LCA, where the analysis determined that significant environmental benefits could be gained by establishing future 2G biorefineries, with potential opportunities arising from process optimization. Among the by-products investigated were a combination of xylitol and itaconic acid or polyester resins. The European project "BIOCORE" recommended multiple by-products based on environmental sustainability assessments of the biorefinery concepts from lignocellulosic biomass (LCB) (O'Donohue 2014).

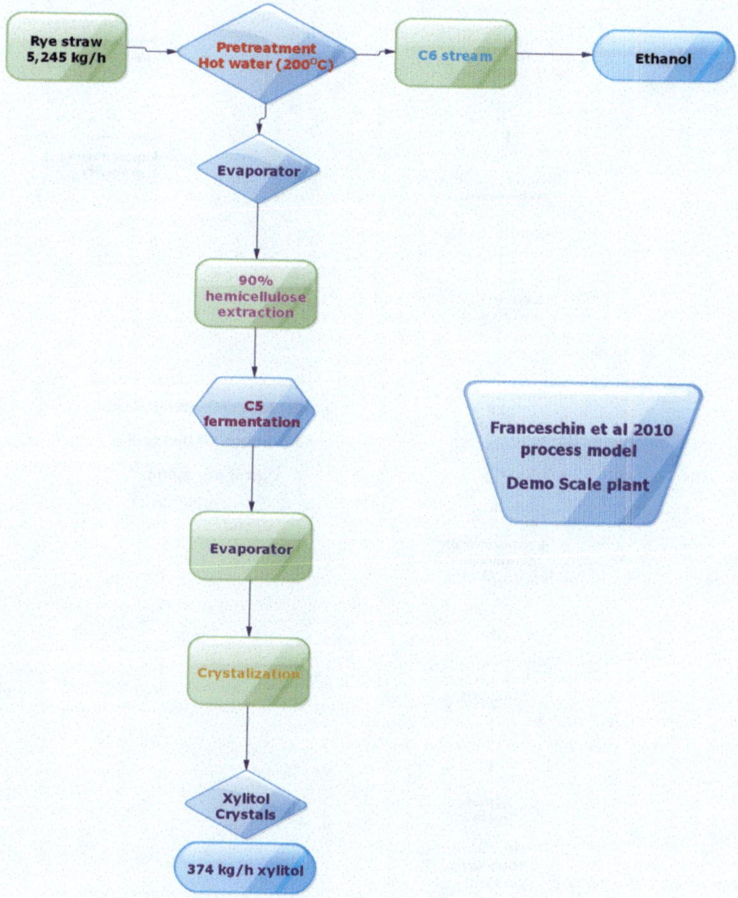

Fig. 9.5 Xylitol production through demonstration plant (Franceschin et al. 2011)

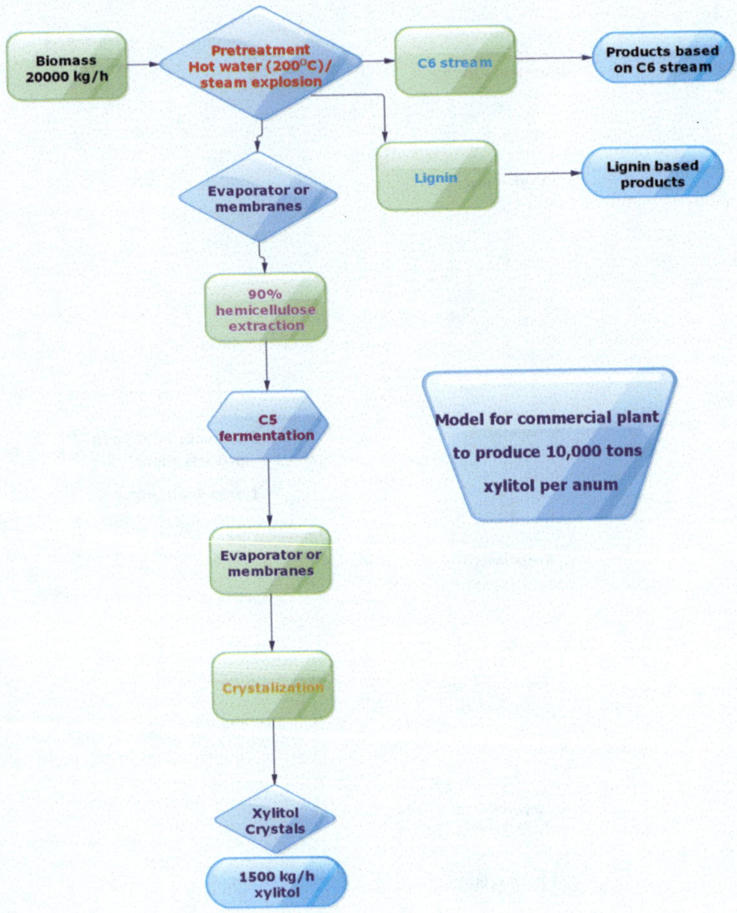

Fig. 9.6 Xylitol production through commercial plant

There have also been lessons learnt about the sustainability of xylitol production from existing commercial enterprises. In China, the commercial xylitol manufacturing process uses corncob as a feedstock for hydrolysis, Biomass Hydrolysis Process (BHP), that generates both water and C6 sugar-based waste, the indiscriminate disposal of which has resulted in environmental pollution. Since 2007, the introduction of new environmental regulations by the Chinese government has resulted in limiting effluent from xylitol production (https://www.foodmanufacture.co.uk/Article/2007/04/27/Still-sweet-on-xylitol#). As a result, these regulations have reduced production capacity of Chinese xylitol producers in order to meet the new wastewater directive.

In contrast, DuPont's (http://www.dupontnutritionandbiosciences.com) Wood Based (DWB) integrated concept to produce xylitol is a more environmentally begin process as the carbon footprint of DWB Xylitol is 90% lower than when production is based on BHP. As with the BIOCORE project, the environmental impacts of the processes and products were assessed. The DWB method requires significantly less energy (85% lower), has less impact on toxicity for both land (94% less) and water (99% less). In the DWB process, the xylose producing facility is integrated with a pulp and paper plant. Moreover, the feedstock originates from sustainably managed forests with the benefit of DuPont's XIVIA™ being more sustainable compared with xylitol manufactured by BHP. These assessments were based on 15 different parameters while evaluating both processes. The major differences were calculated based on Kg CO_2 eq released during the manufacturing of xylitol from wood compared to corncobs.

The IEA Bioenergy Task 42 "Biorefining in a Circular Economy", concluded that to achieve a sustainable biobased and circular economy, several points need to be addressed including reducing fossil fuel dependency, limiting greenhouse gas emissions, designing new processes, developing new technologies, recycling chemicals, and mainly the deployment of new biorefineries in rural locations to develop these areas.

Use of lignocellulosic biomass in integrated biorefineries will improve the environmental sustainability of the green bioeconomy. The above-mentioned BHP and DWB methods produce xylose which is converted to xylitol. During the transition toward production of biobased fuels and chemicals, integrated biorefining technologies enabling C5 stream use for IB production of xylitol and other commodities will improve economic and environmental impact (Fig. 9.7).

9.9 Conclusions and Perspectives

In summary, a xylitol plant requires development of a sustainable process that can be applied at commercial scale, ideally integrated with production of other lignocellulose derived products. The following have been identified as the critical points to be addressed:

Fig. 9.7 Key challenges and possible solutions for developing biotech process for xylitol production

Preliminary process development:

1. Continuity and accessibility of feedstock supply
2. Pretreatment that minimises co-production of fermentation inhibitors
3. Concentration of hydrolysate
4. Selection of micro-organism
5. Fermentation process that maximise rate, titre and yield
6. Purification of product
7. Integration of process steps to maximise efficiency
8. Process validation at both pilot and demonstration-scale need consideration from the beginning
9. Protecting intellectual property and securing investment

Process development needs:

1. Assess biomass availability (>100,000 dry tons per year), transport, supply, biomass size reduction including shredding of bales, chopping, and storage of biomass.

2. Steam explosion-based pretreatment is a high CAPEX process for lignocellulosic biorefineries that may be offset by producing higher value products. Additionally SE or other pretreatment processes should be able to process a variety of feedstock.
3. Availability of large quantity of hydrolysates for large scale fermentation developmental runs is a challenge that is reliant on pretreament interdependency.
4. IB processes are dynamic based on physical, chemical and biological parameters. Fermentation process dynamics will change during scale up and therefore need robus validation at pilot-scale before progressing to demonstration trials.
5. Currently during downstream processing of fermentation broth xylitol results in losses of around 30 to 50%, optimization and step-wise improvements will increase economic viability.

Commercialization needs:

1. De-risk venture capital and commercial failure by evaluating the total process at demonstration scale to enable investor confidence in high CAPEX builds as biorefineries are multimillion-dollar outlays.
2. Pretreatment equipment installation takes time and investment, i.e. the SE rigs need to be ordered, built, and installed based on plant specifications, and feedstock handling specifications. Manufacturing and installations take time as these systems are not able to be purchased readymade.
3. Develop sustainable processes that address the circular economy by producing multiple and higher value products in lignocellulosic biorefineries with standardised LCA low carbon supply chains.
4. Determination to commercialize a new technology.

Despite the above mentioned challenges and barriers, fermentative xylitol production is a promising technology that can be part of lignocellulosic biorefineries, as exemplified in the world's first commercial IB xylitol process started in China. In turn, this should galvanise sugarcane producing countries, such as Brazil and India to capitalise on their in-house sugar mill facilities, feedstock and infrastructure for integrated IB xylitol production (Hernández-Pérez et al. 2019).

Acknowledgements We thank WEFO for BEACON pilot scale biorefining centre and BEACON project (80851), BBSRC/DBT India and INNOVATE UK for Newton Bhabha Funded BIOREVIEW (BB/S011994/1) project.

References

Akgul O, Shah N, Papageorgiou LG (2012) Economic optimisation of a UK advanced biofuel supply chain. Biomass Bioenerg 41:248–255. https://doi.org/10.1016/j.biombioe.2012.01.040

Alexander P, Moran D, Rounsevell MDA (2015) Evaluating potential policies for the UK perennial energy crop market to achieve carbon abatement and deliver a source of low carbon electricity. Biomass Bioenerg 82:3–12. https://doi.org/10.1016/j.biombioe.2015.04.025

BCC research report on Biorefinery Products: Global Markets 2020. https://www.bccresearch.com/
https://www.theccc.org.uk, Biomass in a low-carbon economy, Committee on Climate Change, November 2018
Biorefinery Outlook (http://www.e4tech.com)
Chandel AK, Garlapati VK, Kumar SPJ et al (2020) The role of renewable chemicals and biofuels in building a bioeconomy. Biofuels Bioprod Biorefin 14:830–844
Dahmen N, Lewandowski I, Zibek S et al (2019) Integrated lignocellulosic value chains in a growing bioeconomy: status quo and perspectives. GCB Bioenergy 11:107–117. https://doi.org/10.1111/gcbb.12586
Dale BE (2017) Feeding a sustainable chemical industry: do we have the bioproducts cart before the feedstock horse? Faraday Discuss 202:11–30. https://doi.org/10.1039/c7fd00173h
Felipe Hernández-Pérez A, de Arruda PV, Sene L, da Silva SS, Kumar Chandel A, de Almeida Felipe MDG (2019) Xylitol bioproduction: state-of-the-art, industrial paradigm shift, and opportunities for integrated biorefineries. Crit Rev Biotechnol 39(7):924–943. https://doi.org/10.1080/07388551.2019.1640658. Epub 2019 Jul 16. PMID: 31311338
Franceschin G, Sudiro M, Ingram T et al (2011) Conversion of rye straw into fuel and xylitol: a technical and economical assessment based on experimental data. Chem Eng Res Des 89(6):631–640
Hassan S, Gwilym A Williams, GA, Jaiswal AK (2019) Lignocellulosic Biorefineries in Europe. Current State Prospect Trends Biotechnol 37:231–234
Hernández-Pérez AF, de Arruda PV, Sene L et al (2019) Xylitol bioproduction: state-of-the-art, industrial paradigm shift and opportunities for integrated biorefineries. Crit Rev Biotechnol 39:924–943
Hodgson E et al (2016) Horizon scanning the European bio-based economy: a novel approach to the identification of barriers and key policy interventions from stakeholders in multiple sectors and regions. Biofuels, Bioprod Biorefin 10:508–522. https://doi.org/10.1002/bbb.1665
http://www.beaconwales.org
http://www.bioreviewproject.org/
http://www.dupontnutritionandbiosciences.com/content/dam/dupont/amer/us/en/products/nutrition-health/documents/2e_XIVIA_White_Paper.pdf
http://www.e4tech.com/biorefinery-outlook.php
http://www.energy.gov/eere/bioenergy/poet-dsm-project-liberty
http://www.etipbioenergy.eu/value-chains/conversion-technologies/advanced-technologies/sugar-to-alcohols/commercial-cellulosic-ethanol-plants-in-brazil
http://www.fazergroup.com
http://www.greencarcongress.com/2019/07/20190722-poet.html
http://www.imarcgroup.com
http://www.novapangaea.com
http://www.novasep.com/
http://www.task42.ieabioenergy.com/
https://www.arcitekbio.co.uk/
https://www.bbnet-nibb.co.uk/wp-content/uploads/2019/08/LBNet-UKBioChem10_UK-Top-Bio-based-Chemicals-Opportunities_Dec2017.pdf
https://www.ec.europa.eu/info/sites/default/files/24_axens_julie_maillot.pdf
https://www.eciu.net/netzerotracker
https://www.fazergroup.com/media/news/new-innovation-from-oats--fazer-builds-40-million-euro-xylitol-factory-in-lahti/
https://www.ieabioenergy.com/wp-content/uploads/2020/02/Bio-based-chemicals-a-2020-update-final-200213.pdf
https://www.mondelezinternational.com
https://www.sweet-appeal.com
https://www.valmet.com
https://www.xilinat.com/

https://www.zuchem.dcc.site/news/
http://www.sappi.com
http://www.cambi.com
Kim J-H, Han K-C, Koh Y-H et al (2002) Optimization of fed-batch fermentation for xylitol production by Candida tropicalis. J Ind Microbiol Biotechnol 29:16–19. https://doi.org/10.1038/sj.jim.7000257
Ko BS, Kim J, Kim JH (2006) Production of xylitol from D-xylose by a xylitol dehydrogenase gene-disrupted mutant of Candida tropicalis. Appl Environ Microbiol 72(6):4207–4213. https://doi.org/10.1128/AEM.02699-05. PMID: 16751533; PMCID: PMC1489653
Lopes TF, Duarte LC, Carvalheiro F et al (2022) Production of hemicellulosic sugars from residual lignocellulosic biomass in an integrated small-scale biorefinery: techno-economic and life cycle assessments. In: Brienzo M (ed) Hemicellulose biorefinery: a sustainable solution for value, addition to bio-based products and bioenergy, clean energy production technologies, pp 71–109. https://doi.org/10.1007/978-981-16-3682-0_3
Lüders M, Christian W, Georg V et al (2020) Increase of Miscanthus cultivation with new roles in materials production. Rev Agronom 10:1–21. https://doi.org/10.3390/agronomy10020308
Makinen KK (2000) The rocky road of xylitol to its clinical application. J Den Res 79:1352–1355
Martínez N et al (2007) The potential for hydrogen-enriched biogas production from crops: Scenarios in the UK. Biomass Bioenerg 31:95–104. https://doi.org/10.1016/j.biombioe.2006.07.003
Meyer R (2017) Bioeconomy strategies: contexts, visions, guiding implementation principles and resulting debates. Sustainability 9:2–32. https://doi.org/10.3390/su9061031
O'Donohue M (2014) BIOCORE Final publishable summary report. Toulouse. http://www.biocore-europe.org/file/BIOCOREfinalreport(1).pdf
Ou L, Dou C, Yu JH, Kim H, Park YC, Park S, Kelley S, Lee EY (2021) Techno-economic analysis of sugar production from lignocellulosic biomass with utilization of hemicellulose and lignin for high-value co-products. Biofuels Bioprod Biorefin 15(2):404–415. https://doi.org/10.1002/bbb.2170
Perlack RD, Stokes BJ (Leads) USDA, 2011. U.S. Billion-Ton Update: Biomass Supply for a Bioenergy and Bioproducts Industry. ORNL/TM-2011/224. Oak Ridge National Laboratory, Oak Ridge, TN. 227p
Platt R, Bauen A, Reumerman P et al. EU Biorefinery Outlook to 2030: European Commission, Directorate-General for Research and Innovation, EU biorefinery outlook to 2030 : studies on support to research and innovation policy in the area of bio-based products and services, Publications Office, 2021. https://www.data.europa.eu/doi/10.2777/103465
Ravella SR, Gallagher J, Fish S et al (2012)Overview on commercial production of xylitol, economic analysis and market trends. In: da Silva SS, Chandel AK (eds) D-xylitol. Springer-Verlag, Berlin, Heidelberg, pp 291–306
Rødsrud G, Lersch M, Sjöde A (2012) History and future of world's most advanced biorefinery in operation. Biomass Bioenerg 46:46–59. https://doi.org/10.1016/j.biombioe.2012.03.028
S2Biom (2016) Vision for 1 billion dry tonnes lignocellulosic biomass as a contribution to biobased economy by 2030 in Europe, S2Biom. https://www.s2biom.wenr.wur.nl/
Schröder T et al (2019) Strategic planning of a multi-product wood-biorefinery production system. J Clean Prod 211:1502–1516. https://doi.org/10.1016/j.jclepro.2018.12.004
Silva SS, Chandel AK (2012) D-xylitol: fermentative production, application and commercialisation. ISBN: 978-3-642-31886-3. Springer Verlag, Heidelberg, Germany
Slupska M, Bushong D (2019) Lessons from commercialization of cellulosic ethanol—A POET perspective. Biofuels Bioprod Biorefin 13:857–859. https://doi.org/10.1002/bbb.2033
Somani A, Smith D, Hegarty M et al (2018) Draft genome assemblies of xylose-utilizing *Candida tropicalis* and *Candida boidinii* with potential application in biochemical and biofuel production. Genome Announc 6:e01594-e1617. https://doi.org/10.1128/genomeA.01594-17

Walker DJ, Gallagher J, Winters A et al (2018) Process optimization of steam explosion parameters on multiple lignocellulosic biomass using taguchi method—a critical appraisal. Front Energy Res 6:46. https://doi.org/10.3389/fenrg.2018.0004

Werpy T, Petersen G (2004) Top value added chemicals from biomass: Volume I—Results of screening for potential candidates from sugars and synthesis gas; Report No. DOE/GO-102004-1992; US Department of Energy Office of Scientific and Technical Information, Oak Ridge, TN, USA, 2004. http://www.nrel.gov/docs/fy04osti/35523.pdf

West TP (2021) Xylitol production by Candida Species from hydrolysates of agricultural residues and grasses. Fermentation 7:1–11. https://doi.org/10.3390/fermentation7040243

Williams L, Gallagher J, Bryant D et al (2017) Anaerobic digestion and the use of pre-treatments on lignocellulosic feedstock to improve biogas production and process economics. In: Singh LK, Chaudhary G (eds) Advances in biofeedstock and biofuels: biofeedstock and their Processing, vol 1. Wiley, pp 121–147

Yoon BH, Jeon WY, Shim WY et al (2011) L-arabinose pathway engineering for arabitol-free xylitol production in Candida tropicalis. Biotechnol Lett 33:747–753. https://doi.org/10.1007/s10529-010-0487-2

Zaimes GG, Vora N, Chopra SS et al (2015) Design of sustainable biofuel processes and supply chains: challenges and opportunities. Processes 3:634–663. https://doi.org/10.3390/pr3030634

Open Access This chapter is licensed under the terms of the Creative Commons Attribution 4.0 International License (http://creativecommons.org/licenses/by/4.0/), which permits use, sharing, adaptation, distribution and reproduction in any medium or format, as long as you give appropriate credit to the original author(s) and the source, provide a link to the Creative Commons license and indicate if changes were made.

The images or other third party material in this chapter are included in the chapter's Creative Commons license, unless indicated otherwise in a credit line to the material. If material is not included in the chapter's Creative Commons license and your intended use is not permitted by statutory regulation or exceeds the permitted use, you will need to obtain permission directly from the copyright holder.

Chapter 10
Applications of Xylitol in Food, Material, Health, and Medical Sector

Priscila Vaz de Arruda, Thais Suzane Milessi, Júnia Alves-Ferreira, Luciane Sene, Florbela Carvalheiro, Luís C. Duarte, and Maria das Graças de Almeida Felipe

Abstract Xylitol is considered as one of the top relevant biorefinery sugar-derived products due to its interesting properties and its use has already been approved in more than forty countries. This polyol is mainly recognized in the food sector due to its sweetness equivalent to sucrose but with a significantly lower glycemic index. A great variety of products formulated with xylitol can be found in the market, being most industrially applied in chewing gum production. However, xylitol has others interesting properties, with several clinical applications, acting both on the prevention and treatment of diseases, such as cardiovascular diseases, otitis and diabetes. This chapter comprehensively presents the established and in development xylitol applications, from food to medical sector, including recently discovered COVID-19 treatment potential, and chemical industries applications, exploring xylitol-based materials and its use on tissue-engineering. In addition, xylitol safety and side effects are explored, including the main policies regarding its use.

Keywords Xylitol · Food sector · Nutritional properties · Medical applications · Safety policies

P. V. de Arruda (✉)
Department of Bioprocess Engineering and Biotechnology—COEBB/TD, Federal University of Technology—Paraná (UTFPR), Rua Cristo Rei, 19, Vila Becker, Toledo, PR 85902-490, Brazil
e-mail: priscilaarruda@utfpr.edu.br

T. S. Milessi
Institute of Natural Resources, Federal University of Itajubá, Av. BPS 1300, Itajubá, MG 37500-903, Brazil

J. Alves-Ferreira · F. Carvalheiro · L. C. Duarte
LNEG-Laboratório Nacional de Energia e Geologia, Unidade de Bioenergia e Biorrefinarias, Estrada Do Paço Do Lumiar, 22, 1649-038 Lisboa, Portugal

L. Sene
Medical and Pharmaceutical Sciences Center, State University of Western Paraná, Rua Universitária, 1619 Jardim Universitário, Cascavel, PR 85819-110, Brazil

M. das G. de Almeida Felipe
Department of Biotechnology, Engineering School of Lorena, University of São Paulo, Estrada Municipal Do Campinho s/n, Lorena, SP 12602-810, Brazil

© The Author(s), under exclusive license to Springer Nature Switzerland AG 2022

M. das G. de Almeida Felipe and A. K. Chandel (eds.), *Current Advances in Biotechnological Production of Xylitol*,
https://doi.org/10.1007/978-3-031-04942-2_10

10.1 Introduction

Xylitol is a known acyclic polyhydroxy alcohol or pentitol, comprising five hydroxyl groups (OH), each attached to a carbon atom. It has the molecular formula $C_5H_{12}O_5$, or more explicit as (2*S*, 4*R*)-pentane-1, 2, 3, 4, 5-pentol.

It has several clinical applications (Mäkinen 2000), but it is mainly recognized in the food sector as a relevant additive, especially as a sweetener that can be tolerated by diabetics, but also as an emulsifier, humectant, stabilizer, and thickener (Rice et al. 2020). Table 10.1 presents xylitol properties as compared to other similar additives in the polyols group.

The xylitol molecule has a strong affinity for water and is only slightly soluble in alcohol (1.2 g/100 g solution of 96% ethanol, and 6.0 g/100 g of 96% methanol) (Mussatto 2012), which have to be considered for its formulations. However, xylitol is highly stable to pH variations, which increases its versatility to be used across many different applications.

There are many classes of food products, some chemical products and some health applications in which xylitol has a significant role as shown in Fig. 10.1 and these will be discussed in detail in the following sections.

10.2 Xylitol Applications

10.2.1 Oral and Dental Care

The use of xylitol in oral and dental care is the most known and studied application of this polyol (Fraga et al. 2020; Mäkinen et al. 2013; Salli et al. 2019). Its benefits to oral health are related to four main effects: (i) it prevents caries incidence and

Table 10.1 Xylitol properties as compared to the other polyol additives (Compiled Mitchell 2007; Tiefenbacher 2017; Livesey 2003)

	Mannitol	Sorbitol	Xylitol	Erythritol	Isomalt	Maltitol	Lactitol
Relative sweetness in relation to sucrose = 1	0.6–0.7	0.5–0.6	1	0.5–0.6	0.5–0.6	0.8	0.3–0.4
Glycemic index (glucose = 100)	0	9	13	0	9	35	6
Hygroscopicity	very low	High	Medium	Very low	Low	Low	Medium
Melting point (°C)	165	97	94	121	145 –150	150	122
pH stability in food processing	2–10	2–10	2–10	2–10	2–10	2–10	3–10
Heat stability in food processing (°C)	>160	>160	>160	>160	>160	>160	>160

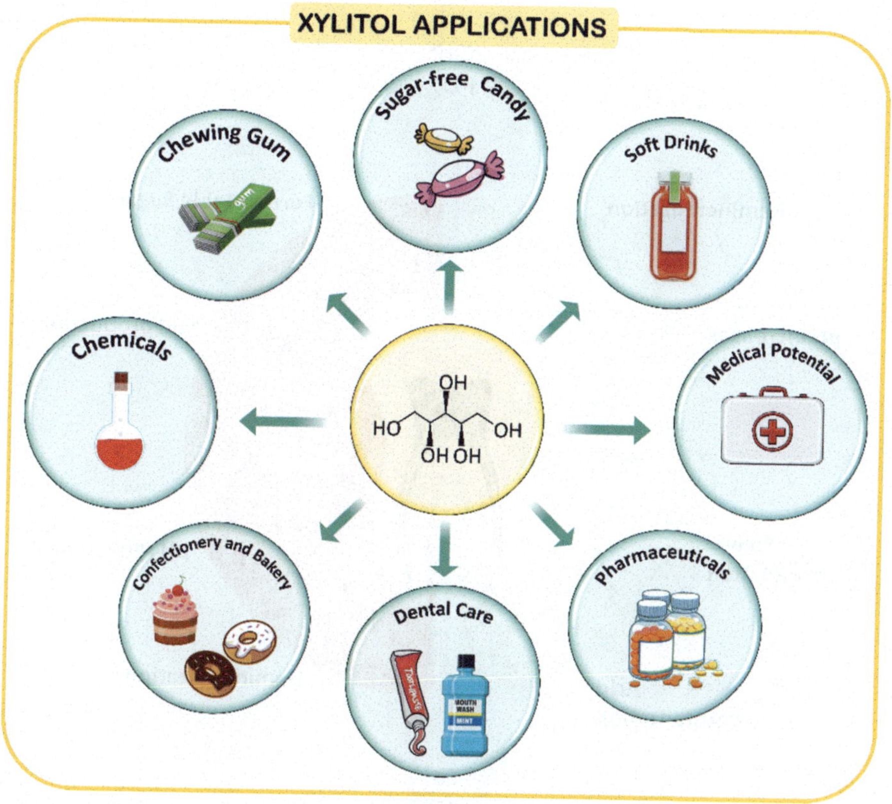

Fig. 10.1 Main classes of food, chemicals and pharmaceutical care products in which xylitol has a significant presence

stabilize the existing ones; (ii) it can promote the dental enamel remineralization; (iii) it reduces the growth of oral pathogenic bacteria preventing plaque formation; and (iv) it buffers the mouth pH preventing the saliva acidification (Mussatto 2012).

Dental caries is caused due to the infection process by oral bacteria, mainly from the genus *Streptococcus*, with compose the oral flora and are found in dental plaque (Ahuja et al. 2020). These bacteria produce mainly lactic acid and extracellular polysaccharides through the fermentation of sugars derived from consumed food, which stimulates the formation of more dental plaque and causes the reduction of the mouth pH (Benahmed et al. 2020), which leads to teeth enamel demineralization and the formation of dental cavities infected by bacteria (caries) (Fig. 10.2). In this sense, the accumulation of bacteria on the teeth and the frequent ingestion of sugars are key factors for caries development (Fraga et al. 2020). However, oral pathogenic bacteria are not able to ferment xylitol, then the use of xylitol limits its proliferation and as a consequence prevents dental plaque and caries formation (Mussatto 2012).

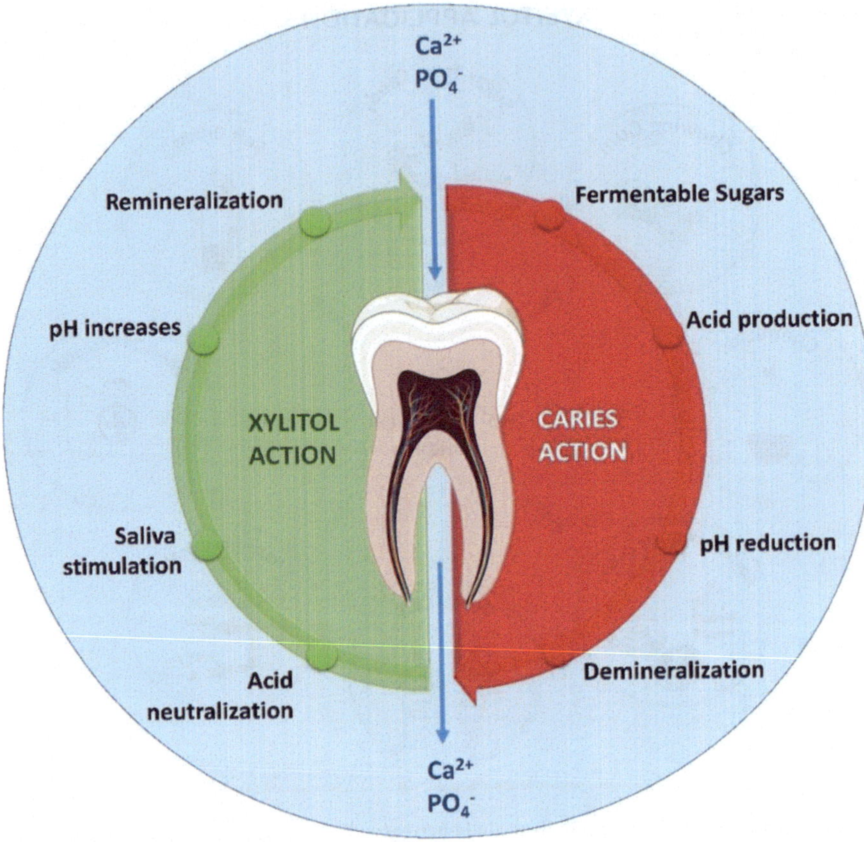

Fig. 10.2 Caries inducing vs caries prevention due to xylitol action

The mechanism for decreasing tooth decay by xylitol occurs through the inhibition of glucosyltransferase, which in turn, inhibits the growth of glucose-fermenting bacteria (i.e., *Streptococcus mutans*), and their adhesion to the tooth surface (Janket et al. 2019). Also, xylitol can bind to calcium ions and stimulate salivary flow (Benahmed et al. 2020), leading to consequent remineralization of teeth enamel (Fig. 10.2).

More recently, the xylitol application combined with other bioactive molecules has been explored, as studies revealed potential synergistic effects and improved remineralization (Ahuja et al. 2020). For instance, a study conducted by Riley et al. (2015) revealed that fluoride toothpaste enriched with 10% of xylitol was superior by 13% in reducing caries when compared to a pure fluoride toothpaste. Cagetti et al. (2020) evaluated the administration of sugar-free chewing gum containing Magnolia bark extract and xylitol and observed that caries lesions, gingival bleeding, dental plaque and *Streptococcus sp.* load were significantly lower compared to the subject group that used chewing-gum containing xylitol only. Gargouri et al. (2018), on the other hand, studied the efficiency of xylitol chewing gum enriched with hydroxyapatite and phosphopeptide-amorphous calcium phosphate and observed enhancement of the eroded teeth remineralization. However, the antimicrobial action decreased when compared with pure xylitol, showing the importance of the two studies to define the best formulation of xylitol application for each oral health objective.

10.2.2 Xylitol as Health Promoter and Applications in the Medical Sector

10.2.2.1 Cardiovascular Diseases

The fermenting bacteria that populate the oral region may spread through the bloodstream and get involved in the pathophysiology of several inflammatory-infectious cascades. Examples are endocarditis, and the raised risk for cardiac and cerebrovascular events since there is evidence that this dissemination may intensify the inflammatory activity in atherosclerotic lesions (Hegde et al. 2012). The direct bacteriostatic effect of xylitol on oral bacteria described above thus can have a positive effect on the prevention of chronic diseases, but there are evidences that its effects can go beyond that. For instances, the fermentation of xylitol by the colon microbiota raises the production of butyrate, a short-chain fatty acid correlated with improved responses to oxidative stress and reduction of lipid peroxidation, favoring the preservation of the endothelial function and prevention of atherosclerosis (Wölnerhanssen et al. 2019).

10.2.2.2 Respiratory Tract Infection

The human respiratory tract is naturally colonized by a microbiota which is responsible to hamper the establishment of exogenous and pathogenic microbes (de Steenhuijsen Piters et al. 2015). The respiratory microbiota is composed mainly of *Staphylococcus* spp, *Moraxella* spp, *Corynebacterium* spp, etc., and its composition varies along the respiratory tract, such as nares, nasopharyngeal and oropharynx (Salli et al. 2019). These microorganisms can spread into the sinus cavity and cause respiratory infections, especially during a viral respiratory infection. The inner surface of human lungs, on the other hand, is coated with a thin layer of antimicrobial substances able to remove inhaled bacteria, preventing pulmonary infections (Mussatto 2012). However, the increase in the salt concentration of this lung layer inhibits the antimicrobial action and leads to chronic infection such as pneumonia (Zabner et al. 2000).

In this sense, xylitol efficiency in the treatment of respiratory infections is related to its low transepithelial permeability and its antimicrobial properties, inhibiting the growth of pathogenic bacteria. Once it is not consumed by most the bacteria and can also decrease salt concentration in the lungs inner layer, stimulating the natural antibiotic activity of the lungs (Ahuja et al. 2020; Mussatto 2012). Most of the studies suggest the application of xylitol in the form of nasal spray to reduce the bacterial load and increase the local defense. Zabner et al. (2000) in a double-blind, randomized, crossover study, sprayed xylitol for 4 days into each nostril of volunteers with and without cystic fibrosis, a human condition that leads to a high propensity for airway bacterial infections, and observed a reduction in the salt concentration of the lung´s layer on healthy subjects. Most remarkably, for volunteers with cystic fibrosis, which generally have the double salt concentration in the lung's layer, xylitol reduced salts up to the levels of healthy individuals. Additionally, the authors observed that using xylitol significantly decreased the number of nasal *Staphylococcus* bacteria compared with a saline solution control group. Also concerning cystic fibrosis, Sajjan et al. (2004) studied xylitol effect on *Burkholderia cepacia,* which is a common pulmonary pathogen during lung transplantation, especially in patients with cystic fibrosis. These authors concluded that the treatment of the human airway explants with xylitol (60–80 mg/mL) inhibited *B. cepacia* growth up to 65%.

Anti-Viral Effects

Xylitol can also be applied to control viral infections. Yin et al. (2014) evaluated the effect of a combination of red ginseng and xylitol in mices infected with H1N1 and observed an increase both in the survival rate and on the soften in influenza infection. Most recently, in the context of the COVID-19 pandemic where the entire world united efforts to develop effective medicines, xylitol is receiving attention as a potential compound to be used in the treatment of SARS-CoV-2. According to Stathis et al. (2021), a combination of xylitol and iota-carrageenan was able to significantly reduce the viral load of SARS-CoV-2 in vitro, in a few seconds. Ciprandi et al. (2021) found that hypertonic nasal saline solution (3%) with xylitol (5%) and hyaluronate (0.2%) could shorten the viral shedding in asymptomatic COVID-19 positive subjects. Varricchio et al. (2021) studied the administration of the same hypertonic nasal saline solution with xylitol and hyaluronate in patients infected by COVID-19 and concluded that this treatment could restore smell and taste in a shorter time, however, the lack of a control group in this study, implies that this effect still requires further investigation. However, the potential of xylitol application for respiratory diseases is clear and these findings bring even more importance to this polyol.

10.2.2.3 Otitis

Acute otitis media (AOM) is a bacterial infection that occurs mainly in childhood between the ages of 6–24 months, with a total incidence of around 80–90% (Danishyar and Ashurst 2021). Among other types of otitis, AOM presents ear pain, middle ear discharge, irritability, and fever, and the use of painkillers and antibiotics is common for its treatment (Uhari et al. 2000). For its prevention, many interventions and environmental factors can reduce its incidence, such as exclusive breastfeeding until 6 months, and no exposure to tobacco. Food supplementation, among others, with zinc and vitamin A and D, as well as the use of xylitol (Danishyar and Ashurst 2021; Dhooge 2020), have also proved beneficial.

The bacterium *Streptococcus pneumoniae* is the main cause of sinusitis and middle ear infections (Kontiokari et al. 1995) and xylitol can prevent or combat the growth of this microorganism (Dhooge 2020). Xylitol has been studied as an AOM-preventive compound since it reduces microbial adhesion to nasopharyngeal cells and alters the expression of *S. pneumoniae* and *Haemophilus influenzae* bacteria (Intakorn et al. 2014). These bacterial species are unable to assimilate xylitol, since it is phosphorylated to xylitol-5-phosphate during the metabolism and the intracellular accumulation of this compound becomes toxic, causing inhibition of glycolytic enzymes and of the bacteria growth, whose survival time is reduced (Kontiokari et al. 1995). The efficacy of xylitol in preventing AOM episodes suggests a dose of 5 g three times a day, however, higher doses and frequency may be necessary (Azarpazhooh et al. 2016; Vernacchio et al. 2014).

10.2.2.4 Diabetes and Weight Management

The use of xylitol as an alternative and low-caloric sweetener for diabetes is explored since the XX century's sixties, once xylitol does not require insulin to be metabolized by humans; being considered as a good energy source for people with diabetes (Benahmed et al. 2020). However, in the last years, this polyol has also gained popularity among overweight people and diabetic patients due to its hypoglycemic, anti-hyperglycemic, anti-diabetic and anti-obesogenic properties (Wölnerhanssen et al. 2019). After taking xylitol, a small increase in blood glucose is observed together with a rapid release of insulin, leading to the control of glucose levels on diabetic and healthy individuals due to the release of insulin (Benahmed et al. 2020).

As described above, xylitol has the same sweetening power as sucrose, but with a significantly lower glycemic index (around 13) and calorific value (2.4 vs. 3.87 kcal/g, for sucrose) (Moriconi et al. 2020; Salli et al. 2019). Xylitol can be absorbed in humans through two main routes: liver and gut, being the former more efficient and permeable to this polyol with a fast absorption (Mussatto 2012). During xylitol assimilation, it is not completely absorbed and the unabsorbed xylitol can be fermented by bacterial gut flora, as described in its prebiotic activities section. The glucose derived from xylitol metabolism is stored as glycogen, being released gradually which do not cause significant changes in blood glucose levels as conventional sugars (Cocate et al. 2011).

Islam (2011) studied the effect of 10%-xylitol solution administration on glycemic control of non-diabetic rats and observed that after 3 weeks the glucose tolerance was improved in the xylitol-consuming group when compared to the control group. Wolnerhanssen et al. (2016) studied the administration of a single oral dose of 50 g xylitol in non-diabetic and obese glucose-intolerant humans and observed a delayed gastric emptying and a small rise in plasma glucose and insulin. Most recently, Argiana et al. (2020) investigated the effect of low-glycemic-index/load desserts prepared with xylitol as a sweetener and high fiber content in patients with type-2 *diabetes mellitus* and observed improvement in terms of glucose and insulin responses when compared to conventional desserts. The authors also observed that xylitol-based desserts seem to suppress hunger and induce satiety. In this sense, xylitol properties can also play an important role in weight management by reducing energy intake and inducing satiety. King et al. (2005) observed that an association between xylitol and polydextrose can be effective as appetite control and appetite suppression.

These results suggest that xylitol application is an appetite-suppressing sugar for overweight and obese humans. However, besides being a good and secure choice of sweetener for patients with diabetes or people who are trying to lose or maintain weight, the long-term data of xylitol influence on weight loss and satiety is inconclusive and need more investigation (Salli et al. 2019).

10.2.2.5 Skin Diseases

Serratia marcescens is a pathogenic Gram-negative bacterium that can cause many of the hospital-acquired infections, wound, invasive burn wound, and soft tissue infections. It is responsible for more than 10% of surgical wound infections related to open burns. It is also able to form biofilms either in tissues or in medical instruments. Unfortunately, it also presents a multidrug-resistant nature. Recently, xylitol was demonstrated to inhibit *S. marcescens* growth (Khayyat et al. 2021). Furthermore, sub-inhibitory concentrations could also affect several processes related to pathogenicity, namely (i) inhibit biofilm formation, (ii) reduce prodigiosin production, (iii) completely block protease activity, (iv) decreased swimming motility, (v) swarming motility, and (vi) increase the sensitivity to hydrogen peroxide. The usefulness of xylitol against *S. marcescens* was also confirmed in in vivo mice survival tests.

Xylitol, in combination with lactoferrin, was also demonstrated to inhibit the growth of established biofilms of a *P. aeruginosa* clinical wound- isolate (Ammons et al. 2011). The prepared lactoferrin/xylitol hydrogel (in combination with silver wound dressings) showed a stronger reduction of the biofilm viability than a commercial wound hydrogel both for *P. aeruginosa* and MRSA (methicillin-resistant *Staphylococcus aureus)* (Ammons et al. 2011). This is particularly relevant, as *P. aeruginosa* and MRSA are the main responsible for skin and soft tissue infections (SSTIs), especially MRSA that accounts for 50% of all SSTIs.

S. aureus also has a significant role in atopic dermatitis, especially for dry skin patients, to whom it is considered one of the most relevant factors to worsen atopic dermatitis, once its toxins and enzymes cause skin damage. Furthermore, it forms a fibrin and glycocalyx biofilm that not only facilitates the adhesion to the skin, but also increases its antibiotic's resistance (Harris 2015). Two in vitro studies showed that a mixture of xylitol and farnesol is an effective agent against *S. aureus*, as xylitol inhibits the formation of the glycocalyx, and farnesol dissolves the fibrin and suppresses bacterial growth. These effects were accomplished without affecting *S. epidermidis*, which is considered as a normal constituent of the human flora (Masako et al. 2005a), although it can induce infection in immune-depressed patients or people with catheters. In addition, this same mixture induced a significant decrease in the *S. aureus* load in other aerobic skin microflora (from 74 to 41%). Furthermore, it improved skin surface hydration. This was shown in a double-blind and randomized study of 17 patients suffering mild to moderate atopic dermatitis on their arms (Masako et al. 2005b).

10.2.2.6 Hemolytic Anemia

Hemolytic anemia is an autoimmune disease characterized by the production of antibodies that react against red blood cells, caused by deficiency of the cytoplasmic enzyme glucose-6-phosphate dehydrogenase (G6PDH), which is responsible for maintaining the appropriate level of the coenzyme NADPH and cell survival (Nkhoma et al. 2009). The hexose monophosphate (HMP) shunt operates in a two-phase process, an oxidative phase, where the NADPH is regenerated by G6PDH enzymes or a non-oxidative phase, which Pentose Phosphate Pathway (PPP) is responsible for precursor biomolecules formation, as illustrated in Fig. 10.3.

The inability of the G6PDH enzyme to regenerate NADPH leads to a decrease in the production of red blood cells since this is an important cofactor in oxidative reactions (Ahuja and Mamtani 2020; Nkhoma et al. 2009; Ylikahri 1979). The metabolism of xylitol by PPP is not depending on the G6PDH enzyme, thus, it supplies the cell with NADPH through the oxidation of xylulose, keeping the integrity of red blood cells (Ahuja et al. 2020). According to these authors, it is common a deficiency of G6PDH in people with malaria disease, due to the consumption of oxidizing chemicals/antimalarial drugs, and the condition eventually results in hemolytic anemia. Thus, for these people, the PPP is the only defense against oxidative damage by generating NADPH, since the mitochondria are absent in red blood corpuscles.

10.2.2.7 Anti-Cancerous and Anti-Inflammatory Activity

Xylitol has natural anti-inflammatory properties by inhibiting the expression of inflammatory cytokines, being extensively applied in the treatment of periodontitis and destruction of gingival tissues prevention (Benahmed et al. 2020). However, in the last decade studies have shown that this property can also be extended to cancer prevention and treatment.

Cancer is usually defined as the uncontrolled growth of cells and, in some cases, can be caused by microbial infection of chronic inflammation. In addition, angiogenesis is a process involved in tumor growth and metastasis, which share a common signaling pathway with inflammations (Yi and Kim 2013). In this sense, the treatment with anti-inflammatory compounds can be an interesting tool for cancer cases. In fact, xylitol was shown to inhibit the proliferation of different types of cancer

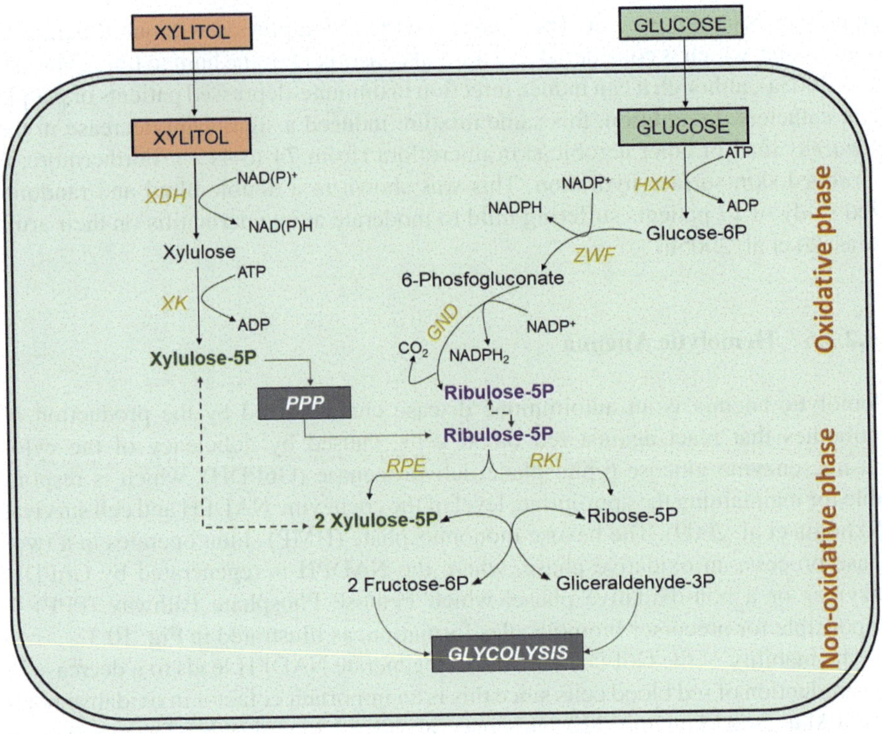

Fig. 10.3 Hexose monophosphate shunt for NADPH regeneration to counter glucose-6-phosphate (G6PDH) deficiency induced hemolytic anemia and xylitol metabolism independent of G6PDH. Abbreviations: HXK = Hexokinase; ZWF = Glucose-6-phosphate-dehydrogenase; GND = 6-phosphogluconate dehydrogenase; RKI = ribose-5-phosphate isomerase; RPE = L-ribulose-5-phosphate 4-epimerase; XDH = xylitol dehydrogenase; XK = xylulokinase; PPP = Pentose phosphate pathway

cells, namely A549, Caki, NCI-H23, HCT-15, HL-60, K562 and SK MEL-2, with higher specificity for human gingival fibroblast cells (Ahuja et al. 2020). Roy and Bhat (2018) investigated the effects of polyols, including xylitol, with an increased number of –OH groups from 2 to 6, on the structure and aggregation of recombinant human γ-Synuclein, which is a model for intrinsically disordered protein reported to play a significant role in neurodegenerative diseases and cancer. Authors found that polyols stabilize the natively unfolded conformation of γ-Syn and delay the structural transition. Qusa et al. (2019) developed a solid dispersion formulation based on olive oil phenolic (−)-oleocanthal and xylitol and observed a high in vivo anti-breast cancer activity by suppressing the human triple-negative breast cancer growth and recurrence after primary tumor surgical excision in nude mice orthotopic xenograft model. According to Tomonobu et al. (2020), the induction of the glutathione-degrading enzyme CHAC1 is the main cause of xylitol-induced apoptotic cell death in cancer cells, once this enzyme induction is required for the endoplasmic reticulum stress that is triggered by xylitol in cancer cells and leads to oxidative stress and eventual apoptotic cell death. According to those authors, a chemotherapeutic combined with xylitol might improve the treatment of cancer patients.

10.2.2.8 Osteoporosis

Osteoporosis is a systemic skeletal disease that causes a decrease in bone mass, increasing bone fragility and the risk of fractures. Xylitol stimulates calcium absorption by the intestine and facilitates its passage from the blood to the bones, increasing the amount of calcium in the bones, preserving the minerals in them and decreasing the weakening of their biomechanical properties and the need for calcium reabsorption. NADH, present during the metabolism of xylitol in the citric acid cycle in the body, is responsible for increasing the transport of calcium ions across the cell membrane and for the synthesis of collagen. Some enzymes in their reduced form can also trigger reactions that preserve the high level of calcium in bones (Mattila et al. 1998a, b, 1996; Sato et al. 2011). Mattila et al. (1998a), described that when animals´ diets were supplemented with an amount of xylitol that varied from 10 to 20% in the formulation of the feed a high peak bone mass could be correlated with higher xylitol doses, which suggest that this molecule is one of the cornerstones in the prevention of osteoporosis. To evaluate whether it is possible to increase bone mass by the means of oral xylitol administration, 3 month old male rats were fed a diet supplemented with xylitol (20% weight/weight) for 8 weeks. Dietary xylitol induced a significant increase in the amounts of calcium and phosphorus in the long bones of the rats, as compared to the controls that were fed the same diet without xylitol (Mattila et al. 1998b).

Despite this important property of xylitol, it was reported that the risk of urolithiasis, could be increased if the intake of this sugar-alcohol occurs for a long time; in that case, metabolic acidosis is the main responsible for that (Janket et al. 2019).

10.3 Xylitol in Pharmaceutical and Medical Industries

Xylitol has high potential as a bioactive compound for pharmaceutical and medical industries due to its properties that provide benefits for oral and dental traits, respiratory system, and prevent or treat important diseases such as diabetes, otitis and, most recently, has shown potential action against Sars-CoV-2 (Fig. 10.4). Due to the great range of pharmaceutical applications, the expected annual growth rate (CAGR) is 7.7% for the pharmaceutical grade xylitol market by 2027 (The Manomet Current 2021). North America is and will continue to be the main consumer of this polyol due to pharmaceutical companies' investments in natural products and the increasing number of dental cases in the region owing to the bad eating habits of consumers,

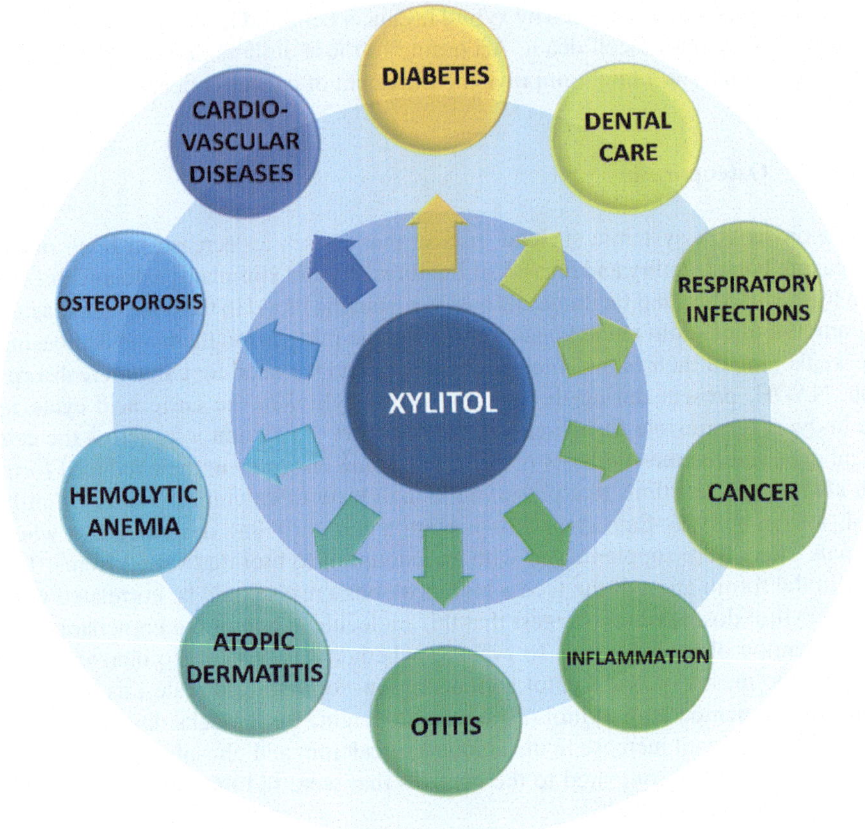

Fig. 10.4 Main applications of xylitol in pharmaceutical and medical industries

which may boost demand for xylitol (Ahuja and Mamtani 2020). The main applications of xylitol in the pharmaceutical and medical industries are fully detailed in the next topics, with special emphasis on the recent advances in each area.

10.3.1 Personal Care and Cosmetics

The use of xylitol in industrial products has already been approved in more than forty countries, for being a non-toxic substance, with high solubility and crystallization capacity, besides providing a fresh feeling (Prakasham et al. 2009). Due to its cariostatic action, it helps in the reduction of oral flora bacteria, as well as increases the level of soluble polysaccharides and makes dental plaque less adherent, facilitating its removal through brushing, being indicated for mouth rinses and toothpaste (Chukwuma and Islam 2018). Due to its antimicrobial potential against skin pathogens, mainly bacteria, including some *Staphylococcus* and *Cutibacterium* acnes it has been used in personal care products (Anglenius and Tiihonen 2020). According to these authors, the concentration level of xylitol influenced the growth-inhibiting and growth-promoting effects on pathogenic microbes, since 1% of xylitol resulted in enhanced *S. epidermidis* growth and inhibition of the other microbes. On the other hand, 5% level inhibited the growth of all pathogenic bacteria evaluated (Anglenius and Tiihonen 2020).

Xylitol has been widely used by the Personal Care industry, especially in moisturizing formulations, due to the several benefits it provides. The main products launched with xylitol for this segment are shampoos, toothpaste, facial care, hair treatments, conditioners, facial cleansers, mouth rinses, among others. The global demand for xylitol in the personal care and cosmetics market was estimated to reach over 9,000 tons by 2025 (Research and Markets 2018).

10.3.2 Xylitol Side Effects

Xylitol is an intermediate product of mammalian metabolism, being present in the human blood in the range between 0.03 and 0.06 mg/100 ml (Mussatto 2012). Besides being well tolerated by humans when ingested in doses of up to 20 g each (with a maximum daily intake of 60 g), it is not digested by human enzymes and approximately 50% of the consumed amount is absorbed by passive diffusion in the intestine (Fraga et al. 2020) or fermented by fecal microbiota. Xylitol can be taken from candies, chewing gums, sweets, and syrups, and for pharmaceutical purposes, it is recommended doses of 5–6 g with a minimum of three exposures per day for a better clinical effect (Benahmed et al. 2020). The ingestion of higher doses results in temporary gastrointestinal disorders, such as diarrhea and flatulence, due to the osmotic pressure increase and imbalance in the large intestine caused by the low rate of xylitol assimilation (Mäkinen 2016; Mussatto 2012). However, this laxative property can also be used to treat constipation, as set before on the subsection of xylitol prebiotic characteristic, being very important its conscious use and the correct dosage. The indiscriminate administration of xylitol can also lead to a high selective pressure in nasal microbiota and the appearance of *S. mutans*, which become xylitol-resistant (Benahmed et al. 2020), highlighting the importance of xylitol correct use for medical purposes.

10.4 Xylitol in Food Industry

10.4.1 *Safety and Legislation*

Xylitol is a permitted food additive recognized by the JECFA (Joint FAO-WHO Expert Committee Report on Food Additives) since 1999 (GSFA 2021). It is listed in the Joint FAO/WHO Codex Alimentarius Commission (2014) in its Table 3 (Additives Permitted for Use in Food in General, Unless Otherwise Specified, in Accordance with GMP) under the INS No 967.[1] Besides FAO, several regional Food Safety authorities issued similar decisions, e.g. FDA (Food Additive Status List—FDA 2019) and EFSA (EFSA—Trusted science for safe food). For instance, in the EU market, xylitol bears the number E-967 and it is classed within Group I and IV: Polyols additives.

Both JECFA and EFSA have given xylitol an Acceptable Daily Intake (ADI[2]) "not specified" which is the highest safety rating that can be given to any food additive. This translates into provisions for use of GMP (Good Manufacturing Practice) and "*quantum satis*" (no maximum level indicated), and as such it can be used as long there is a reasonable technological need.

Nevertheless, this classification is a dynamic endeavor and Food Safety authorities regularly evaluate this and other similar compounds, e.g. the safety of polyols is being re-evaluated by EFSA according to a programme established in (Commission Regulation (EU) 257/2010). For this purpose, EFSA has launched recently (June 2017 and January 2018) calls for the technical and toxicological data on sweeteners authorized as food additives in the EU and for the usage level and/or concentration data in food and beverages intended for human consumption. The current status of a given additive can be continuously monitored through the EU database.[3] Currently authorized uses for xylitol include usage both in solid and liquid form, and its applications range from confectionery including breath refreshing microsweets to Spirit drinks and other alcoholic drinks including mixtures of alcoholic drinks with non-alcoholic drinks and spirits with less than 15% of alcohol, flavored drinks, dried fruit and vegetables, unprocessed mollusks and crustaceans, dry pasta and potato gnocchi (Commission Regulation (EU) 1129/2011; Commission Regulation (EU) 2018/1497).

Besides this legal authorization regarding its safe use, there is a growing concern related to religious/ethical reasons. In this respect, besides the general population, three main groups of consumers are usually considered: on the religious level, the Jewish and Muslim consumers, and on the ethical level, the vegetarians (including, vegan, lacto-vegetarians, ovo-lacto-vegetarians, pescetarians and demi-vegetarians). This translates into the Kosher, Halal and vegetarian food classes, where Kosher is

---

[1] INS No: Additive number in the International Numbering System.

[2] The amount of a food additive that can be consumed in the diet every day throughout life without health risks.

[3] AUTHORISATION OF ADDITIVES (http://www.europa.eu).

the selection and preparation of foods in accordance with traditional Jewish ritual and dietary laws, and Halal is the food conforming to the Islamic (Muslim) dietary laws (Bender 2006). Xylitol is currently listed as Kosher (except for Pesach, e.g. Kosher Search (isitkosherapp.com) and together with other sugar alcohols (namely sorbitol, mannitol, maltitol and erythritol) is generally recognized as Halal, and hence have been approved for use in foods in Muslim countries (Al-Teinaz 2020).

Regarding vegetarian usage, it is also ethically acceptable by the strictest of vegetarians, as a supplement either extracted from plants or prepared (e.g. by fermentation) from other plant sources (Bender 2006).

10.4.2 Food, Confectionery and Bakery Products

Xylitol is present in small amounts in fruits and vegetables, such as plums, strawberries, raspberries, cauliflower, pumpkin, and spinach (Ur-Rehman et al. 2015). Xylitol and other polyols have been applied as sugar replacers, as low-calorie alternatives, in the preparation of food products. However, its application is not restricted to providing sweetness, but also to maintaining other relevant physicochemical properties of sugar, as texture and hardness (Rice et al. 2020).

Xylitol is the sweetest polyol described in the food context, with sweetness equivalent to sucrose. It can be used alone or in combination with other sugar substitutes in the production of several confectionery products that are particularly rich in sugar, such as jellies, hard candies, chewing gum, ice cream and chocolate coatings (Ahuja et al. 2020; Rice et al. 2020). The metabolism of xylitol in the biological system is slower than that of glucose, thus providing an amount of calories 40% lower than sucrose (Kumar et al. 2020).

Xylitol can also be an interesting sugar substitute that enables avoiding browning reactions. Browning/Maillard reactions between reducing sugars and amino acids/proteins are crucial in the chemical stability of foods, giving them a peculiar fragrance and flavor. The absence of aldehyde and ketone groups prevents xylitol from undergoing browning reactions, preventing microbial contamination while acting as a sweetener and preservative. This advantage makes xylitol suitable for the formulation of infant foods, as it does not reduce nutritional value and food quality (Ahuja et al. 2020). Although the addition of xylitol in baked goods provides color and flavor, some browning may occur with the reduction of sugar present in the flour. The color and texture of the cakes also remain similar to the ones made with sucrose. In some cases, cookies prepared with xylitol may have some brown spots, due to the low solubility of xylitol in the dough (Ur-Rehman et al. 2015).

Thus, many works have evaluated the characteristics of products formulated with xylitol and other polyols as compared to sucrose. For example, the addition of xylitol (120 g/L), as a natural sugar substitute in the yoghurt formulation, subjected to 28 days of refrigerated storage (7 °C), showed that there is no harm to the physical-chemical and sensory characteristics of the product compared to sucrose (Costa et al. 2019). The potential of xylitol to be used in confectionery gels at concentrations of 1, 3, 5, 10 and 20% have also been investigated (Cai et al. 2017). The results indicated that the addition of xylitol at low concentrations increases the strength of fish gelatin confectionery gels. However, increased concentration of xylitol (above 3%) can decrease gel network formation, due to increases in the viscosity of the water around the sugar molecules in the continuous phase. Xylitol has also been suggested as a low-calorie bulking agent in the preparation of gluten-free rice chiffon cake, since the texture and sensory properties, in general, were similar to cakes with the addition of sucrose, and even presenting superior traits related to moisture, sweetness and overall appearance (Kim et al. 2014). It was also reported that the use of xylitol in the preparation of sponge cake affected the product quality, since there were changes in the thermal and foaming properties of egg protein, resulting in cakes with smaller volume and harder texture compared to the ones using sucrose (Hao et al. 2016).

10.4.3 Soft Drinks

Consumers are increasingly longing for a healthier lifestyle and well-being and one of the most impacted food sectors is the beverage market, for which drastic changes have been observed since the beginning of the century. Following this consumer trend, beverage manufacturers' interest turned to a market that already existed for quite some time in Japan, the functional drink market (Bawa 2006). Functional drinks are indeed a logical response to consumers' expectations, and popular concepts include boosting the immune system, relieving stress, improving intestinal health, increasing vitality and stamina, controlling cholesterol and body weight, and fighting degenerative diseases (Bawa 2006). As such, the functional beverage market covers many categories, from sports/energy drinks to herbal teas and fruit and vegetable juices. However, the consumer golden rule is that no one is willing to sacrifice taste for health and nutrition benefits. As such, the search for alternative ingredients/additives is still strong. Among the most relevant, polyols, and specifically xylitol, among others, such as erythritol, are particularly well suited for functional beverages, as they are all-natural, non-caloric and non-glycemic bulk sweeteners with taste and mouthfeel enhancing properties, thus helping food designers to ensure a good-tasting quality as well as to provide significant health benefits (Bawa 2006).

The low-energy value of polyols, ranging from 3 kcal/g for xylitol and 0.2 kcal/g for erythritol makes them suitable "candidates" to be used as bulk sweeteners in beverages for the prevention of overweight and/or obesity as well as in weight-management programs (Bawa 2006). Nevertheless, its use in this type of food product is relatively limited as it can cause diarrhea, due to its use in higher amounts than those found in solid foods, and/or its higher potential intake. However, its use in tea and coffee may not cause remarkable gastrointestinal effects, since the intake of these drinks is self-restricting (Mäkinen 2016).

Nevertheless, a recent study conducted in the Spanish market, where the types of low and no-calorie sweeteners are identified on the product labels, showed that the addition of xylitol was not reported in the composition of any drinks and beverages (soft drinks, fruit juices and nectars, sports drinks, energy drinks, vegetable drinks, beverages with mineral salts). The main sugar substitutes present in the overall drinks are acesulfame K and sucralose (Samaniego-Vaesken et al. 2019). The use of xylitol in drinks is reported in countries such as Australia (Lime ReFRESHer, http://www.suncoastlimes.com.au), Korea (Xylitol Citron Tea Drink, http://www.freshjamkorea.com), and Thailand (Scotch Real birds nest, http://www.scotchthailand.co.th).

10.4.4 Chewing Gum

Increasing health awareness in developed and developing markets has resulted in consumers preference for sugar-free chewing gum, thus boosting the market for non-caloric sweeteners such as xylitol (Experts Industries 2014). Chewing gum is by far the main application segment of the xylitol market, surpassing more than 60%. The interest in xylitol for the production of chewing gum is mainly related to its anti-cariogenic activity, together with the high negative solution heat, which generates a strong cooling sensation in the mouth (BeMiller 2019). Xylitol decreases bacterial growth in the mouth compared to products with added sucrose (Ghosh et al. 2019), being considered a non-cariogenic carbohydrate sweetener, as seen above.

A study conducted in a high-caries-risk adult population showed that the long-term use (1 year) of xylitol containing polyol mixture in chewing gum (xylitol, sorbitol, mannitol and maltitol syrup) reduced the risk of caries by 23% as compared to a chewing gum where xylitol was replaced by isomalt (Cocco et al. 2017).

A recently study by Cocco et al. (2020) evaluated the in vitro antibacterial effect and the saliva flow after administration of chewing gums with two different xylitol concentrations (100% and 22% of xylitol). The authors observed that using the concentration of 100% xylitol, saliva levels increased significantly more than the concentration with 22%, showing the relation between this polyol and saliva production stimulation. In addition, cariogenic and periodontal bacteria growth decreased after the use of the two chewing gums, with a greater decrease for the 100% xylitol-based gum. Miake et al. (2003) studying the remineralization of artificially demineralized enamel, immersed in a solution with 20% of xylitol found that after 2 weeks xylitol promoted Ca^{2+} movement and accessibility, inducing deeper layers of remineralization of dental enamel.

Besides its anti-cariogenic ability, xylitol also improves the shelf life, color and flavor of the products; properties such as fast drying and crystallization also make this polyol often used in the coating of pellet forms of sugar-free chewing gum (Ur-Rehman et al. 2015).

The xylitol gum category is divided into two categories: xylitol content above 50% and xylitol content below 50%, the latter being the fastest-growing category. In terms of applications, the global market is segmented especially into breath freshening and tooth protection, with substantial expansion projections for the coming years for the tooth protection segment (forecast 2020–2027) (DATAINTELO 2021). The xylitol gum market is especially dominated by the United States in North America, Brazil in South America, Germany in Europe, China and India in the Asia Pacific. In the Middle East and Africa, the top players are Saudi Arabia, UAE, Egypt, Nigeria, and South Africa (Calibre Research 2021).

10.4.5 Prebiotics and Probiotics Foods

Prebiotics and probiotics[4] play an important role in intestinal health and consequently, consumer awareness and interest in such dietary supplements have been growing for decades (Hutkins et al. 2016). The main groups of prebiotics are the non-digestible fiber and oligosaccharides (Mohanty et al. 2018). Polyols (including xylitol), are also sometimes considered prebiotics (Mohanty et al. 2018).

The effects of low-digestible carbohydrates, such as xylitol, have been investigated in vitro using human fecal cultures (Sato et al. 2016). That study indicated that xylitol stimulates the growth and metabolic activity of *Anaerostipes* spp. in the human colon, promoting the formation of butyrate. This acid can be used in the treatment of colon disease and ulcerative colitis, but its delivery into the colon by oral administration is complicated due to its absorption in the upper part of the gastrointestinal (GI) tract. Butyrate is considered one of the most interesting short-chain fatty acids in the colon mucosa, as it is believed that this compound plays a chemopreventive role in colorectal cancer (Mäkeläinen et al. 2007; Scheppach and Weiler 2004). Thus, ingestion of potential prebiotics such as xylitol can be an alternative to modulate microbial fermentation and increase the production of butyrate in the colon. Studies have indicated which alteration of the metabolic activity of the intestinal microbiota by xylitol affects the metabolism of flavonoids (daidzein) (Tamura et al. 2013), and observed a decrease in the number of *Bacteroides* and a tendency to increase in *Bifidobacterium* and *Prevotella* genus. This was further supported by Uebanso et al. (2017) that showed that xylitol intake of 40 and 194 mg/kg body weight/day significantly alters the intestinal microbiota of mice, leading to an increase in the *Firmicutes* phylum and the genus *Prevotella* and suppressing the growth of the fecal *Bacteroidetes* phylum and the genus *Barnesiella*.

The synergistic effect between xylitol and probiotic bacteria was demonstrated in a study using *Clostridioides difficile*, a microorganism responsible for one of the most prevalent infections in hospitals and nursing homes (Rätsep et al. 2017). The combination of xylitol (5%) with the probiotic *Lactobacillus plantarum* Inducia suppressed the germination of *C. difficile* spores in vitro after 48 h. In addition, pre-feed experiments with a single dose of symbiotic product (*L. plantarum* Inducia 1 ml of 10^{10} CFU/ml and 20% xylitol) or xylitol (20%), increased the hamster

[4] Prebiotics are defined by the International Scientific Association for Probiotics and Prebiotics as a substrate that is selectively used by a host microorganism to produce a health benefit. In general, this effect is mainly associated to the selective growth of bifidobacteria and lactobacilli species in the gastrointestinal tracts *together with the* production of short-chain fatty acids. Probiotics are defined by WHO as "live microorganisms which when administered in adequate amounts confer a health benefit on the host."

survival rates by 78% and 56%, respectively, compared to the 13% survival rate of untreated hamsters. This combination allowed to suppress the germination of spores and outgrowth into vegetative toxin-producing cells of *C. difficile*, as well as declined the pathogenic colonization in the intestine (Rätsep et al. 2017).

Xylitol also has shown to be more efficient than sucralose when added to yoghurts for the survival of the probiotic cultures (*Lactobacillus casei*) in simulated gastrointestinal conditions (Costa et al. 2019).

Nevertheless, and as mentioned before, the high consumption of xylitol and other sugar alcohols can have a laxative effect. Osmotic diarrhea resulting from excessive consumption of these substances is an osmotic response to the presence of carbohydrates that are slowly absorbed into the intestinal lumen, drawing water from surrounding tissues (Mäkinen 2016). On the other hand, when in the large intestine, these compounds can be rapidly fermented into short-chain fatty acids and gases (Nyyssölä et al. 2020). The ability to cause diarrhea is related to the configuration of the molecule. The xylitol molecule is symmetrical and has a lower molar mass, generating less severe gastrointestinal disturbances as compared to molecules of larger size and greater asymmetry (Mäkinen 2016). Thus, people with intestinal disorders, such as irritable bowel syndrome, are advised to avoid polyols ingestion because these can exacerbate gastrointestinal symptoms (Nyyssölä et al. 2020).

The maximum amount of xylitol that can generate gastrointestinal discomfort is quite variable for each individual. Several studies indicate that the maximum dose of xylitol tolerated in adults, after adaptation, is 20 g per dose up to 60 g per day (Culbert et al. 1986). As such, according to the Regulation (EU) 1169/2011, the products with a polyol content starting from 10% should bear a warning statement "excessive consumption may produce laxative effects".

Besides the effect on the GI tract, dietary xylitol plus isoflavonoids can exert a synergistic effect on bone health, with increased production of estrogen metabolites such as equol (Tamura et al. 2013). Unlike glucose, xylitol also can inhibit the growth of *Candida albicans*, and thus decrease the risk of candidiasis and angular cheilitis. In vitro studies have shown that xylitol has an important antifungal effect on *C. albicans* with a minimum inhibitory concentration of 20×10^4 μg/mL and a reduction of 99.95% in colony-forming units at 40×10^4 μg/mL (Talattof et al. 2018). Another study with xylitol-fed mice showed that the growth pattern of *C. albicans* was similar to the control (diet without added carbohydrates), and more than 90% of the animals showed no histological evidence of *C. albicans* invasion of the gastric wall, indicating that xylitol is not used as a substrate by this yeast (Vargas et al. 1993).

Studies demonstrating in situ natural production of polyols in the GI tract represents an emerging field in sugar reduction. Thus, in the appropriate fermentation matrix, polyol producing starter strains such as those mentioned above (Ahuja et al. 2020), can be applied as natural, sugar reducing probiotics assuming they meet regulatory guidelines for use in food (Rice et al. 2020).

10.4.6 *Pet Food and Farm Animals Feed*

Although xylitol is safe for humans, its toxicity has been reported in some animals, in particular dogs. This shows that xylitol assimilation in dogs and humans are quite different. In humans, xylitol assimilation is insulin-independent, but in dogs, it induces a high insulin production (Fawcett et al. 2010). In contrast to dogs, xylitol is much less toxic to cats. Jerzsele et al. (2018) investigated the effect of xylitol in cats using dosages considered toxic for dogs (100, 500 and 1.000 mg/kg body weight), with the potential to cause liver failure or even death. The authors of this study concluded that xylitol at these doses is not toxic to cats, as it did not change the hematological or biochemical parameters analyzed. Only for the highest dose, there was a slight increase in glucose values, which however was within the expected physiological range.

Whereas the intravenous administration of xylitol in rats and horses causes little or no significant increase of insulin in the plasma, in animals such as cows, goats and rabbits it can increase plasma insulin levels similarly or even more markedly than glucose (Kuzuya et al. 1971). According to these authors, xylitol has been reported as a strong stimulator of insulin release for dogs. The contact of pets such as dogs and cats with xylitol, in general, occurs due to the accidental ingestion of xylitol in products used in the domestic environment (Jerzsele et al. 2018). Symptoms like vomiting, anorexia, jaundice, lethargy and weakness were observed in dogs, resulting in hypoglycemia and acute hepatic necrosis. However, the clinical recovery of these domestic animals has been possible after adequate treatment with several drugs (Fawcett et al. 2010).

A detrimental effect was also found to birds. For instance, death from acute hypoglycemia in wild Cape sugarbirds (*Promerops cafer*) species has been reported after ingestion of a homemade solution of xylitol (Gardner and Mitchell 2017). The clinical signs detected, 15–30 min after feeding on the xylitol solution, in most animals were incoordination, weakness, falling from perches, collapse, and death. Gross necropsy performed on some carcasses indicated severe or moderate necrosis or autolysis in various organs and tissues. Cape sugarbirds have very small, rudimentary ceca, and this may indicate low to non-functional capacity for bacterial fermentation of pentose sugars. It is assumed that in these birds, xylitol is rapidly absorbed from the gastrointestinal tract, thus raising insulin levels.

10.5 Xylitol in the Chemical Industry/Xylitol-Based Materials

Since 2004 xylitol is considered as one of the top relevant biorefinery, sugar-derived, pivotal compounds for the chemical industry, as determined by the US Department of Energy (Werpy et al. 2004). This was further strengthened in a second evaluation in 2010 (Bozell and Petersen 2010).

More than a decade has passed, and an updated evaluation of the role of xylitol in the chemical industry is required to realize how these promises were, or not, fulfilled.

Table 10.2 summarizes the four main primary transformation routes for xylitol and its target products: (i) Oxidation, (ii) Dehydration, (iii) Bond cleavage (Hydrogenolysis), and (iv) Direct polymerization. These were considered the base for relevant chemical processes (Werpy et al. 2004).

A first finding is that xylitol has not yet being significantly picked-up by the chemical industry as a major precursor, as its market is still greatly dominated by the food applications described above. Furthermore, earlier estimated market growths, e.g. as reported in (Delgado Arcaño et al. 2020) that xylitol market would value above US$1 Billion by 2022, will also probably not be confirmed (Grand View Research 2021). A careful analysis of the status of the potential xylitol-derived chemical products gives some insights on the reasons for this situation.

For the oxidative route, the main target products are xylaric and xylonic acids, but recent reports describe that both of these acids can be obtained as direct xylose products, which avoids the costs for xylitol production. Xylaric acid production from xylose by the chemical route was demonstrated to be an energy-efficient and low temperature aerobic oxidation process that can be conducted in water using commercial catalysts, e.g. Pt/C. The oxidation occurs preferably at neutral pH and 60 °C, achieving a xylaric acid yield of 64% (Sadula and Saha 2018), which is promising as compared to the performance targets defined before (Werpy et al. 2004). Xylonic acid, can be produced by bioconversion, both by enzymatic conversion from xylan (Lee et al. 2017) and fermentation from xylose (He et al. 2021), and special the later route had gained much attention in recent years. In fact, xylose fermentation already presents fair productivities and yields (He et al. 2021; Liu et al. 2012).

The dehydration of xylitol yields 1,4-anhydroxylitol both in catalyzed (Oltmanns et al. 2013) and non-catalyzed aqueous media (Yamaguchi et al. 2017), a compound, that has many current applications in the pharma, cosmetic and household cleaning products industries.

Table 10.2 Main primary transformation pathways for xylitol, their target products and accomplishments. Based (Werpy et al. 2004)

Primary transformation pathway	Derivative or derivative family	Potential use of derivatives
Oxidations	Xylonic acid Xylaric acid	Concrete dispersant, chelating agent, antibiotic, polyamide, hydrogel modifier Precursor for other chemicals
Dehydration	1, 4-anhydroxylitol	Surfactants, cosmetics, plastic monomers, and raw materials for pharma
Bond cleavage (hydrogenolysis)	Propylene and ethylene glycols	Antifreeze, unsaturated polyester resins
Direct polymerization	Polyesters and nylons	New polymer opportunities

Bond cleavage reactions, most noteworthy, hydrogenolysis can yield glycols as main products, namely propylene glycol and ethylene glycol. These compounds are mainly produced from fossil resources, namely petroleum-based propylene and ethylene via their epoxide intermediates, respectively for propylene glycol and ethylene glycol. As such, their production from renewables is highly desirable, and this was demonstrated at lab-scale (Sun and Liu 2014), where a nearly 100% xylitol conversion and high selectivity of (~61% to almost 70%) to the two glycols, can be achieved under relatively mild conditions (473 K and 4.0 MPa H_2, in the presence of $Ca(OH)_2$). This is a significant advantage over the reported earlier (Werpy et al. 2004) and surpasses the performance requirements. Nevertheless, this process faces competition with processes based on other feedstocks, e.g. glycerol.

Finally, direct polymerization, namely copolymerization with other glycols for the unsaturated polyester resin market has been explored. A thorough review on this subject can be found in (Zhang et al. 2021) and especially below regarding xylitol-based biopolymers.

10.5.1 Xylitol-Based Polymers

Biopolymers are natural polymers entirely produced by living organisms or chemically synthesized from biological material. Biopolymers are composed of similar or repeating units called monomers and have relevant importance in cell biochemistry, playing a structural or functional role, as well acting as storage molecules. Microorganisms synthesize different classes of these biopolymers, such as polysaccharides (composed of sugars and/or sugar acids connected by glycosidic linkages), polyamides (composed of amino acids connected by peptide bonds), polyesters (composed of hydroxy fatty acids linked by ester bonds) and polyphosphates (polyPs; composed of inorganic phosphates linked by anhydride bonds) (Moradali and Rehm 2020).

Many biopolymers have recognized applications in several industrial sectors. Major examples include polysaccharides such as agarose, alginate, carrageenan, cellulose, chitosan, dextran, hyaluronic acid, starch, xanthan gum and others (Di Donato et al. 2014; Shariatinia 2019). Some peptides and polypeptides possess potential applications in drug development, diagnosis, and/or biotechnology (Hayashi et al. 2012; Sable et al. 2017; Vargason et al. 2021).

The use of synthetic biopolymers over naturally occurring materials provides several advantages. Firstly, the method by which they are synthesized allows obtaining polymers with the same composition. Thus, they have unlimited availability and can be produced with a wide range of physical, chemical and mechanical properties that can be modified to improve the properties required for specific applications. Modifications improve heat, moisture resistance, solubility in water, sustainability, flexibility, compatibility, biodegradability and promote their functionalization to superior properties and applications (Dhaliwal and Dosanjh 2018; Ganie et al. 2021).

Under the environmental perspective, polyesters from renewable resources may replace polymers derived from fossil fuel feedstock, and synthetic approaches may render bio-based polyesters with completely innovative properties for novel applications (Vilela et al. 2014). The production of novel biopolymers can also be achieved by synthetic biology for the development of cell factories (Moradali and Rehm 2020).

10.5.2 Xylitol-Based Biopolymers in Tissue Engineering, Wound Healing and Drug Delivery Systems

Bioresorbable polymeric materials are emerging biomaterials to meet demands for biomedical applications such as tissue engineering, drug delivery and wound healing applications (Deepa and Jaisankar 2016).

Sugar alcohols as monomers for polymer synthesis have recently received an appreciable amount of attention from researchers, and have been used as substrates for the synthesis of a wide variety of materials with very different properties and potential industrial applications (Piatek-Hnat et al. 2020).

As polyols are multifunctional alcohols with branched structures whose terminations are -OH groups, these free –OH groups have been utilized to make a variety of polymer structures ranging from cross-linked to linear to star-shaped (Dasgupta et al. 2018).

In this sense, polymers derived from xylitol have gained great interest, especially because xylitol is very accessible, and the resulting biomaterials are biocompatible and biodegradable. Different methodologies for the polymerization of xylitol have been explored using catalysts and solvents (Mosquera et al. 2021).

Many potential medical uses of xylitol-based polyesters have been reported. These include poly(xylitol-co-maleate-co-PEG) (PEG-Polyethylene glycol) hydrogels for cell encapsulation and poly(xylitol sebacate) fiber networks for electrospinning using the core-shell method (Lang et al. 2020).

The most commonly used scaffold polymer in TE is the poly (lactic-co- glycolic acid) (PLGA). PLGA copolymers are made up of various molar ratios of glycolic acid and lactic acid, and their degradative properties are dependent on the molar ratio of poly glycolic acid (PGA) to poly lactic acid (PLA) (Gentile et al. 2014). Xylitol-based hydrogels and elastomers proved to be biocompatible in vitro and in vivo, compared to the prevalent synthetic polymer poly(L-lactic-co-glycolic acid) (PLGA) (Bruggeman et al. 2010, 2008). Anjum et al. (2019) evaluated the antibiofilm activity of xylitol in PLGA/xylitol nanoparticles for the treatment of chronic wound infections. It was demonstrated that PLGA nanoparticles containing xylitol successfully penetrated the EPS (extra polymeric substance) component of the biofilm matrix and potentially overcome the antibiotic resistance associated with the biofilms.

Poly(glycerol sebacate) (PGS) is another soft elastomer suitable for applications in the soft tissue engineering field. However, the rapid degradation kinetics of this polyester has become one of the major drawbacks for applications in tissue engineering. In order to overcome this inconvenience, the development of biodegradable elastomers based on polycondensation reactions of xylitol with sebacic acid referred to as poly(xylitol sebacate) (PXS) elastomers, have recently been the object of several studies (Bruggeman et al. 2010, 2008). In a comparative study on in vitro enzymatic degradation of PGS-and poly(xylitol sebacate) (PXS)-based materials, under static and cyclic flexing conditions, PXS demonstrated significantly slower degradation kinetics than PGS, while having better mechanical flexibility than PGS (Chen et al. 2012).

In another study, dodecanedioic acid (DDA) was added into low modulus poly(xylitol sebacate) (PXS) as the third monomer in order to increase its strength. By varying the monomer rates, a series of novel poly(xylitol sebacate dodecanoate) (PXSD) were synthesized (Sani et al. 2018). When glutamic acid was added as the third monomer into the PXS, the novel l poly (xylitol sebacate glutamate) (PXSG) presented a decreased percent of elongation at break and degradation rate as the glutamic acid ratio increased. Moreover, PXSG had the additional advantage of being synthesized without using any harsh solvents and catalysts (Sani et al. 2021).

A new biocompatible and biodegradable nanocomposite biopolyester PXDDA (poly(xylitol-dodecanedioic acid) has been synthesized based on the chemical reaction between xylitol, DDA and LA as the third element in the polymeric chain. The samples were provided based on the mole fraction of LA and weight percentages of bioglass nanoparticles (n-BG). To meet the desired characteristics in the tissue engineering field, mechanical and hydrophilic properties of PXDDA were controlled by changes in mole ratio between monomers and bioactive glass nanoparticles (n-BG) (Sotoudeh et al. 2021). To increase the effectiveness of growth factors and to improve tissue regeneration, a simple dopamine coating method was developed to load fibroblast growth factor (FGF) on the surface of PXDDA polymeric films. In vitro studies showed that the FGF-polydopamine-PXDDA films have a significant role in supporting the adhesion, spreading, and proliferation of fibroblast cells (Firoozi and Kang 2020).

Two different approaches for obtaining polyesters from xylitol and succinic acid, with and without carboxylic acid activation, were explored by Liz-Andela et al. (2017). While the activation of succinic acid by in situ formation of succinyl chloride yielded a short-chain linear polyester, the direct copolymerization at high temperature produced an amorphous cross-linked material. The thermal properties of the crosslinked materials can be tuned by selecting the appropriate monomer ratio (Liz-Andela et al. 2017). A series of branched polymers instead of cross-linked materials were obtained from the copolymerization of xylitol and succinic acid by rapid microwave synthesis without the use of solvents or catalysts. Although the reaction mechanism is not yet clear, this study raised the possibility of using a rapid and easy polymerization method for the industrial fabrication of materials with a wide range of physical properties (Mosquera et al. 2021).

The success of medical therapy depends on the controlled release of the drugs required for the treatment at the desired anatomical site. Biodegradable polymers can act as carriers of therapeutic agents, as a consequence of the interplay between the diffusion process and the polymer degradation (Macha et al. 2019).

Zhang et al. (2015) developed a drug delivery system consisting of an injectable photocurable composite hydrogel based on methacrylated poly(ethylene glycol)-co-poly(xylitol sebacate) (PEGXS-M) and acrylamidomethyl-b-cyclodextrin (b-CD-NMA) for both hydrophilic and hydrophobic drug release, theophylline and phenethyl alcohol, respectively. These hydrogels demonstrated good biocompatibility with adipose-derived mesenchymal stem cells (ADMSCs). The results suggested that the PEGXS-M-CD composite hydrogels have a great potential in drug delivery for the sustained release of both hydrophilic and hydrophobic drugs. The addition of adamantanamine hydrochloride enhanced the release of the drugs (Zhang et al. 2015).

Due to the great similarity with bone tissue, a bioactive nanohydroxyapatite (HAP)/poly xylitol sebacic adibate (HAP/PXSA)-based composite scaffold was developed for the application of bone tissue engineering with the release of vitamin K (VK). The study revealed improved biocompatibility and more favorable changes in bone regeneration for the VK-loaded HAP/PXSA composite (Dai et al. 2019).

Xylitol based triblock polyesters—poly (xylitol citrate suberate) (PXCSu) were synthesized by the catalyst free melt polycondensation technique and evaluated for antidiabetic drug release by UV-visible spectrophotometer. In vitro drug, antihyperglycemic drug (glimepiride) loading and releasing efficiency results revealed 80% and 70% of drug loading and releasing efficiency, respectively, which indicates a good potential as drug implants for local delivery (Deepa and Jaisankar 2018).

A novel, aspirin-loaded, linear poly (anhydride ester) with xylitol in the backbone was synthesized and designed as poly(2, 3, 4-triaspirin xylitol succinate) (PXAS). Aspirin was incorporated at the non-terminal –OH groups of xylitol and the terminal –OH group was polymerized with a diacid anhydride. This procedure allowed a significant high drug loading (53%) as compared to other systems when the drug is grafted on the polymer and controlled release kinetics of aspirin. Other advantages of this xylitol-based polymer comprise good cytocompatibility and anti-inflammatory properties with potential applications as injectable or an implantable bioactive material (Dasgupta et al. 2017).

10.6 Conclusion

The contribution of scientific research as a basis for creating opportunities for innovations in biomolecules to add value to different industrial sectors is undoubtedly unquestionable and, following the sweeteners, xylitol, one of the top 12 renewable added-value chemicals stands out for its characteristics peculiar and diverse. The demand for this sweetener is growing as well as the possibilities of its applications such as healthcare, food and materials. The characteristics highlighted in health and oral hygiene are the particularity of prevention and reduction of caries inherent to its metabolism, in respiratory diseases by stimulating natural antibiotic activity, in antiviral activity such as reducing the viral load of SARS-COV-2, in the control of blood glucose and weight-related to energy consumption and satiety. In skin diseases by different mechanisms such as bacterial biofilm formation, hemolytic anemia related to the regeneration of enzymatic cofactors, inhibition of different types of cancer cells through the induction of degrading enzymes in these cells, as well as in osteoporosis through stimulation of calcium absorption. In the food industry, due to its physicochemical properties, it has application on different continents as an additive in food and beverages and a growing market including functional foods for a diverse group of consumers. As for its assimilation, this is differentiated in humans when compared to animals such as dogs and rats related to the release of insulin. Also, the characteristics of xylitol as a bioactive compound and observing the recommendations for its use expands the possibilities of its application in the pharmaceutical and medical industries, with a growing market projection in the coming years. On the other hand, advances in research increasingly include xylitol as an input in the chemical industry for the production of a variety of compounds in a biorefinery context, such as biopolymers. Innovations in medical therapy can be made possible by the participation of xylitol in the composition of drugs, which further expands market opportunities with consequent social, environmental and economic contributions considering its peculiar characteristics and production based on renewable sources such as plant biomass rich in xylan.

References

Ahuja K, Mamtani K (2020) Xylitol market size by application (chewing Gum, confectionary, food, personal care, pharmaceuticals, nutraceuticals), downstream application potential (xylaric acid, ethylene glycol, propylene glycol), industry analysis report, regional outlook, industry

Ahuja V, Macho M, Ewe D, Singh M, Saha S, Saurav K (2020) Biological and pharmacological potential of xylitol: a molecular insight of unique metabolism. Foods 9:1–24. https://doi.org/10.3390/foods9111592

Al-Teinaz YR (2020) Halal ingredients in food processing and food additives. Halal Food Handb 149–167. https://doi.org/10.1002/9781118823026.ch10

Ammons MCB, Ward LS, James GA (2011) Anti-biofilm efficacy of a lactoferrin/xylitol wound hydrogel used in combination with silver wound dressings. Int Wound J 8:268–273. https://doi.org/10.1111/j.1742-481X.2011.00781.x

Anglenius H, Tiihonen K (2020) Evaluation of xylitol as an agent that controls the growth of skin microbes: *Staphylococcus aureus*, *Staphylococcus epidermidis*, and *Cutibacterium acnes*. Korean J Microbiol 56:54–58. https://doi.org/10.7845/kjm.2020.0001

Anjum A, Chung PY, Ng SF (2019) PLGA/xylitol nanoparticles enhance antibiofilm activity: via penetration into biofilm extracellular polymeric substances. RSC Adv 9:14198–14208. https://doi.org/10.1039/c9ra00125e

Argiana V, Kanellos PT, Eleftheriadou I, Tsitsinakis G, Perrea D, Tentolouris NK (2020) Low-glycemic-index/load desserts decrease glycemic and insulinemic response in patients with type 2 diabetes mellitus. Nutrients 12:1–9. https://doi.org/10.3390/nu12072153

Azarpazhooh A, Lawrence HP, Shah PS (2016) Xylitol for preventing acute otitis media in children up to 12 years of age. Cochrane Database Syst Rev 2016. https://doi.org/10.1002/14651858.CD007095.pub3

Bawa S (2006) The role of functional beverages with non-caloric and low-calorie sweeteners in the prevention of non-communicable (diet-related) diseases. Agro Food Ind Hi Tech 17:43–47
BeMiller JN (2019) Carbohydrate and noncarbohydrate sweeteners. Carbohydrate Chemistry for Food Scientists. Elsevier Inc., London, Third Edit. https://doi.org/10.1016/b978-0-12-812069-9.00019-4
Benahmed AG, Gasmi A, Arshad M, Shanaida M, Lysiuk R, Peana M, Pshyk-Titko I, Adamiv S, Shanaida Y, Bjørklund G, Org B (2020) Health benefits of xylitol. Appl Microbiol Biotechnol 104:7225–7237. https://doi.org/10.1007/s00253-020-10708-7/Published
Bender DA (2006) Benders' dictionary of nutrition and food technology. Woodhead Publishing Limited, Eighth Edi
Bozell JJ, Petersen GR (2010) Technology development for the production of biobased products from biorefinery carbohydrates—the US Department of Energy's "top 10" revisited. Green Chem 12:539–555. https://doi.org/10.1039/b922014c
Bruggeman JP, Bettinger CJ, Langer R (2010) Biodegradable xylitol-based elastomers: in vivo behavior and biocompatibility. J Biomed Mater Res Part A 95:92–104. https://doi.org/10.1002/jbm.a.32733
Bruggeman JP, Bettinger CJ, Nijst CLE, Kohane DS, Langer R (2008) Biodegradable xylitol-based polymers. Adv Mater 20:1922–1927. https://doi.org/10.1002/adma.200702377
Cagetti MG, Cocco F, Carta G, Maspero C, Campus G (2020) Long-term efficacy of magnolia bark extract and xylitol administered through chewing gums on caries in adults: a 2-year randomized controlled intervention trial. J Funct Foods 68. https://doi.org/10.1016/j.jff.2020.103891
Cai L, Feng J, Regenstein J, Lv Y, Li J (2017) Confectionery gels: effects of low calorie sweeteners on the rheological properties and microstructure of fish gelatin. Food Hydrocoll 67:157–165. https://doi.org/10.1016/j.foodhyd.2016.12.031
Calibre Research (2021) Global Xylitol Chewing Gum Market , Forecast to 2027
Chen Q, Yang X, Li Y (2012) A comparative study on in vitro enzymatic degradation of poly(glycerol sebacate) and poly(xylitol sebacate). RSC Adv 2:4125–4134. https://doi.org/10.1039/c2ra20113e
Chukwuma CI, Islam MS (2018) Xylitol: one name, numerous benefits. Phytochem 409–435. https://doi.org/10.1007/978-3-319-27027-2_33
Ciprandi G, Mantia I La, Brunese FP, Varricchio Al, Varricchio At (2021) Hypertonic saline with xylitol and hyaluronate may shorten the viral shedding duration in asymptomatic COVID-19 positive subjects: a pilot study. J Biol Regul Homeost Agents 35:1151–1154
Cocate PG, Pereira LG, Marins JC, Cecon PR, Bressan J, Alfenas RC (2011) Metabolic responses to high glycemic index and low glycemic index meals: a controlled crossover clinical trial. Nutr J 10. https://doi.org/10.1186/1475-2891-10-1
Cocco F, Cagetti MG, Majdub O, Campus G (2020) Concentration in saliva and antibacterial effect of xylitol chewing gum: in vivo and in vitro study. Appl Sci 10. https://doi.org/10.3390/APP10082900
Cocco F, Carta G, Cagetti MG, Strohmenger L, Lingström P, Campus G (2017) The caries preventive effect of 1-year use of low-dose xylitol chewing gum. A randomized placebo-controlled clinical trial in high-caries-risk adults. Clin Oral Investig 21:2733–2740. https://doi.org/10.1007/s00784-017-2075-5
Commission Regulation (EU) 1129/2011 of 11 November 2011 amending Annex II to Regulation (EC) No 1333/2008 of the European Parliament and of the Council by establishing a Union list of food additives. Off J Eur Union 1–177
Commission Regulation (EU) 2018/1497 of 8 October 2018 amending Annex II to Regulation (EC) No 1333/2008 of the European Parliament and of the Council as regards food category 17 and the use of food additives in food supplements. Off J Eur Union
Commission Regulation (EU) 257/2010 setting up a programme for the re-evaluation of approved food additives in accordance with Regulation (EC) No 1333/2008 of the European parliament and of the Council on food additivies. Off J Eur Union 19–27
Costa GM, Paula MM, Barão CE, Klososki SJ, Bonafé EG, Visentainer JV, Cruz AG, Pimentel TC (2019) Yoghurt added with *Lactobacillus casei* and sweetened with natural sweeteners and/or

prebiotics: implications on quality parameters and probiotic survival. Int Dairy J 97:139–148. https://doi.org/10.1016/j.idairyj.2019.05.007
Culbert SJ, Wang YM, Fritsche HA, Carr D, Lantin E, van Eys J (1986) Oral xylitol in American adults. Nutr Res 6:913–922. https://doi.org/10.1016/S0271-5317(86)80066-5
Dai Z, Dang M, Zhang W, Murugan S, Teh SW, Pan H (2019) Biomimetic hydroxyapatite/poly xylitol sebacic adibate/vitamin K nanocomposite for enhancing bone regeneration. Artif Cells Nanomed Biotechnol 47:1898–1907. https://doi.org/10.1080/21691401.2019.1573183
Danishyar A, Ashurst JV (2021) Acute otitis media. https://www.ncbi.nlm.nih.gov/books/NBK470332/. Accessed 3 June 2021
Dasgupta Q, Madras G, Chatterjee K (2018) Biodegradable polyol-based polymers for biomedical applications. Int Mater Rev 64:288–309. https://doi.org/10.1080/09506608.2018.1505066
Dasgupta Q, Movva S, Chatterjee K, Madras G (2017) Controlled release from aspirin based linear biodegradable poly(anhydride esters) for anti-inflammatory activity. Int J Pharm 528:732–740. https://doi.org/10.1016/j.ijpharm.2017.06.065
DATAINTELO (2021) Xylitol chewing gum market 2020 | Global industry forecast to 2027. https://www.dataintelo.com/report/xylitol-chewing-gum-market/. Accessed 5 Oct 2021
de Steenhuijsen Piters WAA, Sanders EAM, Bogaert D (2015) The role of the local microbial ecosystem in respiratory health and disease. Philos Trans R Soc B Biol Sci. https://doi.org/10.1098/rstb.2014.0294
Deepa K, Jaisankar V (2018) Synthesis and characterisation of antihyperglycemic drug delivery applications of certain xylitol based copolyester. Trends Biomater Artif Organs 32:105–110
Deepa K, Jaisankar V (2016) Synthesis and characterisation of certain biodegradable xylitol based polyesters. Asian J Res Chem 9:679. https://doi.org/10.5958/0974-4150.2016.00094.8
Delgado Arcaño Y, Valmaña García OD, Mandelli D, Carvalho WA, Magalhães Pontes LA (2020) Xylitol: a review on the progress and challenges of its production by chemical route. Catal Today 344:2–14. https://doi.org/10.1016/j.cattod.2018.07.060
Dhaliwal K, Dosanjh P (2018) Biodegradable polymers and their role in drug delivery systems. Biomed J Sci Tech Res 11:8315–8320. https://doi.org/10.26717/bjstr.2018.11.002056
Dhooge I (2020) Acute otitis media in children. BMJ 371:399–411
Di Donato P, Poli A, Taurisano V, Nicolaus B (2014) Polysaccharides: applications in biology and biotechnology/ polysaccharides from bioagro-waste new biomolecules-life. Polysaccharides 1–29. https://doi.org/10.1007/978-3-319-03751-6
EFSA—Trusted science for safe food. https://www.efsa.europa.eu/en. Accessed 19 Oct 2021
Experts Industries (2014) Xylitol—a global market overview. https://www.marketresearch.com/product/sample-8164119.pdf. Accessed 20 Sept 2021
Fawcett A, Phillips A, Malik R (2010) Hypoglycaemia and acute hepatic failure associated with accidental xylitol ingestion in a dog. Aust Vet Pract 40:142–147
Firoozi N, Kang Y (2020) Immobilization of FGF on poly(xylitol dodecanedioic acid) polymer for tissue regeneration. Sci Rep 10:1–8. https://doi.org/10.1038/s41598-020-67261-6
Food Additive Status List—FDA (2019). https://www.fda.gov/food/food-additives-petitions/food-additive-status-list#ftnX. Accessed 19 Oct 2021
de Fraga EG, S., Campos, F.M.T., Cavalcante, M.P. da S., Martins, L.F.B., Neto, E.M.R., Mimica, M.J., (2020) Xylitol, a promising allied for oral heath. J Young Pharm 12:197–200. https://doi.org/10.5530/jyp.2020.12.60
Ganie SA, Ali A, Mir TA, Li Q (2021) Physical and chemical modification of biopolymers and biocomposites. Adv Green Mater Fabr Charact Appl Biopolym Biocompos Woodhead Publ Mater 359–377. https://doi.org/10.1016/B978-0-12-819988-6.00016-1
Gardner BR, Mitchell EP (2017) Acute, fatal, presumptive xylitol toxicosis in Cape sugarbirds (*Promerops cafer*). J Avian Med Surg 31:356–358. https://doi.org/10.1647/2016-234
Gargouri W, Zmantar T, Kammoun R, Kechaou N, Ghoul-Mazgar S (2018) Coupling xylitol with remineralizing agents improves tooth protection against demineralization but reduces antibiofilm effect. Microb Pathog 123:177–182. https://doi.org/10.1016/j.micpath.2018.06.038

Gentile P, Chiono V, Carmagnola I, Hatton PV (2014) An overview of poly(lactic-co-glycolic) Acid (PLGA)-based biomaterials for bone tissue engineering. Int J Mol Sci 15:3640–3659. https://doi.org/10.3390/ijms15033640

Ghosh N, Das A, Sen CK (2019) Nutritional supplements and functional foods: Functional significance and global regulations. In: Bagchi D (ed) Nutraceutical and functional food regulations in the united states and around the world, 3rd edn. Academic Press, pp 13–35. https://doi.org/10.1016/B978-0-12-816467-9.00002-2

Grand View Research (2021) Xylitol market size worth $738.10 Million By 2028 | CAGR: 6.4% https://www.grandviewresearch.com/press-release/global-xylitol-market. Accessed 10 Dec 2021

GSFA (2021) Food additive details—xylitol. https://www.fao.org/gsfaonline/additives/details.html?id=146. Accessed 20 Oct 2021

Hao Y, Wang F, Huang W, Tang X, Zou Q, Li Z, Ogawa A (2016) Sucrose substitution by polyols in sponge cake and their effects on the foaming and thermal properties of egg protein. Food Hydrocoll 57:153–159. https://doi.org/10.1016/j.foodhyd.2016.01.006

Harris R (2015) Phytotherapeutic uses of essential oils. In: Baer K, Buchbauer G (eds) Handbook of essential oils, science, technology, and applications, 2nd edn. Routledge, pp 381–431. https://doi.org/10.1201/b19393-14

Hayashi MAF, Ducancel F, Konno K (2012) Natural peptides with potential applications in drug development, diagnosis, and/or biotechnology. Int J Pept 2012:2012–2014. https://doi.org/10.1155/2012/757838

He T, Xu C, Ding C, Liu X, Gu X (2021) Optimization of specific productivity for xylonic acid production by *Gluconobacter oxydans* using response surface methodology. Front Bioeng Biotechnol 9:1–9. https://doi.org/10.3389/fbioe.2021.729988

Hegde AM, Kavita R, Sushma KS, Shetty S (2012) Salivary sialic acid levels and dental health in children with congenital heart disease. J Clin Pediatr Dent 36:293–296. https://doi.org/10.17796/jcpd.36.3.t585vt7v125630q4

Hutkins RW, Krumbeck JA, Bindels LB, Cani PD, Fahey G, Goh YJ, Hamaker B, Martens EC, Mills DA, Rastal RA, Vaughan E, Sanders ME (2016) Prebiotics: why definitions matter. Curr Opin Biotechnol 37:1–7. https://doi.org/10.1016/j.copbio.2015.09.001

Intakorn P, Sonsuwan N, Noknu S, Moungthong G, Pirçon JY, Liu Y, Van Dyke MK, Hausdorff WP (2014) *Haemophilus influenzae* type b as an important cause of culture-positive acute otitis media in young children in Thailand: a tympanocentesis-based, multi-center, cross-sectional study. BMC Pediatr 14:1–9. https://doi.org/10.1186/1471-2431-14-157

Islam MS (2011) Effects of xylitol as a sugar substitute on diabetes-related parameters in nondiabetic rats. J Med Food 14:505–511. https://doi.org/10.1089/jmf.2010.0015

Janket SJ, Benwait J, Isaac P, Ackerson LK, Meurman JH (2019) Oral and systemic effects of xylitol consumption. Caries Res 53:491–501. https://doi.org/10.1159/000499194

Jerzsele K, Z., Pászti-Gere, E., Sterczer, Bersényi, A., Fodor, K., Szabó, D., Vajdovich, P., (2018) Effects of p.o. administered xylitol in cats. J Vet Pharmacol Ther 41:409–414. https://doi.org/10.1111/jvp.12479

Joint FAO/WHO Codex Alimentarius Commission (2014) Codex Alimentarius : food additives. J Chem Inf Model 53:1689–1699

Khayyat AN, Hegazy WAH, Shaldam MA, Mosbah R, Almalki AJ, Ibrahim TS, Khayat MT, Khafagy ES, Soliman WE, Abbas HA (2021) Xylitol inhibits growth and blocks virulence in *Serratia marcescens*. Microorganisms 9:1–14. https://doi.org/10.3390/microorganisms9051083

Kim J-N, Park S, Shin W (2014) Textural and sensory characteristics of rice chiffon cake formulated with sugar alcohols instead of sucrose. J Food Qual 37:281–290. https://doi.org/10.1111/jfq.12083

King NA, Craig SAS, Pepper T, Blundell JE (2005) Evaluation of the independent and combined effects of xylitol and polydextrose consumed as a snack on hunger and energy intake over 10 d. Br J Nutr 93:911–915. https://doi.org/10.1079/bjn20051431

Kontiokari T, Uhari M, Koskela M (1995) Effect of xylitol on growth of nasopharyngeal bacteria in vitro. Antimicrob Agents Chemother 39:1820–1823. https://doi.org/10.1128/AAC.39.8.1820

Kumar V, Nagar S, Sharma P (2020) Opportunity of plant oligosaccharides and polysaccharides in drug development, carbohydrates in drug discovery and development. Elsevier Inc. https://doi.org/10.1016/b978-0-12-816675-8.00015-4

Kuzuya T, Kanazawa Y, Hayashi M, Kikuchi M, Ide T (1971) Species difference in plasma insulin responses to intravenous xylitol in man and several mammals. Endocrinol Jpn 18:309–320

Lang K, Sánchez-Leija RJ, Gross RA, Linhardt RJ (2020) Review on the impact of polyols on the properties of bio-based polyesters. Polymers (basel) 12:1–25. https://doi.org/10.3390/polym12122969

Lee CC, Kibblewhite RE, Paavola CD, Orts WJ, Wagschal K (2017) Production of D-xylonic acid from hemicellulose using artificial enzyme complexes. J Microbiol Biotechnol 27:77–83. https://doi.org/10.4014/jmb.1606.06041

Liu H, Valdehuesa KNGN, Nisola GM, Ramos KRM, Chung WJ (2012) High yield production of d-xylonic acid from d-xylose using engineered *Escherichia coli*. Bioresour Technol Biocatal 115:244–248. https://doi.org/10.1016/j.biortech.2011.08.065

Livesey G (2003) Health potential of polyols as sugar replacers, with emphasis on low glycaemic properties. Nutr Res Rev 16:163–191. https://doi.org/10.1079/nrr200371

Liz-Andela NI, Benitez R, Martin-Franco J, Rojas G (2017) Tunable polyesterification of xylitol: from linear to crosslinked structures. Polym Int 66:532–539. https://doi.org/10.1002/pi.5301

Macha IJ, Ben-Nissan B, Vilchevskaya EN, Morozova AS, Abali BE, Müller WH, Rickert W (2019) Drug delivery from polymer-based nanopharmaceuticals-an experimental study complemented by simulations of selected diffusion processes. Front Bioeng Biotechnol 7:1–14. https://doi.org/10.3389/fbioe.2019.00037

Mäkeläinen HS, Mäkivuokko HA, Salminen SJ, Rautonen NE, Ouwehand AC (2007) The effects of polydextrose and xylitol on microbial community and activity in a 4-stage colon simulator. J Food Sci 72:153–159. https://doi.org/10.1111/j.1750-3841.2007.00350.x

Mäkinen KK (2016) Gastrointestinal disturbances associated with the consumption of sugar alcohols with special consideration of xylitol: scientific review and instructions for dentists and other health-care professionals. Int J Dent 2016. https://doi.org/10.1155/2016/5967907

Mäkinen KK (2000) The rocky road of xylitol to its clinical application. J Dent Res 79:1352–1355. https://doi.org/10.1177/00220345000790060101

Mäkinen KK, Järvinen KL, Anttila CH, Luntamo LM, Vahlberg T (2013) Topical xylitol administration by parents for the promotion of oral health in infants: a caries prevention experiment at a finnish public health centre. Int Dent J 63:210–224. https://doi.org/10.1111/idj.12038

Masako K, Hideyuki I, Shigeyuki O, Zenro I (2005a) A novel method to control the balance of skin microflora: Part 1. attack on biofilm of *Staphylococcus aureus* without antibiotics. J Dermatol Sci 38:197–205. https://doi.org/10.1016/j.jdermsci.2005.01.006

Masako K, Yusuke K, Hideyuki I, Atsuko M, Yoshiki M, Kayoko M, Makoto K (2005b) A novel method to control the balance of skin microflora: part 2. A study to assess the effect of a cream containing farnesol and xylitol on atopic dry skin. J Dermatol Sci 38:207–213. https://doi.org/10.1016/j.jdermsci.2005.01.003

Mattila PT, Knuuttila MLE, Svanberg MJ (1998a) Dietary xylitol supplementation prevents osteoporotic changes in streptozotocin-diabetic rats. Metabolism 47:578–583. https://doi.org/10.1016/S0026-0495(98)90243-8

Mattila PT, Svanberg MJ, Knuuttila MLE (1998b) Dietary xylitol protects against osseal changes experimental osteoporosis. In: Burckhardt P, Dawson-Hughes B, Heaney RP (eds) Nutritional aspects of osteoporosis. Springer-Verlag, New York, pp 157–162

Mattila PT, Svanberg MJ, Mäkinen KK, Knuuttila MLE (1996) Dietary xylitol, sorbitol and D-mannitol but not erythritol retard bone resorption in rats. J Nutr 126:1865–1870. https://doi.org/10.1093/jn/126.7.1865

Miake Y, Saeki Y, Takahashi M, Yanagisawa T (2003) Remineralization effects of xylitol on demineralized enamel. J Electron Microsc

Mitchell H (2007) Sweeteners and sugar alternatives in food technology. Blackwell Publishing Ltd. https://doi.org/10.1002/9780470996003

Mohanty D, Misra S, Mohapatra S, Sahu PS (2018) Prebiotics and synbiotics: recent concepts in nutrition. Food Biosci 26:152–160. https://doi.org/10.1016/j.fbio.2018.10.008

Moradali MF, Rehm BHA (2020) Bacterial biopolymers: from pathogenesis to advanced materials. Nat Rev Microbiol 18:195–210. https://doi.org/10.1038/s41579-019-0313-3

Moriconi E, Feraco A, Marzolla V, Infante M, Lombardo M, Fabbri A, Caprio M (2020) Neuroendocrine and metabolic effects of low-calorie and non-calorie sweeteners. Front Endocrinol (lausanne). https://doi.org/10.3389/fendo.2020.00444

Mosquera M, Orozco F, Benítez R, Martin J, Rojas G (2021) Controlled branching by step-growth polymerization of xylitol and succinic acid via microwave irradiation. ACS Omega 6:13987–13994. https://doi.org/10.1021/acsomega.0c05875

Mussatto SI (2012) Application of xylitol in food formulations and benefits for health. In: Da Silva SS, Chandel AK (eds), D-Xylitol: fermentative production, application and commercialization. Springer, Berlin, pp 309–323. https://doi.org/10.1007/978-3-642-31887-0_14

Nkhoma ET, Poole C, Vannappagari V, Hall SA, Beutler E (2009) The global prevalence of glucose-6-phosphate dehydrogenase deficiency: a systematic review and meta-analysis. Blood Cells. Mol Dis 42:267–278. https://doi.org/10.1016/j.bcmd.2008.12.005

Nyyssölä A, Ellilä S, Nordlund E, Poutanen K (2020) Reduction of FODMAP content by bioprocessing. Trends Food Sci Technol 99:257–272. https://doi.org/10.1016/j.tifs.2020.03.004

Oltmanns JU, Palkovits S, Palkovits R (2013) Kinetic investigation of sorbitol and xylitol dehydration catalyzed by silicotungstic acid in water. Appl Catal A Gen 456:168–173. https://doi.org/10.1016/j.apcata.2013.02.023

Piatek-Hnat M, Bomba K, Pęksiński J (2020) Structure and properties of biodegradable poly (xylitol sebacate-co-butylene sebacate) copolyester. Molecules 25. https://doi.org/10.3390/molecules25071541

Prakasham RS, Rao RS, Hobbs PJ (2009) Current trends in biotechnological production of xylitol and future prospects. Curr Trends Biotechnol Pharm 3:8–36

Qusa MH, Siddique AB, Nazzal S, El Sayed KA (2019) Novel olive oil phenolic (−)-oleocanthal (+)-xylitol-based solid dispersion formulations with potent oral anti-breast cancer activities. Int J Pharm 569. https://doi.org/10.1016/j.ijpharm.2019.118596

Rätsep M, Kõljalg S, Sepp E, Smidt I, Truusalu K, Songisepp E, Stsepetova J, Naaber P, Mikelsaar RH, Mikelsaar M (2017) A combination of the probiotic and prebiotic product can prevent the germination of *Clostridium difficile* spores and infection. Anaerobe 47:94–103. https://doi.org/10.1016/j.anaerobe.2017.03.019

Regulation (EU) 1169, 2011 of The European Parliament and of The Council of 25, (October 2011) on the provision of food information to consumers, amending Regulations (EC) No 1924/2006 and (EC) No 1925/2006 of the European Parliament and of the Council, an. Off J Eur Union 18–63. https://doi.org/10.1075/ttwia.27.04ker

Research and Markets (2018) Xylitol in personal care and cosmetics market size, share & trends analysis report by application (skin, hair, oral), by Region (Asia Pacific, North America), and segment forecasts, 2018–2025. https://www.grandviewresearch.com/industry-analysis/xylitol-in-personal-care-cosmetics-market. Accessed 15 Sept 2021

Rice T, Zannini E, Arendt K, E., Coffey, A., (2020) A review of polyols–biotechnological production, food applications, regulation, labeling and health effects. Crit Rev Food Sci Nutr 60:2034–2051. https://doi.org/10.1080/10408398.2019.1625859

Riley P, Moore D, Ahmed F, Sharif MO, Worthington HV (2015) Xylitol-containing products for preventing dental caries in children and adults. Cochrane Database Syst Rev 37–38. https://doi.org/10.1002/14651858.CD010743.pub2

Roy S, Bhat R (2018) Effect of polyols on the structure and aggregation of recombinant human γ-Synuclein, an intrinsically disordered protein. Biochim Biophys Acta Proteins Proteomics 1866:1029–1042. https://doi.org/10.1016/j.bbapap.2018.07.003

Sable R, Parajuli P, Jois S (2017) Peptides, peptidomimetics, and polypeptides from marine sources: a wealth of natural sources for pharmaceutical applications. Mar Drugs 15. https://doi.org/10.3390/md15040124

Sadula S, Saha B (2018) Aerobic oxidation of xylose to xylaric acid in water over pt catalysts. Chemsuschem 11:2124–2129. https://doi.org/10.1002/cssc.201800494

Sajjan U, Moreira J, Liu M, Humar A, Chaparro C, Forstner J, Keshavjee S (2004) A novel model to study bacterial adherence to the transplanted airway: inhibition of *Burkholderia cepacia* adherence to human airway by dextran and xylitol. J Hear Lung Transplant 23:1382–1391. https://doi.org/10.1016/j.healun.2003.09.023

Salli K, Lehtinen MJ, Tiihonen K, Ouwehand AC (2019) Xylitol's health benefits beyond dental health: a comprehensive review. Nutrients. https://doi.org/10.3390/nu11081813

de Samaniego-Vaesken M, L., Partearroyo, T., Cano, A., Urrialde, R., Varela-Moreiras, G., (2019) Novel database of declared low- and no-calorie sweeteners from foods and beverages available in Spain. J Food Compos Anal 82:103234. https://doi.org/10.1016/j.jfca.2019.103234

Sani NFM, Jafri NA, Majid NA, Rosaidi NA, Onn M (2021) Effect of glutamic acid as additional monomer in biodegradable poly(xylitol sebacate glutamate) polymer. Egypt J Chem 64:2783–2787. https://doi.org/10.21608/ejchem.2021.54273.3128

Sani NFM, Wahit MU, Tham WH, Majid NAA (2018) Biodegradable poly(xylitol sebacate dodecanoate)/nano-hydroxyapatite composite for potential used in biomedical applications. AIP Conf Proc 1985. https://doi.org/10.1063/1.5047157

Sato H, Ide Y, Nasu M, Numabe Y (2011) The effects of oral xylitol administration on bone density in rat femur. Odontology 99:28–33. https://doi.org/10.1007/s10266-010-0143-2

Sato T, Kusuhara S, Yokoi W, Ito M, Miyazaki K (2016) Prebiotic potential of L-sorbose and xylitol in promoting the growth and metabolic activity of specific butyrate-producing bacteria in human fecal culture. FEMS Microbiol Ecol 93:5–11. https://doi.org/10.1093/femsec/fiw227

Scheppach W, Weiler F (2004) The butyrate story: old wine in new bottles? Curr Opin Clin Nutr Metab Care 7:563–567. https://doi.org/10.1097/00075197-200409000-00009

Shariatinia Z (2019) Pharmaceutical applications of natural polysaccharides. In: Hasnain MS, Nayak AK (eds) Natural polysaccharides drug delivery and biomedical applications. Academic Press. https://doi.org/10.1016/B978-0-12-817055-7.00002-9

Sotoudeh A, Darbemamieh G, Goodarzi V, Shojaei S, Asefnejad A (2021) Tissue engineering needs new biomaterials: poly(xylitol-dodecanedioic acid)–co-polylactic acid (PXDDA-co-PLA) and its nanocomposites. Eur Polym J 152:110469. https://doi.org/10.1016/j.eurpolymj.2021.110469

Stathis C, Victoria N, Loomis K, Nguyen SA, Eggers M, Septimus E, Safdar N (2021) Review of the use of nasal and oral antiseptics during a global pandemic. Fut Microbiol 16:119–130. https://doi.org/10.2217/fmb-2020-0286

Sun J, Liu H (2014) Selective hydrogenolysis of biomass-derived xylitol to ethylene glycol and propylene glycol on Ni/C and basic oxide-promoted Ni/C catalysts. Catal Today 234:75–82. https://doi.org/10.1016/j.cattod.2013.12.040

Talattof Z, Azad A, Zahed M, Shahradnia N (2018) Antifungal activity of Xylitol against *Candida albicans*: an in vitro study. J Contemp Dent Pract 19:125–129. https://doi.org/10.5005/JP-JOURNALS-10024-2225

Tamura M, Hoshi C, Hori S (2013) Xylitol affects the intestinal microbiota and metabolism of daidzein in adult Male mice. Int J Mol Sci 14:23993–24007. https://doi.org/10.3390/ijms141223993

The Manomet Current (2021) https://www.manometcurrent.com/pharma-grade-xylitol-market-to-experience-astonishing-growth-players-danisco-roquette-futaste-huakang-shandong-longlive-bio-technology-yuxin-xylitol-technology/. Accessed 2 Sept 2021

Tiefenbacher KF (2017) Technology of main ingredients—sweeteners and lipids. In: Tiefenbacher KF (ed) The technology of wafers and waffles I: operational aspects. Academic Press, pp 123—225. https://doi.org/10.1016/b978-0-12-809438-9.00003-x

Tomonobu N, Komalasari NLGY, Sumardika IW, Jiang F, Chen Y, Yamamoto K, ichi, Kinoshita, R., Murata, H., Inoue, Y., Sakaguchi, M., (2020) Xylitol acts as an anticancer monosaccharide to induce selective cancer death via regulation of the glutathione level. Chem Biol Interact 324:109085. https://doi.org/10.1016/j.cbi.2020.109085

Uebanso T, Kano S, Yoshimoto A, Naito C, Shimohata T, Mawatari K, Takahashi A (2017) Effects of consuming xylitol on gut microbiota and lipid metabolism in mice. Nutrients 9. https://doi.org/10.3390/nu9070756

Uhari M, Tapiainen T, Kontiokari T (2000) Xylitol in preventing acute otitis media. Vaccine 19:144–147. https://doi.org/10.1016/S0264-410X(00)00294-2

Ur-Rehman S, Mushtaq Z, Zahoor T, Jamil A, Murtaza MA (2015) Xylitol: a review on bioproduction, application, health benefits, and related safety issues. Crit Rev Food Sci Nutr 55:1514–1528. https://doi.org/10.1080/10408398.2012.702288

Vargas SL, Patrick CC, Ayers GD, Hughes WT (1993) Modulating effect of dietary carbohydrate supplementation on *Candida albicans* colonization and invasion in a neutropenic mouse model. Infect Immun 61:619–626. https://doi.org/10.1128/iai.61.2.619-626.1993

Vargason AM, Anselmo AC, Mitragotri S (2021) The evolution of commercial drug delivery technologies. Nat Biomed Eng 5:951–967. https://doi.org/10.1038/s41551-021-00698-w

Varricchio A, Mantia., I. La, Brunese, F., Ciprandi, G., (2021) Smell recovery in patients with COVID-19: an experience with nebulized nasal treatment. J Biol Regul Homeost Agents 35:63–66

Vernacchio L, Corwin MJ, Vezina RM, Pelton SI, Feldman HA, Coyne-Beasley T, Mitchell AA (2014) Xylitol syrup for the prevention of acute otitis media. Pediatrics 133:289–295. https://doi.org/10.1542/peds.2013-2373

Vilela C, Sousa AF, Fonseca AC, Serra AC, Coelho JFJ, Freire CSR, Silvestre AJD (2014) The quest for sustainable polyesters-insights into the future. Polym Chem 5:3119–3141. https://doi.org/10.1039/c3py01213a

Werpy T, Petersen G, Aden A, Bozell J, Holladay J, White J, Manheim A, Elliot D, Lasure L, Jones S, Gerber M, Ibsen K, Lumberg L, Kelley S (2004) Top value added chemicals from biomass. Volume I—Results of screening for potential candidates from sugars and synthesis gas. U.S. Department of Energy (DOE), Oak Ridge, TN

Wölnerhanssen BK, Cajacob L, Keller N, Doody A, Rehfeld JF, Drewe J, Peterli R, Beglinger C, Meyer-Gerspach AC (2016) Gut hormone secretion, gastric emptying, and glycemic responses to erythritol and xylitol in lean and obese subjects. Am J Physiol Endocrinol Metab 310:1053–1061. https://doi.org/10.1152/ajpendo.00037.2016.-With

Wölnerhanssen BK, Meyer-Gerspach AC, Beglinger C, Islam MS (2019) Metabolic effects of the natural sweeteners xylitol and erythritol: a comprehensive review. Crit Rev Food Sci Nutr. https://doi.org/10.1080/10408398.2019.1623757

Yamaguchi A, Muramatsu N, Mimura N, Shirai M, Sato O (2017) Intramolecular dehydration of biomass-derived sugar alcohols in high-temperature water. Phys Chem Chem Phys 19:2714–2722. https://doi.org/10.1039/c6cp06831f

Yi EY, Kim YJ (2013) Xylitol inhibits in vitro and in vivo angiogenesis by suppressing the NF-κB and Akt signaling pathways. Int J Oncol 43:315–320. https://doi.org/10.3892/ijo.2013.1917

Yin SY, Kim HJ, Kim HJ (2014) Protective effect of dietary xylitol on Influenza a virus infection. PLoS ONE 9. https://doi.org/10.1371/journal.pone.0084633

Ylikahri R (1979) Metabolic and nutritional aspects of xylitol. Adv Food Res 25:159–180. https://doi.org/10.1016/S0065-2628(08)60237-2

Zabner J, Seiler MP, Launspach JL, Karp PH, Kearney WR, Look DC, Smith JJ, Welsh MJ (2000) The osmolyte xylitol reduces the salt concentration of airway surface liquid and may enhance bacterial killing. Proc Natl Acad Sci 97:11614–11619

Zhang M, Wu Y, Zhao X, Gao K, Ma PX, Guo B (2015) Biocompatible degradable injectable hydrogels from methacrylated poly(ethylene glycol)-co-poly(xylitol sebacate) and cyclodextrins for release of hydrophilic and hydrophobic drugs. RSC Adv 5:66965–66974. https://doi.org/10.1039/c5ra11902b

Zhang Q, Song M, Xu Y, Wang W, Wang Z, Zhang L (2021) Bio-based polyesters: recent progress and future prospects. Prog Polym Sci 120:101430. https://doi.org/10.1016/j.progpolymsci.2021.101430

Chapter 11
Market, Global Demand and Consumption Trend of Xylitol

Meenu Hans, Nisha Yadav, Sachin Kumar, and Anuj Kumar Chandel

Abstract Xylitol, with an outstanding potential of replacing table sugar and antibiotics as a healthy alternative, is getting global attention in the present era. Food as well as pharmaceuticals/ medical sectors are adopting xylitol as a key ingredient in their products to be consumed by large population suffering from diabetics, obesity, dental and oral infections due to its insulin-independent digestion and antimicrobial activity. Due to the health consciousness of the society in the modern period, the demand of xylitol has increased consistently in the past few years. Hence, to meet the growing demand, various key manufacturers are supplying xylitol and xylitol-based products at the global platform. Along with the commercial production process of xylitol, various attempts are being carried out for producing the cost-effective and eco-friendly approaches to fulfill the consistently rising demands in the society. This chapter deals with the demand, supply and consumption trends of xylitol in the global market.

Keywords Xylitol · Global market · Food industry · Pharmaceuticals · Antimicrobial

M. Hans (✉) · N. Yadav · S. Kumar (✉)
Biochemical Conversion Division, Sardar Swaran Singh National Institute of Bio-Energy, Kapurthala 144601, India
e-mail: meenu.biotech13@gmail.com

S. Kumar
e-mail: sachin.kumar20@gov.in; sachin.biotech@gmail.com

M. Hans
Department of Microbiology, Guru Nanak Dev University, Amritsar, Punjab 143005, India

N. Yadav
Centre for Energy and Environment, Dr. B R Ambedkar National Institue of Technology, Jalandhar, Punjab 144027, India

A. K. Chandel
Department of Biotechnology, Engineering School of Lorena, University of São Paulo, Estrada Municipal Do Campinho, Lorena, São Paulo 602-810, Brazil

© The Author(s), under exclusive license to Springer Nature Switzerland AG 2022
M. das G. de Almeida Felipe and A. K. Chandel (eds.), *Current Advances in Biotechnological Production of Xylitol*,
https://doi.org/10.1007/978-3-031-04942-2_11

11.1 Introduction

Xylitol is a 5-carbon sugar alcohol, which was first extracted from birch trees in Finland (19th century) therefore, popularly recognized as wood or birch sugar, and naturally exists in many vegetables and fruits including berries, plum, cauliflower and lettuces (Ahuja et al. 2020; Kumar et al. 2015, 2009). Use of xylitol as a sweetener is appropriate for diabetic patients, and as an anticariogenic agent, has provided it a world-wide recognition in the pharmaceuticals and food sectors (de Cássia Lacerda Brambilla Rodrigues et al. 2012). Xylitol shares 12% of polyols in the market with a 3-fold expected growth (Rao Ravella et al. 2012). Usage of xylitol provides the alternative and healthy sugar options to people fighting with diabetics and obesity due to its digestion without insulin. Further, antimicrobial characteristic of xylitol makes it a better choice than the consumption of antibiotics, which mostly prove harmful to the consumers. Xylitol has also been used for infusion therapy after surgeries, treatment of shock and burning cases for last century (Aliakbarian et al. 2012; Mäkinen 1978). Commercially, xylitol is produced chemically from waste generated in the food and paper & pulp industries; however, the increasing demand of xylitol in the global market is seeking attention towards alternative routes and substrates for enhanced and cost-effective production to fulfill the growing needs of the society. This chapter highlights the growing demand, supply and consumption trends of xylitol in the global market.

11.2 Current Scenario of Xylitol Production

Asia produces about 50% of the world's total xylitol, with the rest of the production coming from Europe, the United States of America (U.S.A) and Australia (Rao Ravella et al. 2012). The major producers of xylitol are China and U.S.A utilizing corn cob and hardwood (e.g. birch), respectively as the substrate. Current route of the global xylitol production is the chemical method i.e. catalytic conversion of xylose obtained from hemicellulosic fraction of lignocellulosic feedstocks. Chemical reaction involves the hydrogenation of xylose in the presence of Nickel (Ni) catalyst (Mathew et al. 2018; Barathikannan and Paul 2016). Corn cob is the largely generated waste in the food industry in China, whereas hydrolysate of the birch tree is the key by-product of the paper & pulp industry in U.S.A, thus both these substrates are adequately available for providing xylose fraction for conversion into xylitol. Various companies are key suppliers of xylitol in the global market. Table 11.1 summarizes the major suppliers of xylitol in the global market. Various China-based companies such as Hunan JK International Trade Corporation, Futaste Co., Ltd., Jinjiang Weijia Food Co., Ltd., Qingdao FTZ United International, Inc. and Shandong Longlive Bio Companies generate xylitol using corn cob as a substrate. On the other hand, U.S.A-based companies that are Xylitol USA, Inc., Ingredion and DuPont (Danisco) use hydrolysate of birch tree, sugarcane bagasse and waste stream of paper & pulp

Table 11.1 Major global suppliers of xylitol

S. No	Company	Country	Substrate	Website
1	Salvavidas Pharmaceutical Pvt. Ltd	India	Corn husks, sugarcane bagasse and birch	https://www.salvavidaspharmaceutical.in/
2	Avanscure Life sciences Pvt. Ltd	India	Corn husks, sugarcane bagasse and birch	https://www.avanscure.com/
3	National Analytical Corporation	India	Corn husks, sugarcane bagasse and birch	http://www.nacchemical.com/
4	Hunan JK international trade corporation	United Kingdom	Corn cob	http://www.jk-ingredients.com/
5	Futaste Co., Ltd	China	Corn cob	http://www.futaste.com/
6	JinjiangWeijia Food Co., Ltd	China	Corn cob	http://www.jiujiuwang.com/
7	Qingdao FTZ United International, Inc	China	Corn cob	http://www.unitedint.com/
8	Shandong Longlive Bio	China	Corn cob	http://www.longlivegroup.com/
9	Xylitol USA, Inc	U.S.A	Birch trees	http://www.xylitolusa.com/
10	Ingredion	U.S.A	Sugarcane bagasse	https://www.ingredion.com/
11	DuPont (Danisco)	U.S.A	Waste side stream of pulp and paper industry	https://www.dupontnutritionandbiosciences.com/
12	Tate & Lyle	England	Corn, stevia leaves, tapioca	https://www.tateandlyle.com/
13	Cargill Brazil	Brazil	Sugarcane bagasse	https://www.cargill.com.br/

industry, respectively. Danisco is one of the major suppliers of xylitol worldwide using hardwood substrates through catalytic conversion of xylose. Xylitol XIVIA recently claimed the development of a more sustainable product than that of corn cob, where they have generated it from an integrated approach with paper & pulp industry. Tate & Lyle from England uses waste streams of corn, stevia leaves and tapioca as substrate. Cargill Brazil from Brazil utilizes hydrolysate of sugarcane bagasse, whereas Salvavidas Pharmaceutical Pvt. Ltd., Avanscure Lifesciences Pvt. Ltd. and National Analytical Corporation—Chemical Division from India utilize waste streams of corn husks, sugarcane bagasse and birch as the substrates for xylitol production.

11.3 Market Trend

A variety of advantageous properties of xylitol such as sweetening power and low calorific value (2.4 kcal/g as compared to 4 kcal/g of sucrose), have made it a global attraction since last century (Rao Ravella et al. 2012; Makinen 2000; Granström et al. 2007). The global market of xylitol is gradually increasing with an annual value of over US $537 million. The cost of xylitol has declined significantly for the last 10 years, and recently costs from US $4 to $5 per kg (Chandel et al. 2020). However, the cost of the product varies according to the substrate and transportation cost, which in turn based on the site of manufacturing plant and supply chain of the feedstock (Rao Ravella et al. 2012; Kocoloski et al. 2011; Koutinas et al. 2007). Many China-based xylitol producing companies dropped out of the business due to the increasing cost of the raw material leaving the process too expensive to be economically sustainable. Therefore, the most of the companies manufacture xylitol-based products in addition to the sorbitol to lower the manufacturing costs. However, xylitol manufacturing is still increasing consistently owing to the consumer's need of healthy products. The market size of xylitol was worth of US $0.901 billion in 2020, and it is expecting to turn into more than US $1.15 billion by 2023 due to abundant availability of the feedstocks for the large scale production (https://www.researchandmarkets.com/research/2wbg5g/global_xylitol?w=4). The global market was forecasted to rise with a CAGR of 7.44% during year 2019–2024, and it is further estimated to attain US $1.10 billion by 2028 at a CAGR of 2.59% from 2021 to 2028 (https://www.verifiedmarketresearch.com/product/xylitol-market/#:~:text=What%20is%20the%20projected%20market,2.59%25%20from%202021%20to%202028).

11.4 Global Demand and Supply of Xylitol

Due to a variety of applications, xylitol attains a global attention due to health consciousness in the society. For instance, the sugar-free eatables are in high demand by confectionary, pharmaceuticals and personal-care sectors word-wide due to consistently rising desirability of low-calories intake. Consequently, xylitol is a regular constituent of sugar-free candies, chewing gums, medicines, cough syrups and essential food items. Market witnessed a higher selling rate of sugar-free chewing gum than that of sugared gums in 2017. The major international companies selling sugar free chewing gums are Orbit, Trident, Extra, Eclipse and Dentyne (https://www.mordorintelligence.com/industry-reports/xylitol-market). Dentists claimed to intake xylitol containing chewing gums for reducing the harmful gut bacteria by 25–75%, thus recommending xylitol-based gums causing the commercial growth. Xylitol is a standard food ingredient in more than 50 nations leading to increasing commercial attention and demand in the international market (Rao Ravella et al. 2012). In the pharmaceuticals, xylitol is popular to be used a sweetener in the medicines, cough syrups and nutritional dietary items. Asia Pacific region, Middle East, North America, Latin America, Africa and Europe occupy the major share in the global market.

The key suppliers of xylitol in these regions include DuPont, Cargill, Roquette, Ingredion, Zuchem Inc., Novagreen, Thomson Biotech (Xiamen) and Mitsubishi Shoji Foodtech (https://www.researchandmarkets.com/reports/5331151/xylitol-market-global-industry-trends-share). The increasing global demand due to greater awareness of the useful properties of xylitol has led to the steep rise in its production. Hunan JK International Trade Corporation (Grangetown Cardiff, Wales, UK), Incorporated (Wayzata, MN, USA), Tate & Lyle PLC (London, UK), Xylitol USA, Inc. (Colorado, USA), Danisco (DuPont, Copenhagen, Denmark), E. I. du Pont de Nemours and Company (Wilmington, DE, USA), Jinjiang Weijia Food Co., Ltd. (Fujian Province, China), Novagreen Inc (Edmonton, AB, Canada), Futaste Pharmaceutical (Yucheng, China), Shandong Longlive Bio (Yucheng, Shandong, China), Yucheng Lujian (Shandong, China), Qingdao FTZ United International, Inc. (Shandong Road, Qingdao, China),Hangzhou Shouxing BioTechnology Co. Ltd. (Yuhang, China) and Cargill Brazil (Brazil) are the dominant developers and shareholders for xylitol production. China-based Futaste Pharmaceutical Ltd. supplies approximately 35,000 tons of xylitol, 20,000 tons of xylose and other types of sweeteners from hydrolysate of corn cob each year. Further, Hangzhou Shouxing BioTechnology Co. Ltd., China produces 15,000 tons of xylitol using corn cob as substrate whereas Shanghai just import and export Co., Ltd., China supplies 55,000 tons xylitol per year (Ahuja et al. 2020). Table 11.2 summarizes the leading companies supplying xylose-based products in the global market. These companies belong to UK, North America, China, South Korea, India, and supplies confectionary, pharmaceuticals, cosmetics as well as personal care products contributing in the global market share.

11.5 Consumption of Xylitol

Consumption of the xylitol is associated with the potential applications in food, medical, beverages and pharmaceutical sectors with a continuously rising marketplace estimated to cross US $1.14 billion by 2023 as discussed ahead (Ahuja et al. 2020; Chandel et al. 2020). Figure 11.1 presents the major consuming sectors along with the functions played by xylitol as the key ingredient in consumption of the specific goods.

11.5.1 Food Sector

Xylitol has been approved as a safe additive and low-calorie sweetener by the Food and Drug Administration (FDA) in over 50 countries (Ahuja et al. 2019; Barclay et al. 2014; da Silva et al. 2012). Low-calorie of xylitol made it a sugar supplement of choice in a variety of baked good, peanut butter, candies, jams, gelatins, desserts,

Table 11.2 Xylitol based products manufactured by various companies worldwide

S. No	Company	Country	Xylitol based product/s	Website
1	Peppersmith	United kingdom	Chewing gum Mints Candies	http://www.peppersmith.co.uk/
2	Xlear®	North America	Xylitol nasal spray Sinus irrigation products Xylitol gums Mouth wash Xylitol gems Sweetener	https://www.xlear.com/
3	Zap gum	U.S.A	Mints Candies Syrups Chewing gum	https://tryzapp.com/
4	Yesannongsan Co., Ltd	South Korea	Citron tea	https://yesanjam.tradekorea.com/main.do
5	Trident	U.S.A	Chewing gum	https://www.tridentgum.com/
6	Bioxtra	India	Mouthwashes Toothpastes	https://bioxtra.info/en/
7	The Himalaya Drug Company	India	Toothpaste	https://www.himalayawellness.com/
8	Epic Dental	Utah (U.S.A)	Gum Mints Mouth wash Dental protection kits Sweetener	https://www.epicdental.com/
9	Oracoat	Wisconsin (U.S.A)	Mouth moisturizer Renewing melts	https://www.oracoat.com/
10	Izze	Colorado (U.S.A)	Juice	https://www.izze.com/
11	Hager Pharma	North Carolina (U.S.A)	Chewing gum Dry mouth drops Breath spray Pre-pasted toothbrush	https://www.hager-pharma.us/
12	Liqiang Foodstuff Industry Co. Ltd	China	Xylitol Candies Milk slice Bubble gum	http://www.lycone.cn/cn/index.php

(continued)

Table 11.2 (continued)

S. No	Company	Country	Xylitol based product/s	Website
13	Shantou Taijixing Food Co., Ltd	China	Xylitol gum Candies	http://www.shantoutjxfood.shilinzhongyou.com/
14	JiachemDentbio Co., Ltd	China	Xylitol Xylitol chewing gum Xylitol sweetener	http://www.dentbio.com/
15	Healtang Biotech Co., Ltd	China	D-xylose Xylitol Sweetener	http://www.healtang.com/
16	Thomson Biotech (S) Pte. Ltd	China	Xylitol Additives (for food)	http://www.thomsonbiotech.com/
17	Anhui Fukang Pharmaceutical Co., Ltd	China	Pharmaceutical Xylitol products	www.fukang-group.com

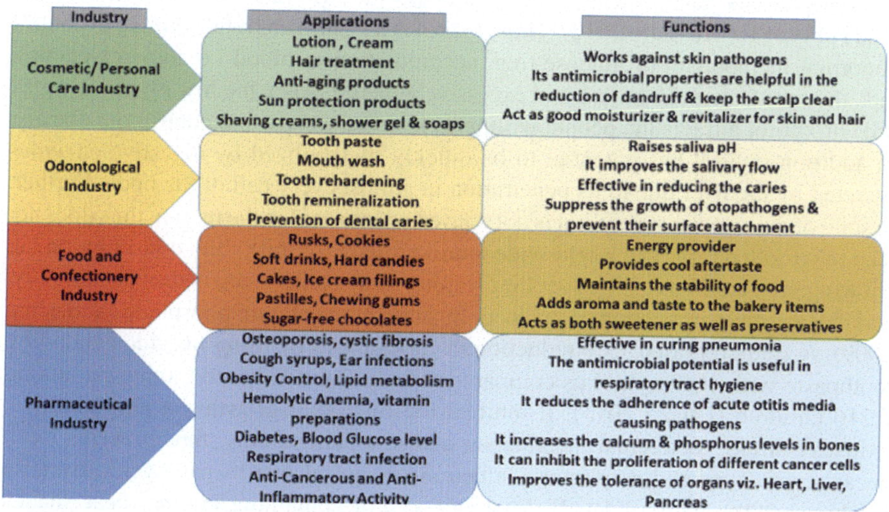

Fig. 11.1 Applications and functions of xylitol in various sectors (Ahuja et al. 2020; Vallejos and Area 2017; Ur-Rehman et al. 2015; Mussatto 2012; Lugani and Sooch 2018; Danhauer et al. 2010)

fruit syrups, drink powders, ketchups, puddings, ice creams and other food items or sprinkled on the food, tea or coffee by majority of the people around the world due to the health consciousness. Xylitol is safe to cook like an ordinary sugar, and does not need to breakdown into liquid form (https://www.bbcgoodfood.com/glossary/xylitol-glossary). In addition, xylitol does not caramelize like sugar, thus does not cause the darkening of the product, which is a common practice in sugary food products (Vallejos and Area 2017). About 70% of the global production of the xylitol contributes for the manufacturing of sugar-free chewing gums and confectionery products with reduced risk of caries in children (Ahuja et al. 2020; Salli et al. 2019; Dasgupta et al. 2017). About 4–10 g of xylitol intake in periods of 3–7 times a day is suggested to give positive results in treating caries (Deshpande and Jadad 2008; Milgrom et al. 2006; Mäkinen et al. 1998; Hujoel et al. 1999). However, an optimal intake amount (4–10 g) in less than 3 times a day does not show any effect (Thaweboon et al. 2004; Rekola 1989; Isokangas 1987). Further, excessive xylitol consumption of about 10–60 g a day does not show any significant reduction but may cause diminished anticariogenicity, abdominal distress and osmotic diarrhea (Giertsen et al. 1999; Uhari et al. 1996; Mäkinen et al. 1995; Salminen et al. 1989).

11.5.2 Medical, Health Sector and Pharmaceuticals

The most important feature of xylitol is its insulin free metabolism as compared to the sucrose, which make it the most advantageous sugar supplement to be taken by the diabetic patients. Thus, it can be digested by people suffering with diabetes mellitus type I or type II (Mussatto 2012). Thus, xylitol is a significant food additive to control abnormal metabolism of glucose (e.g. accumulation in blood i.e. hyperglycemia), fats, thus controls dehydration and excessive appetite. Also, the low glycemic index (GI) of xylitol attracts the people willing to lose the weight and control the obesity. In addition, xylitol has potential to be quickly metabolized by digestive enzymes secreted by the liver due to its penetration in almost every cell of the body. Further, in spite of the sweet taste, xylitol is not converted into acid in the mouth, thus does not lead into tooth decay. The world-wide familiar mode for the xylitol consumption i.e. chewing gum for 20 min improves the cleaning away acid formed in the buccal cavity, and aids in remineralization of the teeth by absorbing calcium phosphate (Soderling 2009). A study reported 13% reduction in caries cases by using fluoride containing toothpaste with 10% xylitol as compared to only fluoride-based toothpaste among 4,216 children (Duane 2015). It inhibits the microbial growth via anti-adhesive, oxidative stress, ineffectual metabolism and low permeability, thus prevents tooth decay, caries, ear and respiratory infections occurred due to pathogenic bacteria (Makinen 2000; Yilikari 1979). Intake of xylitol containing gummy bear snacks helps in reducing *Streptococci mutans* levels in the mouth, which could also be washed off by rinsing the mouth with xylitol that additionally results in the reduced growth of *S. sanguis* and biofilm development (Ganter et al. 2020; Nayak et al. 2014). Likewise, xylitol intake has shown an antimicrobial activity against a variety of pathogens of oral as well as nasopharyngeal systems such as *Pseudomonas aeruginosa* ATCC 9027, *Staphylococcus aureus* ATCC 6538, *Candida albicans*, Lactobacillus, *S. pneumonia*, *S. mitis*, *Haemophilus influenza*, *Moraxella catarrhalis*, *S. epidermidis*, *Porphyromonas gingivalis*, etc. (Zhou et al. 2019; Kozlowska et al. 2015; Marchese et al. 2016; Hajiahmadi et al. 2019; Sakallioglu et al. 2014; Kontiokari et al. 1995; Jain et al. 2016; Han et al. 2005). Thus, Xylitol intake also helps to prevent diseases such as pneumonia, acute otitis media, mastoiditis, meningitis and impaired hearing, hemolytic anemia, metabolic syndrome, atherosclerosis, cardiovascular disease, stroke and even cancer and osteoporosis (Gasmi Benahmed et al. 2020; Ahuja et al. 2020). Xylitol also has profound effect of inhibiting inflammatory reactions and angiogenesis occurring during tumor formation and metastasis, thus exhibits anti-cancerous properties (Yi and Kim 2013).

11.5.3 Other Sectors

Xylitol is also used as an important ingredient of cosmetics and personal care goods such as lotion, cream, hair treatment products, etc. More advanced consumption of xylitol is for the production of highly efficient in terms of mechanical strength and other properties, biodegradable and eco-friendly matrices and fibers e.g. "polyPXBSu" (polymer of xylitol, sebacic acid, butylenes glycol), poly PXBSe (polymer of succinic acid, butylene glycol and xylitol) and xylitol-dicarboxylate-co-butylene dicarboxylate (polymer of dicarboxylic acids: adipic acid, dodecanedioic acid, sebacic acid, succinic acid and suberic acid) (Piątek-Hnat and Bomba 2020). These biodegradable polymers are useful in the healthcare e.g. fibroblast growth factor (FGF) loaded on PXDDA (polymer of xylitol dodecanedioic acid) by dopamine coating method is highly efficient agent to be used in repair and regeneration of tissues as compared to the naturally found less stable FGF with short half-life and rapid inactivated by the enzymes (Firoozi and Kang 2020).

11.6 Future Recommendations

The expanding market of xylitol for the consumption in food, pharmaceuticals/medical sectors and cosmetics/personal care goods leads a thrust to opt the cost-effective and environment-friendly routes using sustainable feedstocks to supply the desired products adequately. Because the conventional chemical process used for xylitol production using corn cob and hardwood hydrolysate is expensive and limited in terms of the technology and substrates, respectively. Integrated biorefinery of lignocellulosic biomass is the one of the best choices to generate xylitol along with other biofuels and renewable chemicals in spite of utilizing individual substrates in a single approach in terms of the energy savings and cost-effectiveness. In place of the extraction of xylose from hemicellulose fraction of biomass through acid or enzymatic hydrolysis, acid pretreated hydrolysate of biomass left as a separate fraction during the production of second generation (2G) biofuels could be a better substrate to be utilized for xylitol production due to enriched source of xylose. Further, traditional expensive chemical process could be replaced by the biological route as a comparatively cheaper and eco-friendly alternative route. Biological route involves the microbial approach via fermentation of xylose by using potential native or genetically engineered xylose-utilizing microbial strains (most commonly yeast e.g. *Pichia sp.*) and enzymatic approach via oxidoreductive process using two enzymes xylose reductase and xylitol dehydrogenase. Both of the suggested methods have their own pros and cons but definitely better choices over expensive and environmentally unfeasible chemical process. These processes are also reported to provide relatively higher product yield and productivity than present commercial process, which is the key requirement for fulfilling the increasing demands of xylitol in the global market.

11.7 Conclusions

Xylitol is paving way in the global market as a key ingredient of interest in the food, pharmaceuticals/medical and personal care/cosmetics sectors due to its unusual properties such as sweet taste with low calories, digestion without requiring insulin, anti-cariogenicity, antimicrobial activity, high mechanical strength and biodegradability. The market size of xylitol is expecting to turn into more than US $1.15 billion by 2023, and further estimated to attain US $1.10 billion by 2028. The global market was estimated to rise with a CAGR of 7.44% during year 2019–2024, and expecting a CAGR of 2.59% from 2021 to 2028. To meet the consistently rising demands of xylitol, cost-effective and sustainable integrated biorefinery of lignocellulosic biomass could be adopted to generate xylitol along with other bio-based chemicals rather than individual process. Further, xylose enriched acid pretreated hydrolysate of biomass used for 2G biofuels production is a better choice of substrate for xylitol production. Further, a commercial expensive chemical process could be replaced by biotechnological route consisting of fermentative or enzymatic approaches due to economic and environmental feasibility of the associated technologies, and also for promising higher yield and productivity of the product.

Acknowledgements Authors gratefully acknowledge the Ministry of New and Renewable Energy, Government of India for providing funds to carry out the research activities. Meenu Hans and Nisha Yadav are thankful to the Sardar Swaran Singh National Institute of Bio-Energy (SSS-NIBE), Kapurthala, India for providing Research Fellowship. Meenu Hans acknowledges Guru Nanak Dev University, Amritsar, India for her Ph.D. registration [Reg. No. 2007.KJ/A.519]. Nisha Yadav also acknowledges Dr. B R Ambedkar National Institute of Technology, Jalandhar, India for her Ph.D. enrollment [Roll No. 21801208].

Anuj K. Chandel thanks the CNPq-Brazil for the scientific productivity program (Process number: 309214/2021-1).

References

Ahuja V, Macho M, Ewe D, Singh M, Saha S, Kumar S (2020) Biological and pharmacological potential of xylitol: a molecular insight of unique metabolism. Foods 9(11):1592

Ahuja V, Sharma A, Rathour RK, Sharma V, Rana N, Bhatt AK (2019) In-vitro and in-silico characterization of xylose reductase from *Emericellani dulans*. Curr Chem Biol 13:159–170
Aliakbarian B, de Faveri D, Perego P, Converti A (2012) An Assessment on xylitol recovery methods. In: da Silva SS, Chandel AK (eds) D-Xylitol. Springer, Berlin, Heidelberg, pp 229–244
Barathikannan K, Paul A (2016) Xylitol: production, optimization and industrial application. Int J Curr Microbiol App Sci 5(9):324–339
Barclay A, Sandall P, Shwide-Slavin C, Brand-Miller J (2014) The ultimate guide to sugars and sweeteners: discover the taste, use, nutrition, science, and lore of everything from agave nectar to xylitol: the experiment, New York, USA
Benahmed AG, Gasmi A, Arshad M, Shanaida M, Lysiuk R et al (2020) Health benefits of xylitol. Appl Microbiol Biotechnol 1–13
De Cássia Lacerda Brambilla Rodrigues R, Canettieri EV, Martinez EA, Canilha L, NapolezSolenzal AI, de Almeida e Silva JB (2012) Statistical approaches for the optimization of parameters for biotechnological production of xylitol. In: da Silva SS, Chandel AK (eds) D-Xylitol. Springer, Berlin, Heidelberg, pp 103–160
Chandel AK, Garlapati VK, Jeevan Kumar S, Hans M, Singh AK, Kumar S (2020) The role of renewable chemicals and biofuels in building a bioeconomy. Biofuel Bioprod Biorefin 14(4):830–844
Danhauer JL, Johnson CE, Corbin NE, Bruccheri KG (2010) Xylitol as a prophylaxis for acute otitis media: systematic review. Int J Audiol 49(10):754–761
da Silva AF, Ferreira AS, da Silva SS, Barbosa Raposo NR (2012) Medical applications of xylitol: an appraisal. In: da Silva SS, Chandel AK (eds) D-Xylitol, Springer-Verlag Berlin Heidelberg, pp 325–342
Dasgupta D, Bandhu S, Adhikari DK, Ghosh D (2017) Challenges and prospects of xylitol production with whole cell bio-catalysis: a review. Microbiol Res 197:9–21
Deshpande A, Jadad AR (2008) The impact of polyol-containing chewing gums on dental caries: a systematic review of original randomized controlled trials and observational studies. J Am Dent Assoc 139(12):1602–1614
Duane B (2015) Xylitol and caries prevention. Evid Based Dent 16(2):37–38
Firoozi N, Kang Y (2020) Immobilization of FGF on poly (xylitol dodecanedioic acid) polymer for tissue regeneration. Sci Rep 10:10419
Ganter J, Hellwig E, Doerken S, Al-Ahmad A (2020) In vitro evaluation of the cariogenic potential of rebaudioside a compared to sucrose and xylitol. Clin Oral Investig 24(1):113–122
Giertsen E, Emberland H, Scheie AA (1999) Effects of mouth rinses with xylitol and fluoride on dental plaque and saliva. Caries Res 33(1):23–31
Granström TB, Izumori K, Leisola M (2007) A rare sugar xylitol. Part I: the biochemistry and biosynthesis of xylitol. Appl Microbiol Biotech 74:277–281
Hajiahmadi M, Yegdaneh A, Homayoni A, Parishani H, Moshkelgosha H, Salari-Moghaddam R (2019) Comparative Evaluation of efficacy of "Green Tea" and "Green Tea with Xylitol" mouthwashes on the salivary Streptococcus mutans and Lactobacillus colony count in children: a randomized clinical trial. J Contemp Dent Pract 20(10):1190–1194
Han SJ, Jeong SY, Nam YJ, Yang KH, Lim HS, Chung J (2005) Xylitol inhibits inflammatory cytokine expression induced by lipopolysaccharide from *Porphyromonas gingivalis*. ClinDiagn Lab Immunol 12(11):1285–1291
Hujoel PP, Mäkinen KK, Bennett CA (1999) The opti-mum time to initiate habitual xylitol gum chewing forobtaining long-term caries prevention. J Dent Res 78(3):797–803
Isokangas P (1987) Xylitol chewing gum in caries prevention. A longitudinal study on Finnish school children. Proc Finn Dent Soc 83:1–117
Jain R, Lee T, Hardcastle T, Biswas K, Radcliff F, Douglas R (2016) The in vitro effect of xylitol on chronic Rhinosinusitis biofilms. Rhinology 54(4):323–328
Kocoloski M, Michael Griffin W, Scott Matthews H (2011) Impacts of facility size and location decisions on ethanol production cost. Energy Policy 39:47–56

Kontiokari T, Uhari M, Koskela M (1995) Effect of xylitol on growth of nasopharyngeal bacteria in vitro. Antimicrob Agents Chemother 39(8):1820–1823

Koutinas AA, Wang RH, Webb C (2007) The biochemurgist- bioconversion of agricultural raw materials for chemical production. Biofuel Bioprod Biorefin 1:24–38

Kozlowska M, Laudy AE, Przybyl J, Ziarno M, Majewska E (2015) Chemical composition and antibacterial activity of some medicinal plants from Lamiaceae family. Acta Pol Pharm 72(4):757–767

Kumar S, Singh SP, Mishra IM, Adhikari DK (2009) Ethanol and xylitol production from glucose and xylose at high temperature by *Kluyveromyces* sp. IIPE453. J Ind Microbiol Biotechnol 36(12):1483

Kumar S, Dheeran P, Singh SP, Mishra IM, Adhikari DK (2015) Bioprocessing of bagasse hydrolysate for ethanol and xylitol production using thermotolerant yeast. Bioproc Biosyst Eng 38(1):39–47

Lugani Y, Sooch BS (2018) Insights into fungal xylose reductases and its application in xylitol production. In: Fungal biorefineries. Springer, Cham, pp 121–144

Mäkinen KK (1978) Biochemical principles of the use of xylitol in medicine and nutrition with special consideration of dental aspects. Experientia Suppl 30:1–160

Makinen KK (2000) The rocky road of xylitol to its clinical application. J Den Res 79:1352–1355

Mäkinen KK, Benett CA, Hujoel PP (1995) Xylitol chewing gums and caries rates: a 40-month cohort study. J Dent Res 74(12):1904–1913

Mäkinen KK, Chiego DJ Jr, Allen P (1998) Physical, chemical, and histologic changes in dentin caries lesions of primary teeth induced by regular use of polyol chewing gums. Acta Odontol Scand 56(3):148–156

Marchese A, Orhan IE, Daglia M, Barbieri R, Di Lorenzo A, Nabavi SF, Gortzi O, Izadi M, Nabavi SM (2016) Antibacterial and antifungal activities of thymol: a brief review of the literature. Food Chem 210:402–414

Mathew AK, Abraham A, Mallapureddy KK, Sukumaran RK (2018) Lignocellulosic biorefinery wastes, or resources. In: Bhaskar T, Pandey A, Venkata Mohan S, Lee D-J, Khanal SK (eds) Waste biorefinery. Elsevier, pp 267–297

Milgrom P, Ly KA, Roberts M, Rothen M, Mueller G, Yamaguchi DK (2006) Mutans Streptococci dose response to xylitol chewing gum. J Dent Res 85(2):177–181

Mussatto SI (2012) Application of xylitol in food formulations and benefits for health. In: Da Silva SS, Chandel AK (eds) D-xylitol. Springer, Berlin, Heidelberg, pp 309–323

Nayak PA, Nayak UA, Khandelwal V (2014) The effect of xylitol on dental caries and oral flora. Clin Cosmet Investig Dent 6:89–94

Piątek-Hnat M, Bomba K (2020) The influence of cross-linking process on the physicochemical properties of new copolyesters containing xylitol. Mater Today Commun 22:100734

Rekola M (1989) Correlation between caries incidence and frequency of chewing gum sweetened with sucrose or xylitol. Proc Finn Dent Soc 85(1):21–24

Russo JR (1977) Xylitol: anti-caries sweetener? Food Eng 79:37–40

Rao Ravella S, Gallagher J, Fish S, Reddy Shetty P (2012) Overview on commercial production of xylitol, economic analysis and market trends. In: da Silva SS, Chandel AK (eds) D-Xylitol, Springer-Verlag Berlin Heidelberg, pp 291–306

Sakallioglu O, Guvenc IA, Cingi C (2014) Xylitol and its usage in ENT practice. J Laryngol Otol 128(7):580–585

Salli K, Lehtinen MJ, Tiihonen K, Ouwehand AC (2019) Xylitols health benefits beyond dental health: a comprehensive review. Nutrients 11:1813

Salminen EK, Salminen SJ, Porkka L, Kwasowski P, Marks V, Koivistoinen PE (1989) Xylitol versus glucose: effect on the rate of gastric emptying and motilin, insulin, and gastric inhibitory polypeptide release. Am J Clin Nutr 49(6):1228–1232

Soderling EM (2009) Xylitol, mutans Streptococci, and dental plaque. Adv Dent Res 21(1):74–78

Thaweboon S, Thaweboon B, Soo-Ampon S (2004) The effect of xylitol chewing gum on mutansStreptococci in saliva and dental plaque. Southeast Asian J Trop Med Public Health 35(4):1024–1027
Uhari M, Kontiokari T, Koskela M, Niemela M (1996) Xylitol chewing gum in prevention of acute otitis media: double blind randomized trial. Brit Med J 313(7066):1180–1184
Ur-Rehman S, Mushtaq Z, Zahoor T, Jamil A, Murtaza MA (2015) Xylitol: a review on bioproduction, application, health benefits, and related safety issues. Crit Rev Food Scinutri 55(11):1514–1528
Vallejos ME, Area MC (2017) Xylitol as bioproduct from the agro and forest biorefinery. In: Food bioconversion. Elsevier Inc., pp 411–432
Xylitol in personal care and cosmetics market size, share & trends analysis report by application (skin, hair, oral), by region (Asia Pacific, North America), and segment forecasts, 2018–2025, Market Analysis Report (2018), Report ID: GVR-2-68038-727-8
Yi EY, Kim YJ (2013) Xylitol inhibits in vitro and in vivo angiogenesis by suppressing the NF-kappa B and Akt signaling pathways. Int J Oncol 43(1):315–320
Yilikari R (1979) Metabolic and nutritional aspects of xylitol. In: Chichester CO, Mrak EM, Stewart G (eds) Advances in food research, vol 25. Academic Press, New York, pp 159–180
Zhou G, Peng H, Wang YS, Huang XM, Xie XB, Shi QS (2019) Enhanced synergistic effects of xylitol and isothia zolones for inhibition of initial biofilm formation by *Pseudomonas aeruginosa* ATCC 9027 and *Staphylococcus aureus* ATCC 6538. J Oral Sci 61(2):255–263
https://www.salvavidaspharmaceutical.in/. Accessed 03 Jan 2022
https://www.avanscure.com/. Accessed 03 Jan 2022
http://www.nacchemical.com/. Accessed 03 Jan 2022
http://www.jk-ingredients.com/. Accessed 03 Jan 2022
http://www.futaste.com/. Accessed 03 Jan 2022
http://www.jiujiuwang.com/. Accessed 03 Jan 2022
http://www.xylitolusa.com/. Accessed 04 Jan 2022
https://www.ingredion.com/. Accessed 04 Jan 2022
https://www.dupontnutritionandbiosciences.com/. Accessed 04 Jan 2022
https://www.tateandlyle.com/. Accessed 04 Jan 2022
https://www.cargill.com.br/. Accessed 04 Jan 2022
http://www.longlivegroup.com/. Accessed 02 Jan 2022
http://www.unitedint.com/. Accessed 02 Jan 2022
http://www.peppersmith.co.uk/. Accessed 30 Dec 2021
https://www.xlear.com/. Accessed 03 Dec 2021
https://tryzapp.com/. Accessed 30 Dec 2021
https://yesanjam.tradekorea.com/main.do. Accessed 30 Dec 2021
https://www.tridentgum.com/. Accessed 29 Dec 2021
https://bioxtra.info/en/. Accessed 30 Dec 2021
https://www.himalayawellness.com/. Accessed 31 Dec 2021
https://www.epicdental.com/. Accessed 31 Dec 2021
https://www.oracoat.com/. Accessed 31 Dec 2021
https://www.izze.com/. Accessed 31 Dec 2021
https://www.hager-pharma.us/. Accessed 01 Jan 2022
http://www.lycone.cn/cn/index.php. Accessed 02 Jan 2022
http://www.shantoutjxfood.shilinzhongyou.com/. Accessed 02 Jan 2022
http://www.dentbio.com/. Accessed 02 Jan 2022
http://www.healtang.com/. Accessed 02 Jan 2022
http://www.thomsonbiotech.com/. Accessed 02 Jan 2022
www.fukang-group.com. Accessed 02 Jan 2022

The manufacturer's authorised representative in the EU is Springer Nature Customer Service Centre GmbH, Europaplatz 3, 69115 Heidelberg, Germany. If you have any concerns regarding our products, please contact ProductSafety@springernature.com

Printed and bound by CPI Group (UK) Ltd, Croydon, CR0 4YY
22/09/2026
02221055-0008